U0210922

避暑山庄万壑松风建筑群北立面图（2012 年）
North elevation of buildings around Wanhe songfeng of Mountain Resort (2012)

避暑山庄月色江声建筑群西立面图（2011 年）
West elevation of buildings around Yuese jiangsheng of Mountain Resort (2011)

避暑山庄如意洲建筑群南立面图（2012 年）
South elevation of buildings around Ruyizhou of Mountain Resort (2012)

避暑山庄烟雨楼建筑群南立面图（2012 年）
South elevation of buildings around Yanyulou of Mountain Resort (2012)

避暑山庄文津阁琉璃瓦屋顶南立面复原图（2013 年）
Restored south elevation of color-glazed tile roof of Wenjinge of Mountain Resort (2013)

避暑山庄文津阁布瓦屋顶南立面测绘图（2012 年）
South elevation survey of buwa roof of Wenjinge of Mountain Resort (2012)

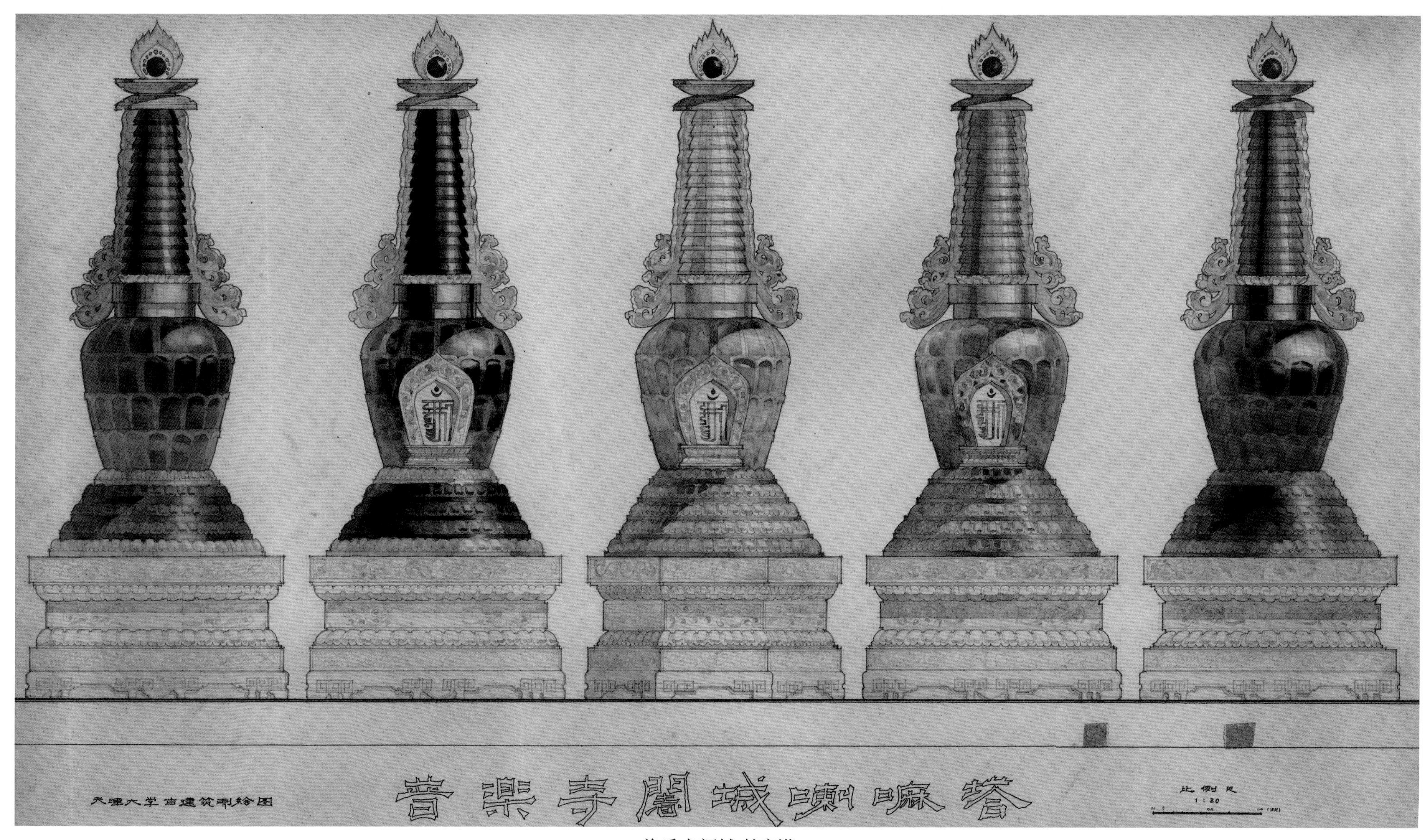

普乐寺阇城喇嘛塔

Stupa of Ducheng of Pulesi

须弥福寿之庙妙高庄严殿天花大样

Ceiling decoration of Miaogao zhuangyandian of Xumi fushou Temple

『十二五』国家重点图书出版规划项目

中国古建筑测绘大系·园林建筑与宗教建筑

承德避暑山庄和外八庙

天津大学建筑学院 承德市文物局 合作编写

杨菁 朱蕾 主编

中国建筑工业出版社

Traditional Chinese Architecture Surveying and Mapping Series:
Garden Architecture & Religious Architecture

BISHU SHANZHUANG AND THE EIGHT OUTER TEMPLES

Compiled by School of Architecture,Tianjin University &
The Relics Bureau of Chengde Municipal
Edited by YANG Jing, ZHU Lei

China Architecture & Building Press

Contents

目录

Introduction

Previously known as Jehol or Rehe (Re River, literally "hot river") after the name of the former province it was located in, Chengde in present-day Hebei province was the summer retreat of the Qing emperors, a complex more famous for its Mountain Resort (Bishu shanzhuang; literally mountain estate for escaping the summer heat) and its Eight Outer Temples (Bawaimiao). Located at a rift basin in the hinterland of the Yan mountain range traversed by Luan River, it was a small, sparsely populated area during the early Qing period when it was known as Upper Camp at Rehe. Since 1681, the twentieth year of Emperor Kangxi's reign, the imperial autumn hunt, an annual rite held at Mulan in Manchuria, spurred the construction of a series of temporary palaces (xinggong) along the route from the capital (in Beijing via Gubeikou Pass) to the hunting grounds. And in case of Rehe, an imperial palace was built in the northeast of the Upper Camp. Construction began in 1703, the forty-second year of emperor Kangxi's reign. Eight years later, in 1711, the palace was officially named Bishu shanzhuang or Mountain Resort. Two years later, in 1713, Kangxi's fifty-second reign year, two Tibetan Buddhist monasteries—Purensi and Pushansi—were built north of the Mountain Resort. Kangxi's successor—the Qianlong emperor—commissioned further large-scale construction inside and outside the ever-expanding compound. Finally, after a period of 90 years and through the patronage of three emperors (Kangxi, Yongzheng and Qianlong), the Eight Outer Temples to the north and east of the Mountain Resort were completed at the end of emperor Qianlong's reign. Originally just the location of the imperial hunting grounds (and a temporary imperial palace built in 1703), the small town of Rehe was named Chengde in 1733 (the eleventh year of Emperor Yongzheng's reign) and made the seat of a government controlling one prefecture (zhou) and five counties (xian) in 1778 (the forty-third reign year of Emperor Qianlong) [1] ,and afterwards, quickly developed into an imperially sponsored garden estate (fig.1).

导言

避暑山庄和外八庙所在的河北省承德市原名『热河』，位于燕山山脉腹地的一处断陷盆地，区内河流大部分属滦河水系。清初此地仅是人烟稀少的小型居民点，称为『热河上营』。康熙二十年（一六八一年）始，清帝每年秋季举行的木兰行围，带动了从北京自古北口外至木兰围场之间沿途一系列行宫的建设活动，热河上营东北部也建立起了热河行宫。康熙四十二年（一七〇三年）行宫初具轮廓，康熙五十年（一七一一年）避暑山庄正式命名，康熙五十二年（一七一三年）在避暑山庄东部兴建了两座藏传佛教寺庙溥仁寺和溥善寺，乾隆年间又对避暑山庄内外进行大规模营建。历经康、雍、乾三代约九十年的不断建设，在乾隆末年形成现今外八庙环绕山庄北部和东部的格局。自清康熙四十二年（一七〇三年）设立热河行宫总管开始，又雍正十一年（一七三三年）以承德之名出现，再于乾隆四十三年（一七七八年）升为承德府，统领一州五县[1]，热河从弹丸之地，随园林经营而迅速发展成一座城市（图1）。

I.The Mountain Resort (Bishu shanzhuang)

Covering an area of 5.6 sq. km, the Mountain Resort faces Wulie River (also known as Re River) on the east and Shizi (Lion) Furrow on the north. Additionally, it borders the Yan mountain range to the west and the modern city of Chengde to the south. The hot spring (Rehe Spring) inside the Mountain Resort flows throughout the year, and its water provides the conditions for a pleasant climate, as described in one of emperor Kangxi's poems ("As the grass and trees flourish, so too do gnats and beetles scatter. The spring water is pristine, and the people are well.")[2]. The Mountain Resort is surrounded by hills: to its north lies a gneiss range with gentle rolling hills and two V-shaped valleys (one bigger than the other). To its east and south are views of the Danxia landform, a geomorphology characterized by red-colored sandstones with steep cliffs. The 59.42-m-high Hammer Peak located east to the Mountain Resort is a local landmark that can be seen from miles away and marks the locality of the Mountain Resort and the Eight Outer Temples.

The unique natural conditions of the Mountain Resort created one of the most beautiful landscapes in the Qing empire (described in one of emperor Kangxi's poems as "as natural looking as nature itself, designed without artificiality").[3] Under emperor Kangxi, the garden design began to follow the principle of elegant simplicity ("colonnaded buildings shall be simple, neither painted not carved, and only then mountain dwellings will be elegant").[4] Today, the 5.6 sq. km large area interspersed with pavilions, terraces, and multi-story buildings and pavilions is the largest Chinese imperial landscape garden still extant, as depicted in *Bishu shanzhuang qishi'er jing* (Seventy-two views of the Mountain Resort) (fig. 2).

(I) The Seventy-two Jing (Scenic Areas) of the Mountain Resort

1.Formation History

When it was still a temporary imperial palace, the area had sixteen scenic spots: Chengbo diecui ("clear ripples with layered greenery"), Zhijing yundi ("zhi [divine mushroom] patch on an embankment to the clouds"), Changhong yinlian ("long rainbow sipping white silk"), Nuanliu xuanbo ("warm currents and balmy ripples"), Shuanghu jiajing ("a pair of lakes like flanking mirrors"), Wanhe songfeng ("pine winds through myriad vales"),

一、避暑山庄

避暑山庄占地面积五百六十万平方米，东临武烈河，北接狮子沟，西枕燕山，南依城市。内有热河泉四季长流，带来『草木茂，绝蚊蝎，泉水佳，人少疾』[二]的宜人小气候。外则群山环抱：北部是片麻岩分布区，山坡平缓，山顶圆滑，大小沟谷呈『V』字形；东部和南部在大自然的鬼斧神工下产生了丰富的丹霞地貌景观，位于山庄正东、高五十九·四二米的磬锤峰，更是避暑山庄及『外八庙』在选址和设计上所依托的重要山水形胜。

得天独厚的自然条件造就了避暑山庄『自然天成地就势，不待人力假虚设』[三]的景观。康熙朝以来的园林建设依照『楹宇守朴，不雘不雕，得山居雅致』[四]的原则，在占地的范围内构筑以亭、台、楼、阁，形成了『避暑山庄七十二景』为代表的我国现存规模最大的皇家园林（图2）。

（一）避暑山庄七十二景

1）形成始末

早在热河行宫时期，山庄内既已形成十六景，是为：澄波叠翠、芝径云堤、长虹饮练、暖流暄波、双湖夹镜、万壑松风、曲水荷香、西岭晨霞、锤峰落照、芳渚临流、南山积雪、金莲映日、梨花伴月、

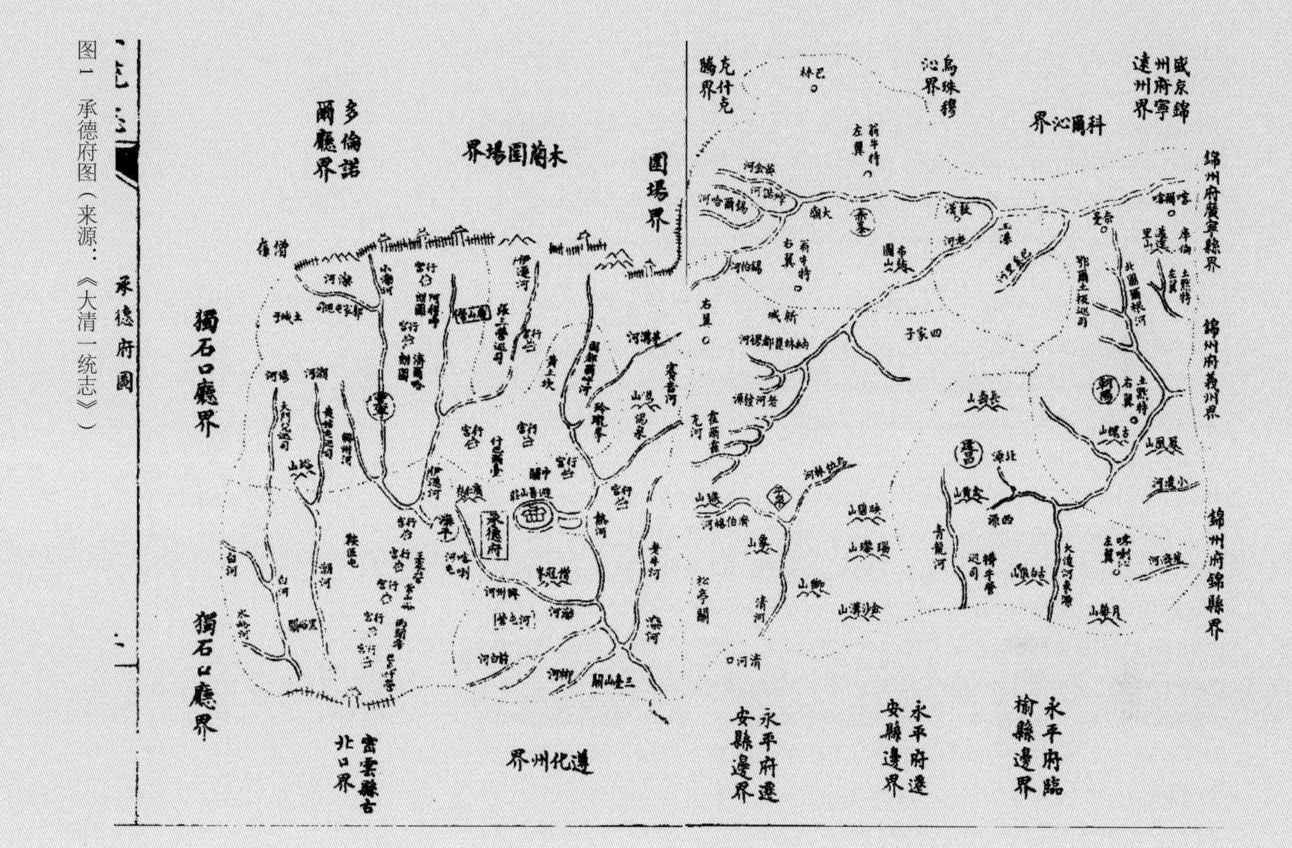

图1 承德府图（来源：《大清一统志》）

图2 避暑山庄总图（来源：《钦定热河志》）

图3 沈嵛《御制避暑山庄诗》中的第十二景图，锤峰落照（来源：《清殿版画汇刊》）

Fig.1 Plan of Chengde prefecture (Source: *Da Qing yitong zhi*)
Fig.2 Layout of the Mountain Resort (Source: *Qinding Rehe zhi*)
Fig.3 Chuifeng luozhao, the twelfth view of Shen Yu's *Yuzhi Bishu shanzhuang shi* (Imperial poems on the Mountain Resort) (Source: *Qingdian banhua huikan*)

莺啭乔木、石矶观鱼、甫田丛樾。[5]

康熙五十年（一七一一年），随着新的宫廷区和金山寺等园林工程初成，避暑山庄正式得名。同年，康熙选取山庄中三十六处景致，以四字命名，它们是：烟波致爽、芝径云堤、无暑清凉、延薰山馆、水芳岩秀、万壑松风、松鹤清樾、云山胜地、四面云山、北枕双峰、西岭晨霞、锤峰落照、南山积雪、梨花伴月、曲水荷香、风泉清听、濠濮间想、天宇咸畅、暖流暄波、泉源石壁、青枫绿屿、莺啭乔木、香远益清、金莲映日、远近泉声、云帆月舫、芳渚临流、云容水态、澄泉绕石、澄波叠翠、石矶观鱼、镜水云岑、双湖夹镜、长虹饮练、甫田丛樾、水流云在。次年，康熙皇帝诗、内阁学士沈嵛绘图、版刻名手朱圭、梅裕凤[6]雕版的《御制避暑山庄诗》刊印成书（图3）。

乾隆六年（一七四一年），避暑山庄又开始大规模的建设活动，乾隆十九年（一七五四年）乾隆皇帝以三字为名，增标了新的三十六景，他在《御制再题避暑山庄三十六景诗序》中指出：『我皇祖，

Qushui hexiang ("scent of lotuses by a winding stream"), Xiling chenxia ("morning mist by the Western Ridge"), Chuifeng luozhao ("sunset at Hammer Peak"), Fangzhu linliu ("fragrant island by flowing waters"), Nanshan jixue ("southern mountains piled with snow"), Jinlian yingri ("golden lotuses reflecting the sun"), Lihua banyue ("pear blossoms accompanied by moon"), Yingzhuan qiaomu ("orioles singing in tall trees"), Shiji guanyu ("observing the fish from a waterside rock"), and Futian congyue ("immense field with shady groves").[5]

In 1711, the fiftieth year of emperor Kangxi's reign, when the landscape around the palace and Jinshan Monastery gradually took shape, the emperor gave his summer residence the name Bishu shanzhuang (Mountain Resort). In the same year, he selected thirty-six scenic spots and gave them each a four-character name: Yanbo zhishuang ("misty ripples bringing brisk air"), Zhijing yundi (see above), Wushu qingliang ("un-summerly clear and cool"), Yanxun shanguan ("inviting the breeze lodge"), Shuifang yanxiu ("fragrant waters and beautiful cliffs"), Wanhe songfeng (see above), Songhe qingyue ("sonorous pines and cranes"), Yunshan shengdi ("scenes of clouds and mountains"), Simian yunshan ("clouds and peaks on all sides"), Beizhen shuangfeng ("nestled in the north between a pair of peaks"), Xiling chenxia (see above), Chuifeng luozhao (see above), Nanshan jixue (see above), Lihua banyue (see above), Qushui hexiang (see above), Fengquan qingting ("clear sounds of a spring in the breeze"), Hao Pu jianxiang ("untrammeled thoughts by the Hao and Pu rivers"), Tianyu xianchang ("exuberant sky"), Nuanliu xuanbo (see above), Quanyuan shibi ("fountainhead in a cliff"), Qingfeng lüyu ("verdant island of green maples"), Yingzhuan qiaomu (see above), Xiangyuan yiqing ("fragrance grows purer in the distance"), Jinlian yingri (see above), Yuanjin quansheng ("sounds of a spring near and far"), Yunfan yuefang ("moon boat with cloud sails"), Fangzhu linliu (see above), Yunrong shuitai ("cloud shapes and figures in the water"), Chengquan raoshi ("clear spring encircling rocks), Chengbo diecui (see above), Shiji guanyu (see above), Jingshui yuncen ("clouds and peaks in mirroring water"), Shuanghu jiajing (see above), Changhong yinlian (see above), Futian congyue (see above), Shuiliu yunzai ("clouds remain as water flows").

In the next year, *Yuzhi Bishu shanzhuang shi* (Imperial poems on the Mountain Resort) was published in a woodblock-print edition containing poems written by the Kangxi emperor and illustrations drawn by Shen Yu, Academician of the Grand Secretariat (or Imperial Household Department; Neibu) (fig. 3).[6] The woodblock cutters were Zhu Gui and Mei Yufeng.

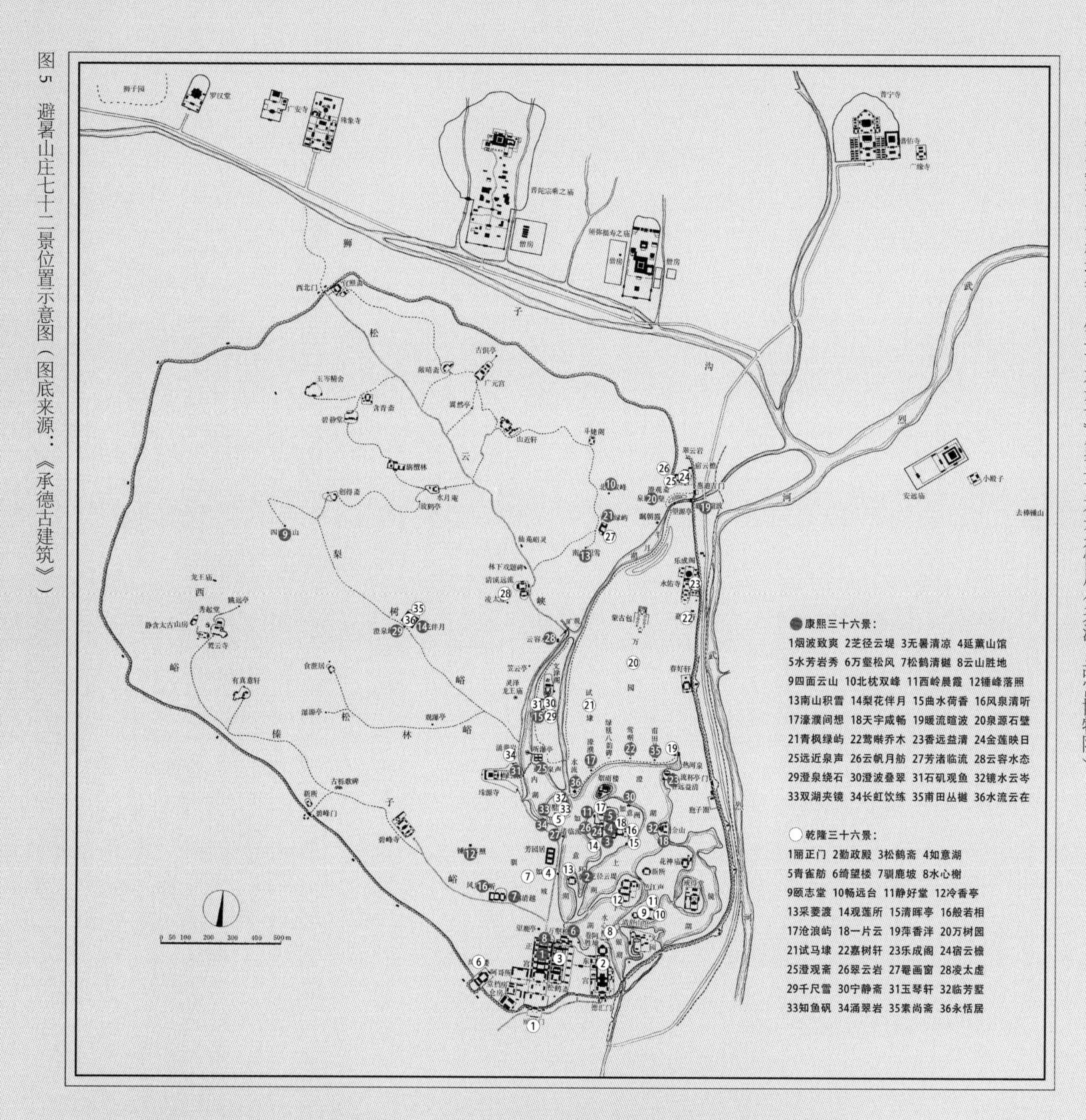

图5 避暑山庄七十二景位置示意图（图底来源：《承德古建筑》）

永恬居
山中境爽氣和致足怡
神悅性既佳風景且樂
阜安引養引恬慎修思
永
皇祖御題齋額意在斯乎
壺中壤島欝懸居景物惟
恬氣象舒已是洞天傳王
簡得教福地續琅書
臣勵宗萬敬錄恭繪

图4 励宗万《御制再题避暑山庄三十六景诗》第三十六景永恬居（来源：故宫博物院）

Fig.4 The thirty-sixth view of Li Zongwan's *Yuzhi zaiti Bishu shanzhuang sanshiliu jing shi* (Later imperial poems on the 36 views of the Mountain Resort) (Source: The Palace Museum)

Fig.5 Location plan of the seventy-two scenic areas in the Mountain Resort (Source: *Chengde gujianzhu*)

In 1741, the sixth year of his reign, emperor Qianlong ordered further large-scale construction at the Mountain Resort. In 1754, his nineteenth reign year, the emperor selected thirty-six additional scenic spots, giving each a three-character name. The last character of the three-character name is often used to designate the type or intended use of a scenic area. In the preface to *Yuzhi zaiti Bishu shanzhuang sanshiliu jing shi* (Later imperial poems on the thirty-six views of the Mountain Resort), emperor Qianlong explains:

"My grandfather, emperor Kangxi, built this imperial garden. He selected thirty-six scenic spots, bestowed a title on them, and ordered them to be painted…I knew that there were more than these thirty-six sights, and so I have chosen thirty-six additional scenic spots."

The thirty-six new scenic spots are: Lizheng Gate (*men*), Qinzheng Hall (*dian*), Songhe Study (*zhai*), Ruyi Lake (*hu*), Qingque Pleasure-boat (*fang*), Qiwang Building (*lou*), Xunlu Hillside (*po*), Shuixin Pavilion (*xie*), Yizhi Hall (*tang*), Changyuan Terrace (*tai*), Jinghao Hall (*tang*), Lengxiang Pavilion (*ting*), Cailing Wharf (*du*), Guanlian Landing (*suo*), Qinghui Pavilion (*ting*), Banruoxiang (*Temple*), Canglang Islet (*yu*), Yipianyun (*Hall*), Pingxiang Landing (*pan*), Wanshu Garden (*yuan*), Shima Racecourse (*dai*), Jiashu Pavilion (*xuan*), Lecheng Pavilion (*ge*), Suyunyan (*Retreat*), Chengguan Landing (*zhai*), Cuiyunyan (*Pavilion*), Yanhuachuang (*Terrace*), Lingtaixu (*Pavilion*), Qianchixue (*Waterfall*), Ningjing Study (*zhai*), Yuqin Terrace (*xuan*), Linfang Lodge (*shu*), Zhiyu Rock (*ji*), Yongcuiyan (*Temple*), Sushang Study (*zhai*), and Yongtian Retreat (*ju*). (Fig 4)

Twenty of Kangxi's thirty-six scenic spots are located in the lake zone (on Ruyi Island and on the lake shores). Qianlong's thirty-six new scenic spots are also concentrated in the lake zone but additionally they extend northward and southward (fig 5). There are in fact even more scenic spots in the Mountain Resort. The seventy-two officially designated sites represent only a small fraction of the scenic landscape. But at the same time, they represent the essence of the imperial garden and, in exemplary fashion, served as a mouthpiece for spreading government authority and the philosophy of the Qing emperors. Through systematic choice of name for each scenic spot and its visual depiction (painting) and textual description (poem), the beauty of the Mountain Resort, a "forbidden" garden-palace reserved only for the physical use by the emperor, became more transparent and accessible to the larger public (at least in theory).

圣祖仁皇帝，肇斯灵囿，标三十六景，题句、绘图垂示册府。朕……乃知三十六景之外，佳胜尚多。萃而录之，复得三十六景。』这三十六景为：丽正门、勤政殿、松鹤斋、如意湖、青雀舫、绮望楼、驯鹿坡、水心榭、颐志堂、畅远台、静好堂、冷香亭、采菱渡、观莲所、清晖亭、般若相、沧浪屿、一片云、萍香泮、万树园、试马埭、嘉树轩、乐成阁、宿云檐、澄观斋、翠云岩、罨画窗、凌太虚、千尺雪、宁静斋、玉琴轩、临芳墅、知鱼矶、涌翠岩、素尚斋、永恬居（图4）。

康熙三十六景中，有二十景分布在如意洲岛和湖区沿岸；乾隆三十六景也以湖区为中心，向南、北方均有拓展（图5）。但七十二景并未涵盖山庄的全部景色，在乾隆的阐释下，七十二景实为避暑山庄的浓缩。通过系列题名，配以图文对应的画册，禁苑美景得到昭示，成为表达清帝治国理念的特殊媒介。

二）后世影响

（甲）对清代皇家园林的影响

避暑山庄七十二景是『景』文化在皇家园林中的反映，中国传统文化中惯常以『景』来命名著名地点或风景名胜，如潇湘八景、西湖十景、燕京八景等，并利用绘画和文学作品对地方『景』文化进

2. Profound Influence on Later Generations

(1) Influence on Imperial Qing-dynasty Gardens

The seventy-two scenic areas in the Mountain Resort built on the traditional Chinese idea of *jing* (view) that was applied to imperial garden design (*jing* referring to a framed view of a beautiful landscape; and more broadly, a sight, scene, scenic spot or area). The word *jing* was usually used to designate beautiful or otherwise significant places and landscapes, for instance, the eight scenes of Xiaoxiang region (depicted in *Eight Views of Xiaoxiang*), the ten scenes around the West Lake in Hangzhou (depicted in *Ten Views of the West Lake*), and the eight scenes of Yanjing ([Beijing]; depicted in *Eight Views of Yanjing*). Painting and poem albums with the same name served to create and spread this sight-centric component of visual culture. Kangxi's thirty-six *jing* and Qianlong's thirty-six *jing*, collectively referred to as the seventy-two scenic areas of the Mountain Resort, were depicted and described in more than ten albums (fig. 6; table 1) that were printed or painted during the reigns of Kangxi and Qianlong.

Table 1: Paintings of the seventy-two scenic areas in the Mountain Resort ⑦

Year	Painter	Title of Painting Album
The 50th year under the reign of Emperor Kangxi (1711)	Shen Yu	*Yuzhi Bishu shanzhuang shi* (Imperial poems on the Mountain Resort) (in Chinese and Manchu language)
The 51st year under the reign of Emperor Kangxi (1712)	Dai Tianrui	*Yuzhi Bishu shanzhuang shi* (Imperial poems on the Mountain Resort)
The 52nd year under the reign of Emperor Kangxi (1713)	Matteo Ripa	*Yuzhi Bishu shanzhuang sanshiliu jing shi* (Imperial poems on the thirty-six views of the Mountain Resort)
The 51st – 54th year under the reign of Emperor Kangxi (1712-1715)	Wang Yuanqi	*Bishu shanzhuang sanshiliu jing* (Thirty-six views of the Mountain Resort)
The 4th year under the reign of Emperor Qianlong (1739)	Zhang Ruoai	*Yuzhi Bishu shanzhuang shi* (Imperial poems on the Mountain Resort)
The 17th year under the reign of Emperor Qianlong (1752)	Zhang Zongcang	*Bishu shanzhuang sanshiliu jing tu* (Imperially produced illustrations on the thirty-six views of the Mountain Resort)
The 17th year under the reign of Emperor Qianlong (1752)	Fang Cong	*Yuzhi Bishu shanzhuang sanshiliu jing shi* (Imperial poems on the thirty-six views of the Mountain Resort)
The 17th year under the reign of Emperor Qianlong (1752)	Li Zongwan	*Yuzhi Bishu shanzhuang shi* (Imperial poems on the Mountain Resort)

行创作和传播。康熙三十六景和乾隆三十六景统称『避暑山庄七十二景』，康乾两朝曾多次书写和绘制相关诗图不下十种（图6、表1）。

表一 避暑山庄七十二景相关绘画⑦

年代	绘制人	图册名称
康熙五十年（一七一一年）	沈嵛	御制避暑山庄诗（汉文、满文两种）
康熙五十一年（一七一二年）	戴天瑞	御制避暑山庄诗
康熙五十二年（一七一三年）	马国贤	御制避暑山庄三十六景诗
康熙五十一年至五十四年（一七一二至一七一五年）	王原祁	避暑山庄三十六景
乾隆四年（一七三九年）	张若霭	御制避暑山庄诗
乾隆十七年（一七五二年）	张宗苍	避暑山庄三十六景图
乾隆十七年（一七五二年）	方琮	御制避暑山庄三十六景诗
乾隆十七年（一七五二年）	励宗万	御制避暑山庄诗

现存版本均以沈嵛版本为圭臬。这些绘画采用了较高的视点，表现了类似于『平远』的效果；建筑物采用了写实的手法，但周围山水的比例根据画面效果而进行了夸张，强调了山环水抱的意象；图中空无一人，也没有任何人类活动的暗示。图面的整体氛围朴素而宁静⑧。

All later editions followed the first album by Shen Yu. Shown from a high viewpoint, building facades are depicted parallel to the picture plane (as if "looking at the front horizontally") and displayed without distortion (only the sides are skewed) at a realistic scale. The surrounding mountains, lakes, and rivers are enlarged to highlight the idea of architecture being embraced by the natural landscape. The atmosphere of painting is unadorned and tranquil, without any trace of human activity. ⑧

The poems and paintings of the Mountain Resort had great impact on Qing imperial garden design, both conceptually and practically. The way by which the sites were named and further poetized and visualized set an example for the creation of the forty scenic areas in Yuanmingyuan (literally, the Garden of Perfect Brightness but more commonly known as the Old Summer Palace), the sights at Jingji shanzhuang, ⑨ the twenty-eight sights in Jingyiyuan, and the sixteen sights in Jingmingyuan, all gardens and mountain resorts that were designed during the reigns of the emperors Yongzheng and Qianlong. The scenic spots were named either with four characters, such as in case of Yuanmingyuan, (the eight sights inside) Jingji shanzhuang, and Jingmingyuan; or with three characters, such as in case of the six new sights, sixteen additional sights, and eight sights outside Jingji shanzhuang as well as the twenty-eight sights of Jingyiyuan.

Painted albums corresponding to these famous sights emerged in large numbers. Many were modeled on the albums of the Mountain Resort, especially *Yuanmingyuan sishi jing* (Forty views of Yuanmingyuan) (fig. 7). The forty paintings, drawn in color on silk with great delicacy by court painters Shen Yuan and Tang Dai and then mounted on a paper folio, were published (in form of two albums) in 1744, the ninth reign year of emperor Qianlong. Each finished painting was placed in a square frame (of figured damask) with calligraphy on the facing half. The corresponding poems were chosen from *Sishi jing duiti shi* (Poems corresponding to the forty views) composed by emperor Qianlong but written by Wang Youdun, a calligrapher and high-rank official in the Ministry of Works. Buildings are realistically painted with fine brushwork; even details like the decorative polychrome painting of architectural members (*caihua*) are clearly distinguishable. The mountains and lakes surrounding the architecture look natural, but in fact, the hills were piled up artificially and the lakes dug out from the flat ground as Yuanmingyuan was an artificial garden.

Other works also adopted the artistic format of the painted album but to varying degrees. The painted albums that developed out of this can be grouped into three types. The first type modifies the arrangement of image and corresponding text and places painting and

这些诗图给清代皇家园林的建设带来了灵感，而其提炼景点、赋诗咏志并制作图咏的方式，也成为在雍正和乾隆朝的皇家园林建设中涌现出来的『圆明园四十景』『静寄山庄诸景』⑨、『静宜园二十八景』和『静明园十六景』等诗图结集作品的范例。这些景点或以四字为名，如『圆明园四十景』『静寄山庄内八景』和『静明园十六景』；或以三字为名，如『静寄山庄外八景』『新六景』和『附列十六景』，以及『静宜园二十八景』。

大量相关绘画作品层出不穷，其中不乏以避暑山庄图册为蓝本的作品，例如乾隆九年（一七四四年）绢本彩绘的《圆明园四十景》（图7）：作品共四十幅，每幅右侧由沈源、唐岱等宫廷画师绘制一景，画面几乎是正方形，工笔彩绘的建筑十分写实，甚至彩画等细节都清晰可辨，建筑仿佛环绕于真山真水之间，但圆明园实为平地叠山理水的人工园林；每幅左侧为乾隆皇帝的《四十景对题诗》，由工部尚书汪由敦书写。亦有更多作品在图册的基础上发展了各种类型的表达方式：董邦达的《田盘胜概图》仍然是图册形式，但没有沿用图文分列两面的方法，而是图文合一，在画面上方是乾隆皇帝御笔的『静寄山庄内外八景诗』；方琮的《静明园十六景图屏》共八面，每面上下排列两景，上幅为长方形抹圆角、下幅为长方形；张若澄的《静宜园二十八景（手卷）》则利用长卷将二十八景巧妙融入整体地势，突显皇家园林磅礴大气、自然天成的韵味。

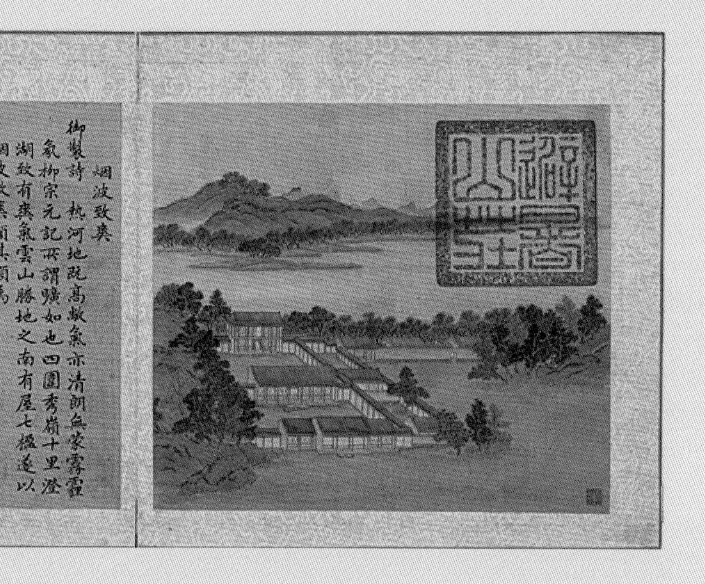
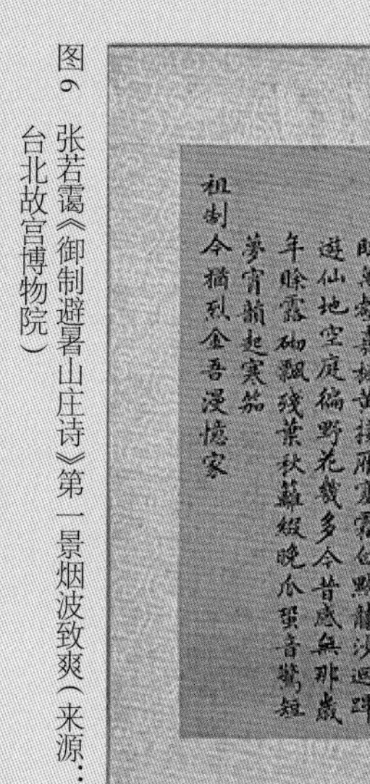

图6 张若霭《御制避暑山庄诗》第一景烟波致爽（来源：台北故宫博物院）

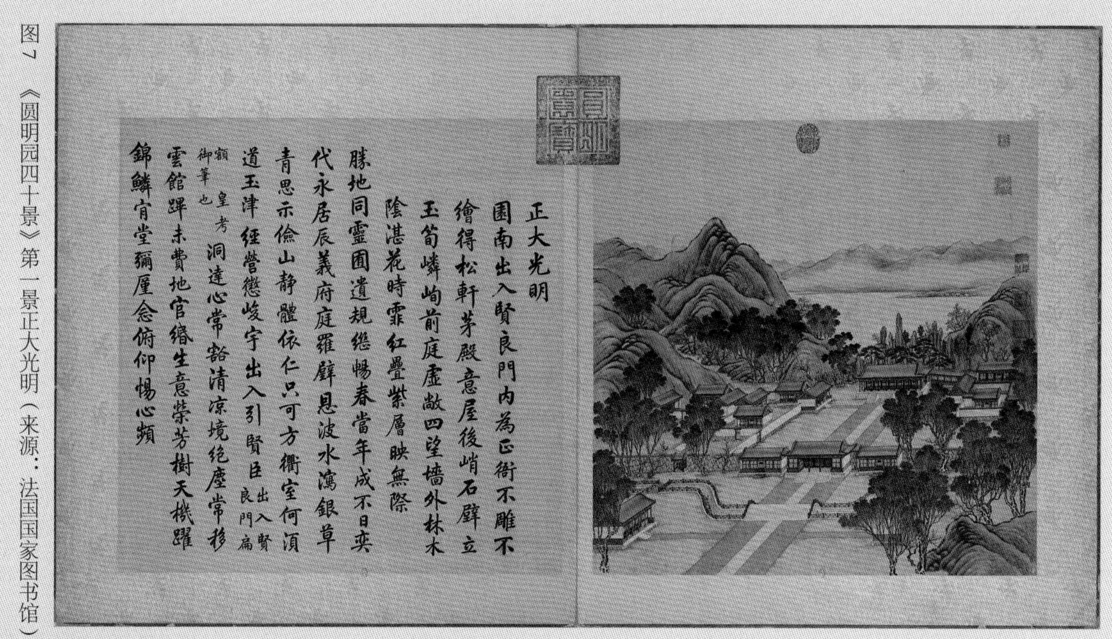

图7 《圆明园四十景》第一景正大光明（来源：法国国家图书馆）

图8 《御制避暑山庄三十六景诗》（来源：台北故宫博物院）

图9 万壑松风建筑群

Fig.6 Yanbo zhishuang, the first view of *Yuzhi Bishu shanzhuang shi* (Source: National Palace Museum in Taipei)

Fig.7 Zhengda guangming, the first view of *Yuanmingyuan sishi jing* (Forty views of Yuanmingyuan) (Source: National Library of France)

Fig.8 *Yuzhi Bishu shanzhuang sanshiliu jing shi* (Imperial poems on the thirty-six views of the Mountain Resort) (Source: National Palace Museum in Taipei)

Fig.9 Buildings at Wanhe songfeng

poem at the same page, such as *Tianpan sheng gaitu* (General drawings of famous sights in Tianpan) by the Qing calligrapher and painter Dong Bangda; or the illustrations of the eight inner and outer sights of Jingji shanzhuang, where emperor Qianlong's poems are shown on the top left or top right of a folio. The second type has two paintings arranged on the same folio—one above the other—such as in case of *Jingmingyuan shiliu jing tuping* (Paintings of the sixteen views of Jingming Garden) by the court painter Fang Cong, an eight-page album where each top painting is put in a rectangular frame with with rounded edges and each bottom painting in a rectangular frame. A good example of the third type is *Jingyiyuan ershiba jing* (Twenty-eight views of Jingyi Garden) by the official Zhang Ruocheng, where all views are laid out on a long hand scroll, with each of the twenty-eight scenic area being skillfully integrated in the surrounding environment to emphasize the garden's majestic atmosphere and accordance with nature.

(2) Influence on European Gardens

In 1713, the fifty-second reign year of emperor Kangxi, Matteo Ripa (Chinese name Guo Maxian, 1682-1746), a Jesuit missionary and artist from Naples, Italy, made copperplate engravings of *Yuzhi Bishu shanzhuang sanshiliu jing shi* (Imperial poems on the thirty-six views of the Mountain Resort) (fig. 8), modeled on Shen Yu's woodblock prints. Under the Kangxi and Qianlong emperors, many Jesuit priests worked at the Qing court and contributed to imperial garden design thanks to their knowledge of Western science and technology. For example, in the Office of Mathematics established by emperor Kangxi at *Mengyangzhai* (Study for the Cultivation of the Youth) in Changchunyuan (one of the three garden sections of the Old Summer Palace [Yuanmingyuan]), they translated Western works on mathematics. Many "efficient Westerners" worked in *Ruyiguan*, also located at Yuanmingyuan. *Wanhe songfeng* in the Mountain Resort at Chengde was also a place where Western priests worked during Kangxi's reign. Today, only the studio in Chengde has survived. It is the only building in Qing-dynasty imperial gardens that still bears witness to this cultural exchange between China and the West.

Although *Wanhe songfeng* (fig. 9) was one of the thirty-six scenic areas in the Mountain Resort designated by emperor Kangxi, it was less an imperial residence than a meeting place for cultural and artistic activities of his entourage. Many renowned Jesuit priests including Matteo Ripa, Joseph-Marie Amiot (Chinese name Qian Deming), Jean-Denis Attiret (Chinese name Wang Zhicheng), Giuseppe Castiglione (Chinese name Lang Shining), and Pierre-Martial Cibot (Chinese name Han Guoying) once worked here. The advanced knowledge in

（乙）对欧洲园林的影响

康熙五十二年（一七一三年），来自意大利那不勒斯的传信部传教士马国贤（Matteo Ripa，一六八二至一七四六）以沈嵛的木版画为蓝本，主持印制了铜版《御制避暑山庄三十六景诗》（图8）。康乾时期，大量精通科学技术的传教士为清廷服务，皇家园林里也留下了他们工作的身影。他们在康熙于畅春园『蒙养斋』设立的算学馆内翻译西方历算著作；圆明园的『如意馆』是大量『效力西洋人』的工作地点；承德避暑山庄的『万壑松风』亦是康熙时期传教士们工作的场所。如今『蒙养斋』和『如意馆』建筑均不复存在，『万壑松风』成为皇家园林中唯一现存的中西文化交流见证地。『万壑松风』建筑群是康熙避暑山庄三十六景之一（图9），在山庄的早期历史中除了宫殿功能外，更是重要的文化、艺术场所。以马国贤、钱德明（Joseph–Marie Amiot）、王志诚（Jean–Denis Attiret）、郎世宁（Giuseppe Castiglione）、韩国英（Pierre–Martial Cibot）等为代表的传教士曾在『万壑松风』中工作，他们将西方先进的制图、绘画手段运用于其作品中，对中国艺术、科技产生了多方面的深远影响。马国贤和郎世宁等传教士遵照康雍乾三代皇帝的旨意，培养了一批能够运用西方透视学、铜版画技艺进行美术创作的中国宫廷画家，郎世宁的学生，满族画家伊兰泰就是其中的代表性人物，他受命创作了著名的圆明园西洋楼铜版画。

cartography and painting that they brought with them had a far-reaching impact on Chinese art and science. At the request of the three emperors Kangxi, Yongzheng, and Qianlong, Ripa and Castiglione trained a group of Chinese court painters in Western perspective drawing and copperplate engraving. Following this tradition, the Manchu painter Yi Lantai, one of the leading figures in the Imperial Academy, produced his famous copperplate engravings of the European buildings (Xiyanglou) at Yuanmingyuan.

Ripa's *Yuzhi Bishu shanzhuang sanshiliu jing shitu* (Imperial poems and illustrations on the thirty-six views of the Mountain Resort) is the first work produced in China that applied the European techniques of object shading in accordance with the rules of perspective. Ripa adopted a Western style for the detailed treatment of mountains, rivers, rocks, plants, and the sky, but still followed the basic Chinese composition laid out by Shen Yu (a year earlier). In late 1723 Ripa left China. After his arrival in England in 1724, he sold replicas of his engravings that were soon copied by local European artists. In 1753, a group of British artists including Robert Sayer published a painting album with twenty leaves (*The Emperor of China's Palace at Pekin, and His Principal Gardens*) based primarily on Ripa's engravings that were modified through additional figures, animals, ships, distant buildings, and even mythical creatures in the porcelain-painting style of the time (fig. 10). British artists also added a new front page ("The emperor of China's palace at Pekin"; actually, the Meridian Gate) and last page ("The Mughal emperor's throne").

Richard Boyle, Third Earl of Burlington (1694-1753) and amateur architect who gathered a group of garden enthusiasts with similar interests, had bought a set of Ripa's copperplate engraving (fig. 11) in 1724. [10] They had profound influence on English garden design (through the discovery of Chinese gardens or what was thought to be a Chinese garden). A telling example is Boyle's estate Chiswick House, a Palladian-style villa with the first English landscape garden designed by William Kent (1685–1748). Kent was a forerunner of the natural-style landscape garden that peaked in England in the second half of the eighteenth century. We can discern elements of Ripa's engravings in his designs that transform the flat two-dimensional shapes of the copperplates into three-dimensional form. The English landscape garden revolutionized Western garden design and challenged the traditional formal (geometric) French garden popular in Europe at that time.

In the mid-nineteenth century, the American landscape architect Andrew Jackson Downing (1815–1852) put forth new ideas on public urban parks based on his study of English landscapes and gardens. Downing influenced Frederick Law Olmsted (1822–1903), the

马国贤的《御制避暑山庄三十六景诗图》是中国第一次出现有阴影的绘画作品，山、水、岩石、植物和天空的画法也采用了西方铜版画的处理方式，但各个元素是严格按照沈嵛版本制作而成。一七二三年底，马国贤离开中国，并于一七二四年途经英国，他出售了几份《御制避暑山庄三十六景诗图》的复制品。马国贤的版本还经历过再创作：一七五三年英国人Robert Sayer等人选出了十八幅马国贤铜版画，仿效流行的外销瓷画风，在其上添加了人物、动植物、船只、远景建筑甚至神话生物等（图10），使其充满了浓郁的东方神秘色彩，并在首页增加了一幅『北京的皇宫』（应为午门），结尾增加一幅『莫卧尔皇帝的宝座』，共二十幅加以出版，该画册名为《中国皇帝的北京宫殿和主要园林》

英国伯灵顿伯爵三世理查德·博伊尔（Richard Boyle, 3rd Earl of Burlington，一六九四至一七五三）身边聚集了一批园林新风格的狂热爱好者。他于一七二四年从马国贤手中购买了避暑山庄铜版画（图11），对英国园林设计产生了巨大的影响[10]。这其中包括主要由威廉·肯特（William Kent，一六八五至一七四八）设计的伯爵本人府邸奇斯威克大宅（Chiswick House）的园林，它是英国最早的自然风景园作品。肯特是英国十八世纪后半期自然风景式园林进入全盛期的先导者，从他的作品中不难看出马国贤铜版画的影响，通过对绘画的观察，肯特将这些场景移植到其园林创作中来。英国自然风景园对西方园林产生了革命性的作用，冲击了传统的几何形园林。

十九世纪中叶，美国景观建筑师安德鲁·杰克逊·唐宁（Andrew Jackson Downing，一八一五至一八五二）受到英国风景园林设计影响，对城市公共园林设计进行了深入的思考，并将自己的设计理念贯穿于纽约中央公园的设计竞赛中。他对后来被誉为美国景观建筑之父的弗雷德里克·劳·奥姆斯

Father of American Landscape Architecture, with whom he had developed a concept that would later win the competition for the design of the new Central Park in New York City. Olmsted created many city parks in the United States, which are not only significant to the history of landscape design, but also to the formation of Western urban lifestyle.

(II) Scenic Areas besides the Designated seventy-two Scenic Areas

1. Miniature Lakes and Mountains

The seventy-two scenic areas are copies (miniatures) on a reduced scale of famous gardens built all over the empire during the reigns of the Kangxi and Qianlong emperors, and they are the epitome of architectural achievement at the Mountain Resort. But there are in fact more beautiful scenic vistas in the Mountain Resort. The area is usually divided into palace, lake, flatland, and mountain districts according to landform and function (fig. 12).[11]

Located at the south of the Mountain Resort and adjacent to the city, the palace district served as temporary residence for the Qing emperors and comprised three major building groups—the Principal Palace, Songhezhai, and the Eastern Palace.

To the north of the palace district is the lake district that comprises eight lakes, each centered around one of eight differently-sized islands, and together with the long lakeside (scenic area of Zhijing yundi) covers an area of about 43 ha.[12] Sixteen major building groups are located here.[13]

The flatland district to the north of the lake district extends to the foot of the mountains in the northwest, and contains two scenic areas on a natural flat terrain—Wanshuyuan and Shimadai. The other scenic areas of the flatland district are arranged around these two broad fields—in the eastern section, Yongyousi, Jiashuxuan, and Chunhaoxuan; to south adjacent to the lake, Shuiliu yunzai, Haopu jianxiang, Yingzhuan qiaomu and Futian congyue; in the west, Wenjinge, Qushui hexiang, Tingpu (Pavilion) and Yuanjin quansheng; in the north around Huidi jimen are Chengguanzhai and Nuanliu xuanbo.

The mountain district located in the northwest of the Mountain Resort covers four fifths of the resort's total area, comprising several hills and small mountains ranging from 20 m to 100 m in height with more than forty building clusters. Taking full advantage of the surrounding topography, the architecture is spread across four valleys—Songyunxia, Lishuyu, Songlinyu, and Zhenziyu.

特德（一八二二至一九〇三）产生了深远的影响。奥姆斯特德设计了全美众多的公共园林，这些园林不仅是园林史上重要的作品，也对西方城市生活方式产生了巨大的影响。

（二）七十二景之外

一）湖山缩影

『七十二景』是避暑山庄的缩影，也是康乾两代帝王园林经营的写照。但是在七十二景之外亦有大量景观存在。人们惯常根据地形和功能将避暑山庄分为宫殿区、湖区、平原区和山区（图12）。[11]

宫殿区是清帝行宫所在，位于山庄南部，毗邻市区，包括正宫、松鹤斋和东宫三个主要组群；湖区位于宫殿区北部，包括洲岛在内约为四十三公顷，共有大小岛屿八座，这些岛屿与长堤把湖面分成八个部分[12]，主要建筑组群约16处[13]。

平原区为湖区再北，直至西北山脚下，有万树园和试马埭两处自然平地，主要建筑则围绕这两块阔野而建——东部有永佑寺、嘉树轩和春好轩，南部临湖有水流云在、濠濮间想、莺啭乔木和莆田丛樾四亭，西部有文津阁、曲水荷香、听瀑和远近泉声等，北部惠迪吉门周围有澄观斋和暖流暄波等组群。

山区面积占避暑山庄总面积的五分之四，位于园内西北部，区内各山总体高度二十至一百米，建筑则围绕松云峡、梨树峪、松林峪和榛子峪四条沟峪依山就势而建，共有组群四十余处。

避暑山庄经过几代清帝的经营，充分实现了乾隆所言『非为一己之豫游，盖贻万世之缔构也』的宏伟规划，其西高东低，包括了湖、平原和山区的布局似乎是中华版图的写照。今所知最早反映避暑山庄

图13 冷枚《避暑山庄图》（来源：故宫博物院）

图14 郎世宁等《万树园赐宴图》（来源：故宫博物院）

图15 郎世宁等《马术图》（来源：故宫博物院）

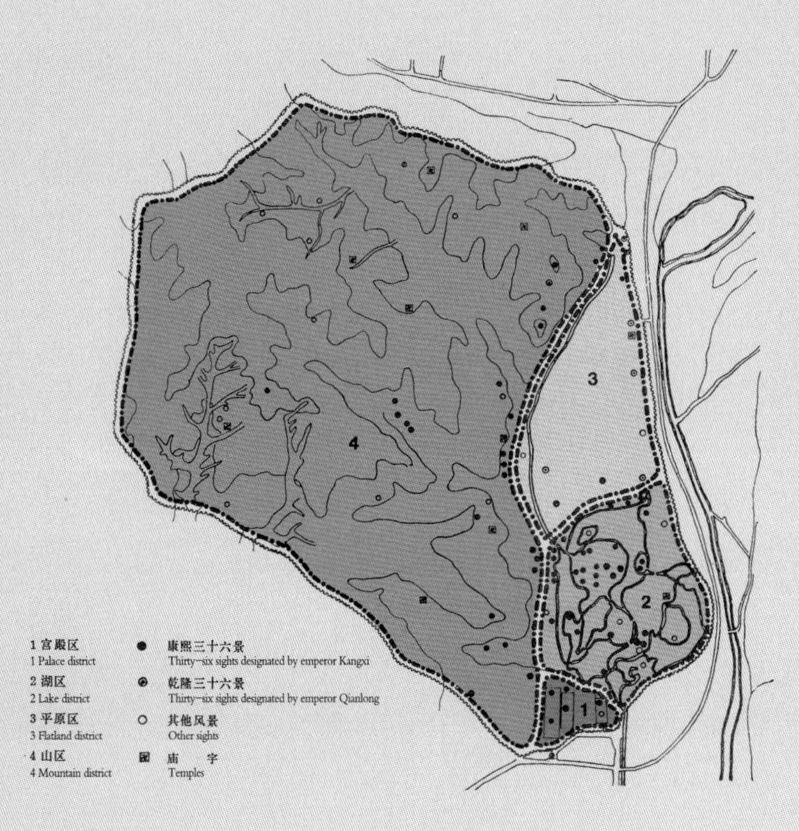

图12 避暑山庄功能分区示意图（图底来源：《承德古建筑》）

图10 《中国皇帝的北京宫殿和主要园林》中的第四幅（来源：大英图书馆）

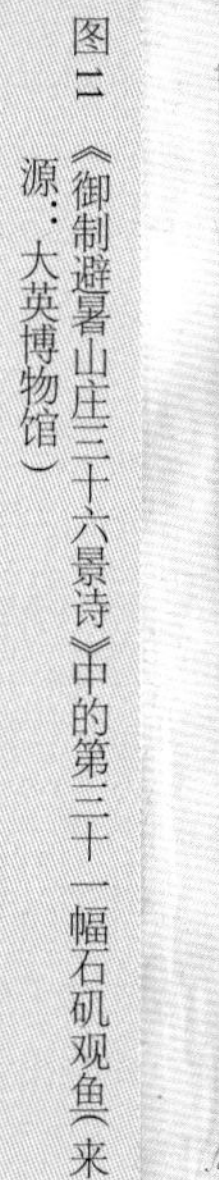

图11 《御制避暑山庄三十六景诗》中的第三十一幅石矶观鱼（来源：大英博物馆）

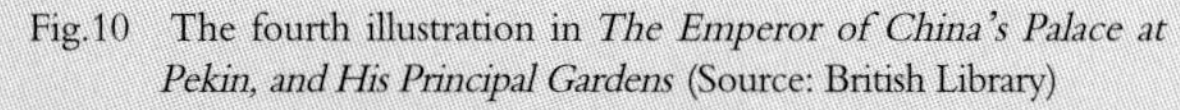

Fig.10 The fourth illustration in *The Emperor of China's Palace at Pekin, and His Principal Gardens* (Source: British Library)
Fig.11 Shiji guanyu, the thirty-first painting of *Yuzhi Bishu shanzhuang sanshiliu jing shi* (Source: British Museum)
Fig.12 Map of different districts in the Mountain Resort (Source: *Chengde Gu Jianzhu* (*The Traditional Chengde Architecture*))
Fig.13 *Bishu shanzhuang tu* (Illustration of the Mountain Resort) by Leng Mei (Source: The Palace Museum)
Fig.14 *Wanshuyuan ciyan tu* (Imperial banquet in the Garden of Ten-thousand Trees) by Giuseppe Castiglione et al. (Source: The Palace Museum)
Fig.15 *Mashu tu* (Horsemanship) by Giuseppe Castiglione et al. (Source: The Palace Museum)

After construction and repair by successive Qing emperors, the Mountain Resort fully realized a goal as expressed by the Qianlong emperor when he described the complex as "created not for my recreation alone, but as a legacy for future generations." Owing to its west-high-east-low layout—with the lake, flatland, and mountain districts reproducing territorial features of the empire— the Mountain Resort at Chengde became a powerful mirror image of the Chinese empire at that time. The earliest depiction of the Mountain Resort is a hanging scroll entitled *Bishu shanzhuang tu* (Painting of the Mountain Resort) (fig. 13) by the court painter Leng Mei. Although the reason for its commission and time of completion are still under debate,[14] it is generally agreed that the painting reflects the original layout of the Mountain Resort before intervention and expansion under emperor Qianlong. In addition to Kangxi's designated thirty-six scenic areas, the painting also depicts various other views next to the lake district, for example Yuese jiangsheng, Huanbi, and Kuangguan. However, many scenic areas built by emperor Kangxi outside the lake district, especially the building clusters in the mountain and palace districts, are not shown, such as Lihua banyue, Danbo jingcheng, and Yanbo zhishuang. Furthermore, neither are Purensi nor Pushansi included, two of the Eight Outer Temples also built under emperor Kangxi. What is shown in the foreground is the scenic area of Wanhe songfeng that connects the palace district with the lake district, while at the center is shown the lake district in its pre-Qianlong condition. The lake district is portrayed as a centerpiece surrounded by jagged peaks and mountains. To its west, is the mountain district, to the east, Wulie River and Hammer Peak Mountains, and to the north, the flatland district with the distant mountains behind Shizi Furrow. That is to say, Leng Mei emphasized the landscape and architecture surrounding the lake district, with most buildings facing the lake, and created a traditional-style landscape (painting) through use of traditional painting techniques characterized by the preference for blue and green colors.

2. Multi-cultural Summer Capital

Although his hanging scroll shows the overall scenery of the Mountain Resort, Leng Mei deliberately seems to have ignored the multi-cultural features that were present, for example, at Wanshuyuan and Shimadai in the flatland district. In his painting, the flatland district is just a transitional zone between the lake district and the mountain district, densely wooded with herds of wandering deers and flocks of flying birds. However, two other paintings drawn by Castiglione during emperor Qianlong's reign offer a better look at the flatland district and give insights into its function.

Covering an area of 1,000+ mu (a unit of area used in imperial China), Wanshuyuan was an

全貌的图像，是宫廷画家冷枚的《避暑山庄图》（图13）。该画的具体年代和绘制原因虽有争议[14]，但基本反映了乾隆皇帝扩建避暑山庄之前的景象。画中除了康熙三十六景外，亦有月色江声、环碧、旷观等湖区附近景点，但湖区以外的很多景点，尤其是康熙年间建设的山区和宫廷区建筑并未画出，如梨花伴月、澹泊敬诚和烟波致爽等，外八庙中建于康熙年间的溥仁寺和溥善寺也未加描绘。画面最前景的是连接宫殿区和湖区的万壑松风组群，而乾隆改造前的湖区则成为画面的中心。湖区周围重峦叠嶂，西面的山区、东面的武烈河与磬锤峰山脉、北面的平原区和狮子沟后的远山，以夸张的形态环绕湖区。冷枚《避暑山庄图》强调了围绕湖区的山水和建筑环境，图中的建筑均临湖而建，结合青绿山水的表现方法，营造出强烈的传统园囿风貌。

（二）多元夏都

冷枚的《避暑山庄图》反映了避暑山庄整体的风貌，但似乎刻意忽略了避暑山庄中的一些多元文化要素，例如北部的平原区的万树园和试马埭。冷枚图上的平原区树木繁密，林间有鹿群闲走，枝头有鸟儿盘旋，是宏大皇家园林中湖区和山区之间的过渡地带。但乾隆年间郎世宁等绘制的两幅宫廷绘画，则近距离表现了平原区在山庄中的特殊功能——万树园占地一千余亩，草地上遍植高大乔木，《万树园赐宴图》（图14）描绘了乾隆十九年（一七五四年）五月，皇帝在此搭建蒙古包和帐篷，宴请投奔清廷的厄鲁特蒙古杜尔伯特部首领的政治事件；后者位于万树园西南，地平草茂，是清帝为了木兰秋狝，挑选蒙古进献的骏马之地，《马术图》（图15）就是表现乾隆十九年十一月，乾隆皇帝与前来归附的准噶尔部阿睦尔撒纳在试马埭附近观看马术的场景。郎世宁的两幅绘画，表现了避暑山庄的另

expansive meadow with tall trees at the edges. Lacking buildings in the conventional sense, there were only a few Mongolian yurts that would be set up temporarily for meetings and banquets. *Wanshuyuan ciyan tu* (Imperial banquet in the Garden of Ten-thousand Trees) (fig. 14) shows a banquet given by the Qianlong emperor for princes of the Dorbets (also written Dörbets), one of the Oirat tribes, in the fifth month of 1754, the nineteenth reign year of emperor Qianlong, to honor the alliance between the tribe and the Qing court. The scene takes place in front of a majestic Mongolian yurt with smaller tents on each side. By contrast, Shimadai, located southwest of Wanshuyuan, was a large flat grassland area where the Qing emperors would appraise fine horses offered by Mongolian tribes for the autumn hunt. Castiglione's second painting—*Mashu tu* (Horsemanship) (fig. 15)—commemorates another banquet (series) given by emperor Qianlong in the eleventh month of the same year (1754), this time for the Dzungar leader Amursana, who had just surrendered to the court. The scene takes place in the scenic area of Shimadai, where they would watch equestrian displays.

Castiglione' two paintings reveal another side of the Mountain Resort. Besides being a summer retreat to escape the heat offered by its favorable geographical location, the Mountain Resort "became an important place for emperors to receive religious leaders and ethnic minorities' leaders, and to conduct military campaigns and control the frontier regions." [15] Out of the many activities that took place at the Mountain Resort, receiving Mongolian princes and dukes was one of the most significant and frequent tasks. Chengde was conveniently located between the Chinese heartland and the regions still under Mongolian rule, and in addition, close to the royal hunting grounds at Mulan. From here, the Qing emperors were able to make inspection tours northwards and to meet Mongolian nobility.

Located (about 70 kilometers) north of the Great Wall, Chengde has a cold temperature climate. The annual hunt at Mulan allowed the emperor to escape the heat of the capital located further south in Beijing and to avoid the risk of contracting smallpox. At that time, Mongolian nobles not suffering from smallpox were in fact afraid to visit the capital and pay tribute to the emperor during the New Year Festival (a ritual known as *nianban*).[16] Instead, after the establishment of the annual autumn hunt in the twentieth year of emperor Kangxi' reign, Mongolian nobles not suffering from smallpox accompanied the emperor to the imperial hunt to show their respect (a ritual known as *weiban*).[17] That is to say, the climatic conditions, geographic location, and topography of Chengde satisfied the demands not only of "escaping the heat" but also of "escaping smallpox". Thus, Chengde became the summer capital of the Qing emperors, and played a significant role as a political center of power from emperor Kangxi's reign until the Second Opium War in 1860.

外一面，即由于承德所处的特殊地理位置，在消夏避暑之外，『成为皇帝接待宗教领袖、接待少数民族首领、指挥边疆战争等边疆治理活动的重要场所』[十五]。

在避暑山庄举行的这些特殊活动中，接待蒙古各部王公是最主要、最频繁的活动之一：承德的地理位置处于内地和蒙古控制区域之间，又紧邻木兰围场，方便皇帝北巡塞外会见蒙古王公。而且承德地处关外，气候寒冷，待到木兰秋狝之时，恰好避开天花的高发期——当时的条件下，未出痘的蒙古贵族惧怕进京『年班』[十六]；康熙二十年（一六八一年）木兰行围制度确立后，未出痘的蒙古贵族每年跟随皇帝到围场狩猎，称为『围班』[十七]。可以说，承德的气候条件、地理位置和地形地貌均满足了『避暑』和『避痘』的需求，成为清代重要的夏都，其作用从康熙皇帝开始一直持续到第二次鸦片战争。

除了作为夏都，担负与蒙古各部定期会见的重任，避暑山庄也接待过众多宗教领袖和各国使节。乾隆四十五年（一七八〇年），六世班禅从后藏日喀则来到承德觐见乾隆皇帝，这也是第一位到达承德的西藏最高宗教领袖。安南战争之后的乾隆五十四年（一七八九年）七月，越南国王阮惠遣其侄阮光显奉表到达避暑山庄，乾隆皇帝在山庄东宫的清音阁大戏楼接见阮光显一行，并命他随同文武大臣和蒙古王公入座赏戏（图16）。英国的马戛尔尼使团在乾隆五十八年（一七九三年）出使大清期间来到承德，随团画家威廉·亚历山大也根据众人的回忆，绘制了多幅反映避暑山庄景象的作品，其中就有乾隆皇帝在万树园帐篷前会见使团成员的画面（图17）。

三）写仿江南

避暑山庄在边防和外交方面的重要地位毋庸置疑，作为其构园主体的园林景观和建筑多以『写仿江南』为手段。例如，康熙三十六景中的『天宇咸畅』和『镜水云岑』写仿了镇江金山寺，山庄烟雨楼仿自嘉兴烟雨楼，文园狮子林原型为苏州狮子林，以及仿照宁波天一阁建造的七座《四库全书》藏书楼之一——避暑山庄文津阁（图18）。

清乾隆三十九年（一七七四年）到乾隆四十八年（一七八三年），北至盛京，南至杭州，七座模仿浙江范氏天一阁的皇家藏书楼逐一建成，又及乾隆五十五年（一七九〇年）《四库全书》在七阁中全部贮藏完毕。至此乾隆皇帝通过一部《四库全书》及以天一阁为原型的皇家七阁藏书楼，将自己的文化伟业推向高峰。这七座藏书楼中的四座建于北方，是文渊阁、文源阁、文津阁和文溯阁，其中除文源阁与圆明园同毁于兵燹外，其他三座基本保持完好。另外在江浙地区还依托南巡行宫建设了三座

In addition to being the imperial summer retreat and seasonal political center where regular meetings with Mongolian tribe leaders took place, the Mountain Resort also served as a meeting place with religious leaders and diplomatic envoys from foreign countries. In 1780, the forty-fifth reign year of emperor Qianlong, the Sixth Panchen Lama came from Shigatse, the traditional residence of the heads of the Gelug School, to Chengde to present himself to emperor Qianlong as the first of highest-ranking religious leaders of Tibetan Buddhism who came here from Tibet. Almost a decade later, in July 1789, the fifty-fourth reign year of emperor Qianlong, after the war between Qing China and Annam (today Vietnam) was over, the Vietnamese king Nguyen Hue sent his nephew with an official letter to the Mountain Resort, agreeing to pay annual tribute to the Qing court (despite his initial military success). Emperor Qianlong received the Vietnamese delegation at Qingyinge, an entertainment area with a three-story theater in the Eastern Palace at Chengde, and ordered the kings' nephew to sit and watch plays with military and civilian Qing officials and Mongolian nobles (fig. 16). In 1793, the fifty-eighth year of emperor Qianlong's reign, a British delegation led by George Macartney arrived in Chengde on the occasion of the emperor's birthday. The artist William Alexander, who accompanied the Macartney Embassy to China, drew several watercolors of the Mountain Resort (after his colleagues' memories since he was not present at the meeting with the Qianlong emperor). One painting shows a scene where emperor Qianlong welcomes the delegation in front of a large Mongolian yurt at Wanshuyuan (fig. 17).

3. In Imitation of Jiangnan Architecture

The Mountain Resort played a key role in frontier defense and diplomatic affairs between China and the outside world, but its scenic areas were modeled after southern Chinese architecture and landscape from the Jiangnan region. For example, Tianyu xianchang and Jingshui yuncen, two of the thirty-six scenic areas designated by emperor Kangxi, were modeled on Jinshan Monastery in Zhenjiang, Jiangsu province; Yanyulou emulated a building with the same name built in Jiaxing, Zhejiang province; Wenyuan shizilin imitated the Lion Grove Garden (Shizilin) in Suzhou; and Wenjinge, one of the seven libraries to safeguard a set of the *Comprehensive Library of the Four Treasuries* (Siku quanshu) took its inspiration from Tianyige in Ningbo (fig. 18).

From the thirty-ninth year (1774) to the forty-eighth year (1783), emperor Qianlong commissioned seven libraries to be built that were all modeled on Tianyige (that housed the private collection of the Ming-dynasty high-ranking military official Fan Qin), The northernmost of these libraries was built in Shenyang and the southernmost in Hangzhou.

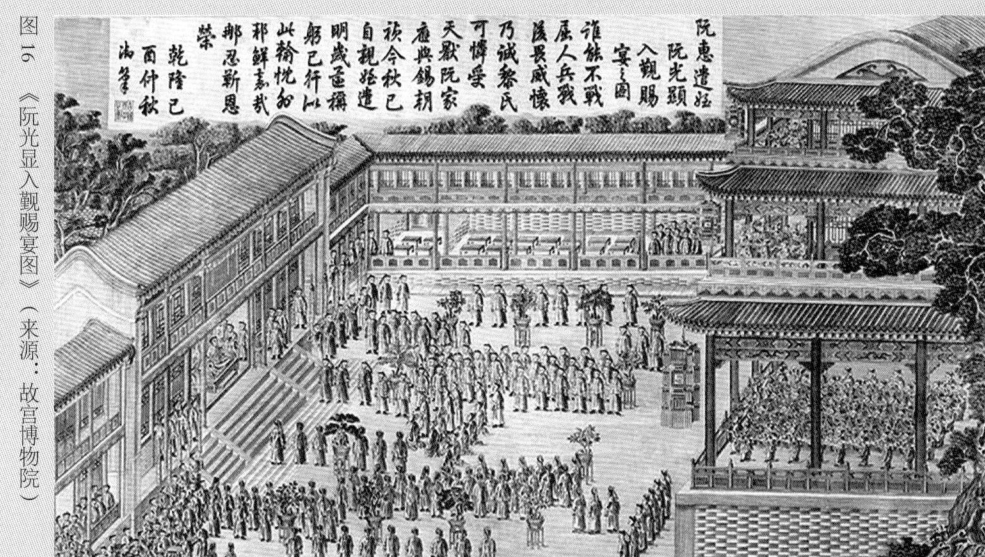

图16 《阮光显入觐赐宴图》（来源：故宫博物院）

图17 威廉·亚历山大《乾隆皇帝在万树园接见英国使臣》（来源：大英博物馆）

Fig.16 *Ruan Guangxian rujin ciyan tu* (Imperial banquet for Nguyen Quang-Hien) (Source: The Palace Museum)

Fig.17 *Emperor Qianlong Meeting British Envoys at Wanshu Garden* by William Alexander (Source: British Museum)

Fig.18 Tianyi Library in Ningbo and Wenjin Library in the Mountain Resort

Fig.19 Aerial Photo of architecture at the scenic area of Wenjinge (Source: Li Zhe, 2012)

图19 文津阁组群航拍（李哲 2012年）

图18 浙江宁波天一阁和承德避暑山庄文津阁

藏书楼——文宗阁、文汇阁和文澜阁，不过均毁于太平天国战火，只有文澜阁在光绪年间得以重建。

避暑山庄的文津阁是四库七阁中最先落成者，也是避暑山庄内最后建成的景点之一。文津阁位于湖区和山区的边界，在千尺雪景区以北，始建于乾隆三十九年（一七七四年）。文津阁周围有曲墙环绕，坐北朝南，三面临水呈半岛状，阁的东北部有水门与山庄水系相通。从南往北，文津阁组群依次为门殿、假山、水池、文津阁和碑亭、花台、假山。文津阁是现存四阁中造园规模最大、气势最佳者。乾隆在《趣亭》中称赞文津阁的造园艺术：『阁外假山堆碧螺，山亭名趣意如何？泉声树影则权置，静对诗书趣更多。』[十八]这种园林氛围的塑造，离不开其选址：文津阁坐北朝南，被山石、水池和树木围绕，最外围还有围墙，围墙外两条水系环绕，使得文津阁组群呈现出『两水夹一岛』的环境意象，也使其相对独立于南部的千尺雪组群自成一体（图19）。

文津阁的布局对天一阁的还原程度最高，乾隆皇帝多次写诗阐释文津阁园林的造园意向：『天一阁前原有池，池南更列假山峙。文津之阁率仿为，故亦叠石成嵔垒。有峰有壑有溪涧，涧水琴音泻池沚。东则月台西西山，又如宝晋斋传米。山亭因以趣为名，林泉引兴诚佳矣。贮书四库其趣多，餍饫优游

In the fifty-fifth year of his reign (1790), the seven copies of the *Comprehensive Library of the Four Treasuries* were sent to them. A monumental task in the cultural history of China—i.e. collecting, reproducing, and housing an immense anthology of manuscript books as the core of the Qing imperial library collection—was finally accomplished, presenting the Qianlong emperor in the best possible light (as a man of letters and wise sovereign). Four libraries were located in the North—Wenyuange in the Forbidden City, Wenyuange at Yuanmingyuan (written with a different Chinese character than the Forbidden City Wenyuange), Wenjinge in Chengde, and Wensuge in Shenyang. The library at Yuanmingyuan burnt down, but fortunately, the other three northern libraries have survived. The remaining three of the seven libraries were located in the South, in Jiangsu and Zhejiang provinces, as part of the temporary palaces built along the route of emperor Qianlong's southward inspection tour—Wenzongge in Zhenjiang, Wenhuige in Hangzhou, and Wenlange in Yangzhou. They were all destroyed during the Taiping Rebellion. Wenlange was rebuilt afterwards during the reign of emperor Guangxu.

Wenjinge in Chengde was the first of the seven libraries but the last of the scenic areas in the Mountain Resort that was completed. Located at the border of the lake district and the mountain district, and north of the scenic area of Qianchixue, construction began in the thirty-ninth year of emperor Qianlong's reign. Enclosed by winding walls, the library building stands in the north of a small peninsula surrounded by water on three sides. A water gate is northeast of Wenjinge that connects to the main water system of the Mountain Resort. The scenic area of Wenjinge comprises, from north to south, a gatehouse, an artificial hill, a pond, the library building, and a stele pavilion. Among the extant four libraries, Wenjinge is the one with the largest garden and the grandest appearance.

In his poem *Quting* (Pleasure Pavilion) emperor Qianlong praised the gardening art of Wenjinge, saying:

"The artificial hill piled up outside the library looks like a green snail, how about if I name this building 'Pleasure [Pavilion]'? I would rather not enjoy the trickling spring or the tree shadow, for I would get more pleasure from quietly reading poetry and literature."[8]

Such an atmosphere benefits from its location. The south-oriented library building is surrounded by rockery, water (from the pool), and trees, and further surrounded by walls and encompassed on two sides by water—the Wenjinge scenic area is referred to as "an island enclosed between two rivers", and really stands apart from the nearby buildings of Qianchixue located to its south (fig. 19).

The layout of Wenjinge is most similar to that of Tianyige, the emperor Qianlong writing several poems to explain this intentional resemblance. For example:

"Originally, there was a pool before Tianyige, and to the south of the pool, stood an artificial hill. In imitation of Tianyige, the garden of Wenjinge also has an artificial hill with stones piled up on top of each other. Peaks, ravines and mountain streams gather here, and gurgling streams rush down to the pool. To its east is the crescent-moon-shaped platform (*yuetai*) and to its west, the West Hill. They both resemble those in the garden of Baojinzhai built by Mi Fu. I name the pavilion on the top of the hill 'Pleasure [Pavilion]', because its trees and its spring bring me enjoyment, and the *Comprehensive Library of the Four Treasuries* stored there gives me great pleasure, for I can read a large number of books."[20]

The imperial poem mentions two specific monuments—the crescent-moon-shaped platform (*yuetai*) and the West Hill—that emulate similar architectural elements and landscapes built at Baojinzhai ("Treasuring-the-Jin Studio"), the study of the great Northern-Song calligrapher and painter Mi Fu. He named his study after the calligraphy of three distinguished Eastern-Jin scholars—Xie An, Wang Xizhi and Wang Xianzhi—he had collected and housed there. Mi Fu's passion for calligraphy and the important role he played in the transmission of classical styles is known through *Baojinzhai fatie*. Through allusion to Mi Fu, the Qianlong emperor who was a passionate art lover and collector himself attached a more profound meaning to the garden of Wenjinge, which reveals his goal to produce a copy that is an improvement on the original—refining and enhancing Jiangnan-style architecture in a creative way.

(III) Summery

The cultural event of "Four Treasuries and Seven Libraries"—establishing a comprehensive imperial library in multiple editions and erecting seven buildings in imitation of Tianyige to store them—demonstrates emperor Qianlong's determination to unify the whole country through learning. But the preference of the Kangxi and Qianlong emperors for Jiangnan-style architecture as a stylistic means of expression has a deeper meaning. The Manchu court did not just yearn for the advanced Han-Chinese culture flourishing in southern China. Even more so, the Manchus aimed to showcase their leading role in the cultural competition with the regions south of the Yangze River (Jiangnan) that was stimulated through their interest.

The Mountain Resort at Chengde is the largest imperial garden complex still extant and

意在此。』[19]除了表明文津阁园林沿用了天一阁的园林结构外，这段御制诗阐述了月台、西山的引入效法了宋代书画家米芾的宝晋斋布局。宝晋斋是米芾的书斋，因收藏了谢安、王羲之和王献之三位晋代名流的书法真迹而得名，有《宝晋斋法帖》传世留名。酷爱收藏的乾隆皇帝借助这个典故，赋予了文津阁园林更加深远的含义，反映了他要超越原作、青出于蓝的意图，而文津阁正是写仿江南的升华和提高之作。

『四库七阁』写仿天一的文化事件，充分体现了乾隆皇帝『汇八方河流如万川归海』的统一天下学问之决心。在这种历史背景下，避暑山庄内康、乾两朝对江南园林大量的写仿实例，并不是单纯的清代统治者对先进汉族文化的孺慕和向往，也是清廷和江南在文化上的相互影响和竞争的结果。

（三）小结

无论是借景言志的『避暑山庄七十二景』，还是体现清帝国版图的湖山布局；无论是复杂的边疆形势带来的多元夏都风貌，还是反映清廷和汉族文化纷繁关系的『写仿江南』造景手法——避暑山庄这座现存规模最大的皇家园林，是清代历史上一页恢宏壮丽的篇章。

随着乾隆朝『外八庙』组群的兴建，承德这座京师以北的夏都，在政治、军事、经济、文化、民族关系和中西交流等方面迸发出更加绚丽多彩的光芒。

二、外八庙

自康熙五十二年（一七一三年）至乾隆四十五年（一七八〇年），承德避暑山庄东北和北部山麓河谷间，陆续修建起十二座藏传佛教寺庙（图20）。其中八座有喇嘛常驻，即溥仁寺、溥善寺、普宁寺、普佑寺、安远庙、普陀宗乘之庙、殊像寺和须弥福寿之庙，并归理藩院喇嘛印务处统管。其余四座寺庙，普乐寺、罗汉堂和广安寺由驻防八旗护卫，广缘寺为喇嘛集资修建。八座寺庙在京设有『八处』，且因地处古北口外，俗称『外八庙』，并成为避暑山庄周围所有十二座藏传佛教寺庙的统称。

（一）营建历史

康熙五十二年（一七一三年），适逢玄烨六十大寿，蒙古科尔沁、土谢图亲王等各部王公至避暑山庄叩祝万寿，并进银二十万两『请建庙于热河』。溥仁寺（图21）和溥善寺即在此背景下建立于武烈河畔，拉开外八庙组群兴建的序幕。

marks a magnificent chapter in the history of the Qing dynasty—both in terms of the multi-cultural atmosphere due to Chengde's position in the frontier region and the architectural appearance in Jiangnan style that reflects the intricate relationship between Manchu (court) and Han-Chinese (people). And it represents the pinnacle of garden design because of the borrowed scenery of the *Seventy-two Views of the Mountain Resort* (by incorporating the surrounding landscape into the garden composition) and the imitative layout of lakes and mountains that emulate the territory of the Qing empire on a reduced scale.

The construction of the Eight Outer Temples during the reign of emperor Qianlong—the subject of the next chapter—further illustrates the important role of Chengde as the seasonal political center north of the capital (Beijing) in the study of the history of politics, military, economy, culture, and ethnic relations as well as for interpreting encounters between China and the West.

II. The Eight Outer Temples

From the fifty-second reign year of emperor Kangxi to the forty-fifth reign year of emperor Qianlong, 12 Tibetan Buddhist complexes were successively built at the foot of the mountains and in the river valleys in the north and northeast of the Mountain Resort in Chengde (fig. 20). Lamas resided in eight of them—Purensi (Universal Benevolence Monastery), Pushansi (Universal Virtue Monastery), Puningsi (Great Buddha Monastery), Puyousi (Universal Blessing Monastery), Anyuanmiao (Ili Temple), Putuo zongcheng Temple (Potaraka Temple of Doctrine Transmission) Temple, Shuxiangsi (Manjusri Statue Monastery), and Xumi fushou (Sumeru Happiness and Longevity) Temple. They were put under Lamaist responsibility, governed directly by Lifanyuan (Board for the Administration of Outlying Regions; or Court of Colonial Affairs) and registered in the Lama Office. Three of the remaining four temples—Pulesi (Round Pavilion Monastery), Luohantang (Arhat Hall), and Guang'ansi (Vast Peace Monastery)—were under control of the Eight Banners, but Guangyuansi (Vast Distant [Lands] Monastery) was also built with money raised by the Lamas. The first eight temples established "Eight Offices" in the city of Beijing, but located north of Gubeikou—the strategically important pass of the Great Wall, separating northeastern China from Beijing region—they were collectively referred to as "Eight Outer Temples" (Waibamiao), a term that ultimately became a collective name for the twelve Tibetan Buddhist temples in Chengde.

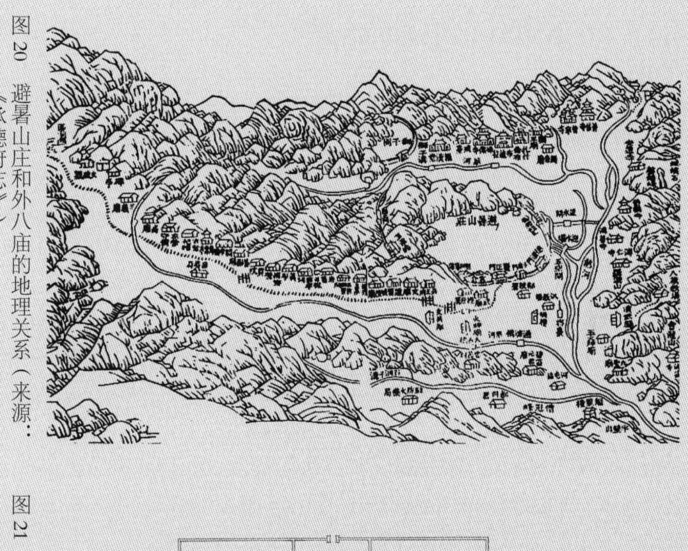

图20 避暑山庄和外八庙的地理关系（来源：《承德府志》）

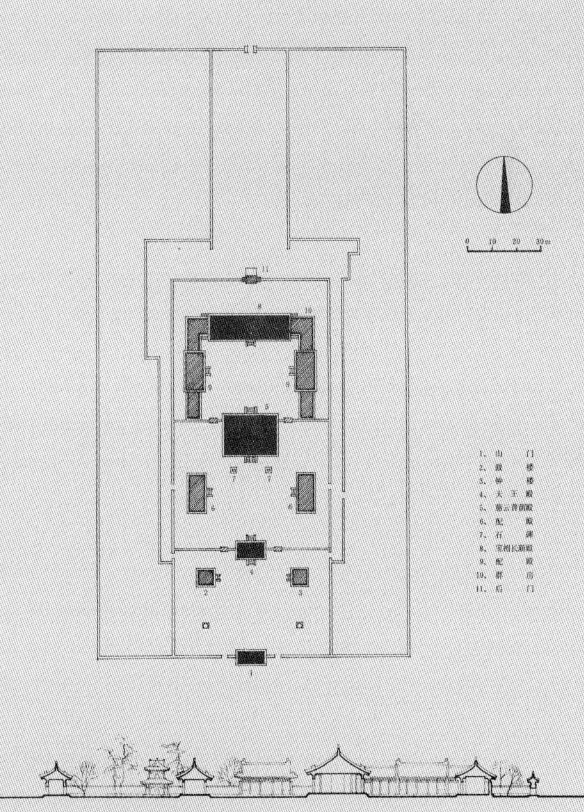

图21 溥仁寺总平面和总剖面图（来源《承德古建筑》）

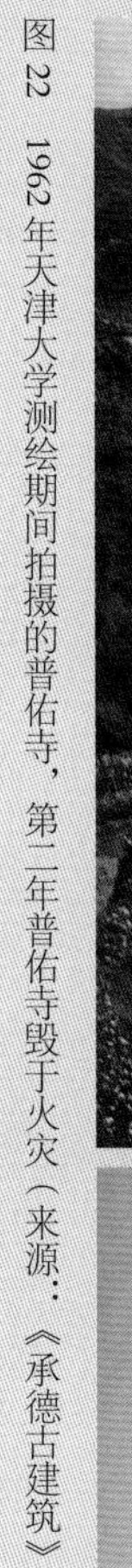

图22 1962年天津大学测绘期间拍摄的普佑寺，第二年普佑寺毁于火灾（来源：《承德古建筑》）

Fig.20 Geographical relation between the Mountain Resort and the Eight Outer Temples (Source: *Chengde fuzhi*)

Fig.21 Site plan and section of Purensi (Source: *Chengde gujianzhu*)

Fig.22 Historical photo of Puyousi (destroyed by fire in 1963) taken during the 1962 survey by Tianjin University (Source: *Chengde gujianzhu*)

(I) Construction History

In 1713, the fifty-second year of emperor Kangxi's reign, Tüsheet princes and nobles of the Khorchins and other Mongolian tribes came to the Mountain Resort. They offered congratulations on the occasion of the emperor's sixtieth birthday and paid tribute of 200,000 *liang* of silver in exchange for the permission to build a temple at Chengde. In the same year, Purensi (fig. 21) and Pushansi were built next to Wulie River, which marked the beginning of the construction of the Eight Outer Temples.

After Qing troops had crushed the rebellion of the Dzungar Oirat in 1755, the twentieth year of Qianlong's reign, several Oirat nobles presented themselves before the emperor at the Mountain Resort. Besides holding banquets and bestowing rewards, the emperor also ordered the construction of Puningsi at Shizi Furrow, five miles northeast of the Mountain Resort as a gesture of his good will and desire to maintain peace among ethnic minorities. Just like the scenic area of Xumi lingjing located at the back hill of Qingyi Garden (today Yiheyuan or more colloquially, the New Summer Palace in Beijing), Puningsi was modeled after the famous Lamaist Samye Monastery in Shannan prefecture, Tibet, and was completed in 1759, the twenty-fourth year of emperor Qianlong's reign. In the next year (1760), the emperor commissioned another temple to the east of Puningsi —Puyousi (fig. 22).

In 1764, the twenty-ninth year of emperor Qianlong's reign, the Dzungar rebellions led by Dawachi and Amursana were crushed. In support of the imperial court, the Dashidava tribe of the Dzungar people decided to move to Rehe. Emperor Qianlong issued an imperial edict to build a temple on a hill northeast of the Mountain Resort—Anyuanmiao also known as Ili Temple (named after the Ili Basin that had come under Qing control a decade earlier)—in imitation of local architecture in Yining (also known as Ghulja) in Xinjiang (formerly Kashgaria). After the victory over the Dzungars, the Qing court also suppressed the revolt of the Khoja brothers.

In 1766, the thirty-first year of emperor Qianlong's reign, construction of Pulesi began as requested by Rolpai Dorje, the Third Changkya and his chief administrative lama and teacher. The complex was not designed as a Lamaist monastery but as a cultural-political venue for holding banquets and audiences with minority leaders (Kazaks and Khalkhas) from the northwest.

In 1771, the thirty-sixth year of emperor Qianlong's reign, when the queen mother turned eighty years old, the largest of the Eight Outer Temples—Putuo zongcheng Temple—was completed after four years of construction. The date of completion of the temple that mirrored the Potala

乾隆二十年（一七五五年）清军平定准噶尔叛乱，乾隆皇帝幸避暑山庄时，厄鲁特蒙古部落来觐。乾隆皇帝赐宴封赏同时，敕令修建普宁寺于避暑山庄东北五里的狮子沟。与北京清漪园（现颐和园）后山须弥灵境同写仿自西藏山南桑耶寺，至乾隆二十四年（一七五九年）初成。乾隆二十五年（一七六〇年）在普宁寺东部敕建普佑寺（图22）。

乾隆二十九年（一七六四年）平定准噶尔达瓦奇和阿睦尔撒纳叛乱。准噶尔的达什达瓦部因支持清政府，决定整个部族迁居热河。乾隆帝下诏依据喀什的固尔扎庙，于避暑山庄东北部山冈上建安远庙（亦称『伊犁庙』）。

清廷在平定准噶尔叛乱之后，又平定了大小和卓木叛乱。乾隆三十一年（一七六六年）在三世章嘉国师的建议下修普乐寺。普乐寺内不设喇嘛，是专供西北边疆地区哈萨克、布鲁特和柯尔克孜等少数民族首领朝见时瞻礼用。

乾隆三十六年（一七七一年）是太后八十大寿，乾隆皇帝仿照西藏拉萨布达拉宫建造普陀宗乘之庙。此庙从乾隆三十二年（一七六七年）起动工，历时四年，是外八庙组群中规模最大一处。寺庙落成时恰逢土尔扈特部东归（图23），乾隆帝下令在普陀宗乘之庙竖起满汉蒙藏四体《土尔扈特全部归顺记》和《优恤土尔扈特部众记》御碑，以纪念这一重大事件。

乾隆三十七年（一七七二年）在普陀宗乘庙以西造广安寺。此庙亦名『戒台寺』，是乾隆皇帝与蒙古王公贵族举行法会之处。

图24 普宁寺

图23 《万法归一图》屏，表现了乾隆皇帝在普陀宗乘之庙万法归一殿会见土尔扈特部首领渥巴锡的场景（来源：故宫博物院）

Fig.23 *Wanfa guiyi tu*, showing emperor Qianlong's reception of a Torghut prince (Ubashi Khan) in Wanfa guiyi Hall at Putuo zongcheng Temple (Source: The Palace Museum)

Fig.24 Puningsi

乾隆三十九年（一七七四年）在普陀宗乘庙和广安寺之间兴建殊像寺。内有宝相阁供文殊菩萨骑狻猊巨像，是仿五台山殊像寺明弘治像而建。落成后的殊像寺为满族喇嘛庙，且贮藏满文大藏经一部。同年，广安寺西侧建罗汉堂。

乾隆四十五年（一七八〇年）乾隆皇帝七十大寿，六世班禅『祝厘来自后藏，上嘉其远至，于山庄建扎什伦布庙居之』。扎什伦布庙即须弥福寿之庙，为外八庙中最后一座。

（二）寺庙布局

外八庙组群中康熙时期建设的溥仁寺和溥善寺坐落在热河泉以东，武烈河东岸的河谷内。虽为黄教寺庙，其寺庙布局为典型伽蓝七堂汉式布局，坐北朝南。

普宁寺是外八庙中乾隆时期建造的第一座。武烈河支流流经其西侧，寺庙总体布局顺应自然山水之势，据冈面南。普宁寺前部为伽蓝七堂式，由牌坊、山门、钟鼓楼、碑亭、天王殿、配殿和大雄宝殿组成。后部仿照西藏山南桑耶寺，主要建筑名为『大乘之阁』，周围环绕自然山体和人工假山，并配以四大部洲、八小部洲、日月殿和四色塔等建筑，模仿佛教须弥山，形成以高阁为中心的藏式曼荼罗布局（图24）。

随后建设的安远庙（图25）和普乐寺（图26）均为前汉后藏式，并在建筑总体布局上呼应了避暑山庄和周围山岭形胜：安远庙中轴线偏向西南，除顺应所在山冈和西北部武烈河流向外，建筑轴线

Palace in Lhasa, Tibet, coincided with the returning of the Torghut people back to China from the east where they had been suppressed by the Russians (fig. 23). To commemorate the event, emperor Qianlong ordered to set up steles (*Torghut quanbu guishun ji* [Record of submission of the whole Torghut clan] and *Youxu Torghut buzhong ji* [Record of pacifying the Torghut people]) inscribed in four languages (Manchu, Chinese, Mongolian, and Tibetan).

In 1772, the thirty-seventh year of emperor Qianlong's reign, Guang'ansi also called Jietaisi was built to the west of Putuo zongcheng Temple as a venue for the Qianlong emperor and Mongolian nobility to hold a dharma assembly. Two years later, in 1774, construction at Shuxiang (Manjusri Statue) Temple began between the Guang'an and Putuo zongcheng temples. It was named after the large statue of Manjusri riding on a lion (*suanni*) that was enshrined in its Baoxiang Pavilion and imitated a similar, well-known statue (dated to the reign of the Ming emperor Hongzhi) standing in another temple with the same name on Mount Wutai. As a monastery for Manchu Lamas, it also housed a copy of the tripitaka written in Manchu. In the same year, Luohan Hall was built to the west of Guang'ansi.

In 1780, the forty-fifth year of emperor Qianlong's reign, when he was seventy years old, the Sixth Panchen Lama came from rear Tibet to offer birthday congratulations to the emperor who, on that occasion, ordered to build a temple in the Mountain Resort that imitates the Panchen Lama's residence in Tibet's Shigatse. The Tashilhunpo Monastery in Chengde is also called Xumi fushou (Sumeru Happiness and Longevity) Temple and is the last one of the Eight Outer Temples built at the Mountain Resort.

(II) Layout of the Eight Outer Temples

Two of the Eight Outer Temples were built under emperor Kangxi—Purensi and Pushansi—and are situated to the east of Rehe spring in a valley on the east bank of Wulie River. Although affiliated with the Gelug School or Yellow Hat sect of Tibetan Buddhism, the temple still has a Han-Chinese Buddhist layout with a south-oriented compound consisting of seven halls (*qielan qitang*).

Puningsi is the first of the Eight Outer Temples built under emperor Qianlong. A branch of Wulie River runs east of the compound that fits into the surrounding natural terrain. The front part of Puningsi follows the seven-hall (*qielan qitang*) design, consisting of an archway, gatehouse, bell tower and drum tower, stele pavilion, Hall of Heavenly Kings (Tianwangdian), side halls, and Treasure Hall of the Great Hero (Daxiongbaodian). The rear part is modeled on the Tibetan Samye Monastery with a main building—Mahayana (Dacheng) Pavilion (Dacheng zhige)—surrounded by natural and artificial hills and accompanied by Sidabuzhou, Baxiaobuzhou, Riyue halls, and the Sise Pagoda. The towering central pavilion is a reference to Mount Sumeru, and seen as a whole, the rear section is constructed to resemble a Tibetan Mandala (fig. 24).

Anyuanmiao (fig. 25) and Pulesi (fig. 26) were subsequently built, both with a Chinese-style front part and a Tibetan-style rear part. The structure echoes the Mountain Resort and the surrounding landscape. The buildings are aligned along the central axis, which is slightly inclined southwestwards so as to follow the direction of the surrounding hills and the flow of Wulie River and to point toward the pagoda at Yongyousi. By contrast, the architecture of Pulesi is oriented westwards, with its west side facing the Mountain Resort and its east side Hammer Peak. The main buildings of the two temples—the square Pudu Hall and the circular Xuguang Pavilion—are the most significant scenic spots in the eastern part of the Mountain Resort.

Shizi Furrow, a valley formed by a branch of Wulie River in the north of the Mountain Resort, runs almost parallel to the resort's north wall. To its north—where foothills protect the Mountain Resort like screens are arranged from east to west—exist the Xumi fushou, Putuo zongcheng, Shuxiang, and Guang'an temples as well as Luohan Hall. The Puto zongcheng (fig. 27) and Xumi fushou (fig. 28) temples were modeled after the Potala Palace in Lhasa and the Tashilhunpo Monastery in Shigatse, and are those among the Eight Outer Temples that most closely resemble Tibetan architecture. They are laid out in a unique way and consist of a "Great Red Platform" with a tall ring-shaped podium building (*qunlou*) atop that encloses a Chinese-style Buddha hall. This arrangement of architecture in concentric squares takes the shape of the Chinese character hui and is thus known as the *hui*-shape plan.

(III) Architectural Achievements

While preserving their own characteristics, the Eight Outer Temples share common features in terms of layout and technology, and this shows the high level of imperial construction achieved during the Qing dynasty.

The *Qinding Rehe zhi* (Imperial record of Rehe) completed in 1781, the forty-sixth reign year of emperor Qianlong, describes the Eight Outer Temples after their completion as

指向山庄内的永佑寺塔；普乐寺布局坐东朝西，西面朝向山庄，东面与磬锤峰呈对景关系。这两座寺庙主体建筑 普度殿和旭光阁 亦一方一圆，与磬锤峰一起形成山庄东部主要的景观标志物。

避暑山庄北部武烈河支流所形成的河谷名为『狮子沟』，沟与山庄北墙走向基本一致。狮子沟北的山麓如屏风拱卫避暑山庄，由东向西面向山庄排列着 须弥福寿之庙、普陀宗乘之庙、殊像寺、广安寺和罗汉堂。普陀宗乘（图27）和须弥福寿（图28）二庙分别写仿拉萨布达拉宫和日喀则扎什伦布寺，是外八庙组群中最具藏式特色的寺庙。这两座寺庙均以大红台为中心，形成『回』字形裙楼环绕汉式中心佛殿的特殊布局。

（三）建筑成就

承德外八庙在建筑布局和技术上，既统一又各具特色，是清代皇家建筑最高水平的体现。成书于乾隆四十六年（一七八一年）的《钦定热河志》这样形容外八庙组群整体建成后的情景：『寺庙……在山庄外者，觐扬祖烈，轮奂如新，诸藩虔构精庐，延洪锡羡……琳宫屹峙，概益增洵乎！会八部龙天而赞诵，遍百神河岳以皈依矣。』这里的『八部龙天』为佛教吸收其他宗教神祇充当护法神的二十诸天神以及龙神等，暗喻外八庙作为护法神祇拱卫于山庄东北山麓。总体来说，外八庙和山庄的拱卫关系形成于乾隆朝。康熙朝奠定了避暑山庄总体格局，但溥仁寺和溥善寺均沿袭传统寺庙坐北朝南布局，与山庄之间并无特殊对应。乾隆朝在建设中逐渐形成了外八庙和山庄相呼应的整体格局，多数建筑轴线朝向山庄（图29）。且单体寺庙布局有三个特点：第一，选址以山地为主，顺应自然山势和河流走向。第二，在建筑布局上将藏式『都纲法式』和汉式『伽蓝七堂』的布局方式相结合，突出轴线序列，即前部空间主要由汉式的山门、碑亭、大雄宝殿等建筑组成，后部采用中心构图的曼荼罗式布局。第三，融合文人园的创作意象，形成灵活又变通的皇家寺庙园林布局，如殊像寺后部随山势经营的假山石洞和点缀其间的亭、台、楼、阁等（图30），普陀宗乘庙中部轴线上的古树与叠石，以及须弥福寿庙大红台基部的叠石、侧面的堆山。

承德外八庙在建筑技术上最突出的成就之一，是熟练运用木结构建造技术建造巨大空间。外八庙组群中的木结构多层建筑代表了清代楼阁技术与艺术的高峰，普宁寺大乘阁、安远庙普度殿、普乐寺旭光阁、普陀宗乘之庙万法归一殿和须弥福寿之庙妙高庄严殿是最突出实例。如普宁寺大乘阁通高三十七点四〇米，内奉二十余米高的千手千眼观音像（图31）；普乐寺旭光阁制仿北京天坛祈年殿，重檐攒尖圆殿，内部建有立体坛城供奉上乐王佛；普陀宗乘之庙万法归一殿曾被瑞典探险家斯文·赫定（Sven Hedin）赞为『最漂亮、装饰最华丽的建筑』，并将其1:1复制（图32），在一九三三年芝加哥『百年进步』世界博览会上作为『中国金亭』进行展出，引起轰动[十]。这些楼阁式建筑均体量巨大，

follows: “The monasteries and temples…situated outside the Mountain Resort, built to represent my ancestors’ achievements, still looking new despite repair on several occasions, exquisite structures reverently built by various (leaders of) outlying regions… magnificent palaces rising high, the grandness of the Mountain Resort has truly increased! The eight groups of spiritual beings gather here to chant, together with various river and mountain gods and immortals.” The phrase Babu tianlong refers to a group of beings present during Mahayana sutra convocations, and traditionally includes deva, *naga, yaksa, gandharva, asura, garuda, kimnara,* and *mahoraga*. They are thought to have originated in other religious traditions, but were subsequently incorporated into Buddhist cosmology where they were viewed as protectors of the Dharma. Such wording suggests that the Eight Outer Temples, situated at the foothills northeast of the Mountain Resort, were intended as guardians that surrounded and protected the Mountain Resort. However, the Eight Outer Temples took up their protective role only under emperor Qianlong (fig. 29). Although the Mountain Resort was completed with a well thought-out design by the time of emperor Kangxi, the Puren and Pushan temples were not yet responding to the architecture and landscape in the resort, still following the traditional south-facing orientation of Chinese monasteries and temples. But by the time of emperor Qianlong, the Eight Outer Temples and the Mountain Resort gradually came to be in dialogue with one another, with the axis of most of the temples now corresponding to the main axis of the Mountain Resort.

Furthermore, each of the Eight Outer Temples shares important design features. First, in terms of site selection, hilly areas were given priority, and the selected locations were to match the natural terrain of the hills and the flow of surrounding rivers. Second, in terms of architectural layout, the temples combine concentric Tibetan design (*dugang fashi*) with axial Han-Chinese design (*qielan qitang*). That is to say, their front sections are aligned one after the other a Chinese-style gatehouse, a stele pavilion, and a Treasure Hall of the Great Hero (Daxiongbaodian), while in the rear part a main building is positioned in the center as embodiment of a Tibetan mandala. Third, Chinese scholars’ gardens were a source of inspiration, and the integration of their creative design led to a flexible and versatile layout of imperial temples and gardens. A good example is the artificial hills and stone caverns of Shuxiang Temple that fit well into the hilly terrain in the rear part of the temple interspersed with pavilions, terraces, multi-story buildings and towers (fig. 30). Another good example is the ancient trees and rock formations that are aligned along the central axis of Xumi fushou Temple or the rockery at the foot of the Great Red Platform and the artificial hills on its side.

图26 普乐寺

图25 安远庙

Fig.25 Anyuanmiao
Fig.26 Pulesi

图28 须弥福寿之庙

图27 普陀宗乘之庙

Fig.27 Putuo zongcheng Temple
Fig.28 Xumi fushou Temple

图29 避暑山庄和周围寺庙全图（来源：美国国会图书馆）

图30 殊像寺假山与建筑剖面复原示意图

Fig.29 General map of the Mountain Resort and its surrounding temples (Source: American Library of Congress)
Fig.30 Reconstruction of section, artificial hill and buildings at Shuxiangsi

One of the most remarkable achievements in the construction of the Eight Outer Temples is the skillful use of timber-frame technology to create a large space. The tall wooden buildings of the Eight Outer Temples represent the technological and artistic pinnacle in the construction of multi-story architecture in the Qing dynasty. The most striking examples are Mahayana Pavilion at Puningsi, Pudu Hall at Anyuanmiao, Xuguang Pavilion at Pulesi, Wanfa guiyi Hall at Putuo zongcheng Temple, and Miaogao zhuangyan Hall at Xumi fushou Temple. Among them, Mahayana Pavilion rises to a total height of 37.4 m in order to house a statue of the Thousand-handed Thousand-eyed Guanyin (Avalokiteśvara) statue more than 20-m tall (fig. 31). Xuguang Pavilion, in imitation of Qinian (Prayer for Good Harvest) Hall at the Altar of Heaven in Beijing, is a circular structure with double-eaved pyramidal roof that was built as a three-dimensional mandala to enshrine Cakrasamvara. Furthermore, Wanfa guiyi Hall was praised by the Swedish explorer Sven Hedin as "the most beautiful and the most magnificently decorated building". Hedin even made a 1-to-1 replica of the hall for the 1933 World's Fair in Chicago (themed "A Century of Progress"), where the so-called "Chinese Golden Pavilion" stirred the attraction of the visitors (fig. 32).[20] The core of these massive multi-story structures is a skillfully made vertical shaft opening that uses tall columns (*tongzhu*) spanning the distance from the ground floor to the roof, which is a technique characteristic for traditional Chinese architecture in the final stage of development.

Anther remarkable contribution that the Eight Outer Temples made is the fusion of the building traditions of different ethnic groups into a new spatial whole, and the use of common materials and familiar techniques to imitate 'foreign' non-local styles. For example, the Eight Outer Temples enhanced the Tibetan flat-roof house (*diaofang*) through combination with Han-Chinese timber-frame and stone-arch techniques. Furthermore, the Great Red Platform of the Putuo zongcheng and Xumi fushou temples represents an advanced expression of spatial design and religious symbolism of Han and Tibetan Buddhist cultures and moreover, an innovative treatment of massive-volume buildings in hilly areas. The basement of each Great Red Platform is a solid stone structure pierced by arched passageways, while the upper floors belong to a ring-shaped podium building (*qunlou*) in a mixed construction style combining brick and wood that encloses a courtyard (hui-shape plan). The tall square pavilion in the middle of the courtyard symbolizes the core of the Tibetan "*dugang* pattern" that is pulled aloft vertically. In this way, the massive building volume was divided into three parts—podium building, *hui*-shape courtyard, and central Buddha hall—which improved both lighting and ventilation.

However, such a bold attempt in structure brings some problems. For example, after

内部为中央数层通高处理，多用通柱，熟练运用木材的包镶技术，体现了中国传统建筑发展到晚期的技术特征。

外八庙在建筑技术上另一个突出贡献，是不同民族建筑结构结合形成特有空间，以及运用熟悉的材料技术模仿其他民族的建筑艺术表现。前一方面，外八庙建筑发展了源自西藏的『平顶碉房』结构，并将其与汉族的木结构体系和石拱技术相结合。突出典型是普陀宗乘和须弥福寿二庙最核心的『大红台』部分：它们底层均以石材砌筑，且留有拱券结构的甬道以供通行；石结构上都采用了『回』字形裙楼围合庭院，为砖木结合结构；庭院中间建方形高阁象征西藏『都纲法式』的中空拔起部分。这种做法把一个巨大体量分成了裙楼、『回』字形庭院和中央佛殿，加强了建筑的采光、通风，充分显示了空间表达和宗教象征方面的艺术成就，也为大体量山地建筑的处理提供了一种新手法。

不过，结构方面的大胆尝试也产生了问题，如普陀宗乘之庙大红台建成后经多次维修、改建，其内部仍于20世纪初期彻底倒塌，损坏的主要症结恰恰就是汉藏结合的部位——平屋顶的屋面承重处，以及砖石结构和木结构的连接处。后一方面，即利用汉地材料技术来模仿蒙藏地区特殊的装饰构件与艺术形象，如：应用彩色琉璃模仿蒙藏建筑檐下金属装饰『边坚』；采用鱼鳞镀金铜瓦模仿蒙藏建筑的『铜镀金』屋顶（图33）；使用汉地传统的 灰背 代替 蒙藏地区的『阿嘎土』铺筑平屋顶屋面；利用 木过梁 取代蒙藏地区梯形窗上的砖石过梁等。

图32 芝加哥的中国金亭（来源：*Building a Century of Progress The Architecture of Chicago's 1933–1934 World's Fair*）

图31 普宁寺大乘之阁千手千眼观音像

图33 避暑山庄内展出的镀金四联铜瓦

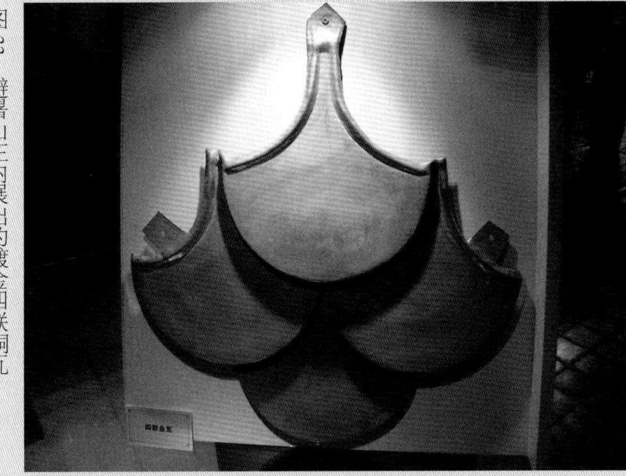

Fig.31 Thousand-handed Thousand-eyed Guanyin statue inside Mahayana Pavilion at Puningsi

Fig.32 The "Chinese Golden Pavilion" at the 1933 World's Fair in Chicago (Source: *Building a Century of Progress: The Architecture of Chicago's 1933–1934 World's Fair*)

Fig.33 Quadruple gold-plated copper tile in the Mountain Resort collection

repeated repair and rebuilding, the inside of the Great Red Platform at Putuo zongcheng Temple fell down completely in the early twentieth century. The main reason for the structural failure lies in the fusion of Chinese-style and Tibetan-style construction and happened in the fringe area where the load-bearing part of the flat roof was joined to the mixed brick-timber walls. That is to say, the builders used Han-Chinese materials and techniques to imitate structural and decorative elements typical for Mongolian and Tibetan people. They used the following: colored glazed tiles to imitate the decorative metal (*bianjian*) installed under the eaves of Mongolian and Tibetan buildings; scale-shaped gilded copper tiles to imitate the Mongolian and Tibetan "gold-plated copper" roof; simple Han-Chinese lime-covered roofs to replace the Mongolian and Tibetan flat roof traditionally built with "Aga soil" (fig. 33); and a wood lintel (guoliang) to replace the stone lintel placed above a Mongolian or Tibetan trapezoid window.

(IV) Summary

The Eight Outer Temples in Chengde represent the largest religious center of Tibetan Buddhism in North China, and also the region with the most Tibetan Buddhist buildings built for the Qing court outside of the capital (Beijing). Emperor Kangxi indicates this in his *Yizhi Purensi beiwen* (Stele inscription for Puren Temple): "I think about the way to rule a country… applying soft tactics toward frontier people has always been a challenge since ancient times. Our ancestors accomplished extraordinary achievements in this respect. Many distant peoples came and paid allegiance, and their grace and kindness took root in peoples' hearts. The Mongolian tribes that the Three Sovereigns and Five Emperors failed to subjugate are today as peaceful as the Middle Kingdom." After succession to the throne, the Qianlong emperor established a national policy of "promoting the Gelug School, pacifying Mongolia". As a consequence, temple construction of several of the Eight Outer Temples became tied to significant events in Qing minority politics, which highlights the peaceful political (re)unification of domestic and frontier regions, as well as the important role of religion, especially Tibetan Buddhism, in the cultural unification of the multi-national Qing empire.

（四）小结

承德外八庙是中国北方规模最大的藏传佛教中心，也是京畿之外清代皇家藏传佛教建筑最密集的区域。康熙皇帝在《御制溥仁寺碑文》中写道：『朕思治天下之道……柔远能迩，自古难之。我朝祖功宗德，远服要荒；深仁厚泽，沦及骨髓。蒙古部落，三皇不治，五帝不服，今已中外无别矣』。乾隆皇帝即位以来，也秉承这种『兴黄安蒙』的基本国策。因此外八庙几乎每座寺庙的修建都紧密联系当时重大的民族政治事务，也是清代内陆边疆地区矛盾纷争与和平统一进程具体而微的缩影，反映了宗教尤其是藏传佛教在清代形成大一统多民族国家进程中的重要作用。

注释

（一）即平泉州、滦平县、丰宁县、隆化县、朝阳县、赤峰县。

（二）康熙三十六景诗之二，《芝径云堤》，康熙五十年（一七一一年）。

（三）同上。

（四）康熙三十六景诗之四，《延薰山馆》，康熙五十年（一七一一年）。

（五）张玉书《扈从赐游记》，康熙四十七年（一七〇八年）。

（六）朱圭和梅裕凤是内廷雕刻高手，朱圭善绘，梅裕凤善临刻康熙御书。他们在合作《御制避暑山庄诗》之前最著名的作品是康熙三十五年（一六九六年）刊行，玄烨撰、焦秉贞绘的《御制耕织图诗》。

（七）表格基于《避暑山庄七十二景》编委会，避暑山庄七十二景. 北京：地质出版社，1993 中总结的版本增补；《钦定热河志》中亦有多个版本七十二景相关绘画。

（八）美国学者 Whiteman 在 *Translating the Landscape* 一文中探讨了沈嵛三十六景图的艺术特点，认为其是不同于明代私家园林图咏的，反映康熙皇帝治国理念的新形式。

（九）静寄山庄有内八景、外八景、新六景和附列十六景，共三十八景。

（十）Harry F. Mallgrave. *Modern Architectural Theory: A Historical Survey, 1673–1968*. Cambridge University Press,2009. P54

（十一）这种划分方法未见于清代文献；一九五六年卢绳的《承德避暑山庄》一文中根据功能将山庄分为行宫、苑景两部分；一九六〇年周维权在《避暑山庄的园林艺术》一文中将宫苑分置的分法细化，提出了苑包括平原、湖泊和山岳三部分；一九八二年天津大学建筑系和承德市文物局编著的《承德古建筑》中『把全园划分为四个具有不同特色的区域』——宫殿区、湖区、平原区和山区。

（十二）如意湖、澄湖、上湖、下湖、镜湖、银湖、半月湖和内湖（西湖）。

（十三）湖区东部有水心榭、清舒山馆、文园狮子林、戒得堂、汇万总春之庙、金山、香远益清等；湖区中部有月色江声、采菱渡、如意洲、烟雨楼等；湖区西部有芳园居、芳渚临流、长虹饮练、临芳墅、知鱼矶等。

（十四）杨伯达在《冷枚及其「避暑山庄图」》中提出此图绘于康熙五十二年（一七一三年）皇帝六旬万寿之前；《承德古建筑》一书则认为绘于康熙四十七年（一七〇八年）或四十八年（一七〇九年）；亦有学者考证该图绘于乾隆初年避暑山庄扩建之前。

（十五）丁海斌，于新. 夏都的设立及其在元、清两代边疆治理中的特殊作用 [J]. 中央民族大学学报（哲学社会科学版），2015（2）：79–83.

（十六）『年班』是清朝入关后为了进一步巩固与蒙古各部之间的政治联盟，实行的制度，即蒙古王公每年要到京师朝见皇帝。

（十七）高勇. 清朝天花的防治和影响 [D]. 内蒙古大学，2005：28–30.

（十八）和珅、梁国治主编. 热河志·文津阁·趣亭·庚子. 乾隆四十六年（一七八一年）.

（十九）和珅、梁国治主编. 热河志·文津阁·趣亭·庚子. 乾隆四十六年（一七八一年）.

（二十）刘瑜，杨菁，王英妮. 芝加哥世博会的中国金亭——兼论早期世博会的中式建筑 [J]. 世界建筑，2013（08）：116–119.

Notes

① Namely Pingquan prefecture and the Luanping, Fengning, Longhua, Zhaoyang, and Chifeng counties.

② Emperor Kangxi's second poem ("*Zhijing yundi*") in *Yuzhi Bishu shanzhuang shi* (Imperial poems on the Mountain Resort), 1711.

③ ibidem

④ Emperor Kangxi's fourth poem *Yanxun Shanguan* in *Yuzhi Bishu shanzhuang shi*, 1711.

⑤ Zhang Yushu, *Hucong ciyou ji*, the forty-seventh year of emperor Kangxi's reign.

⑥ Both Zhu Gui and Mei Yufeng were masters in engraving in the imperial palace, the former good at painting while the latter good at engraving based on the imitation of Emperor Kangxi's handwriting. Before the advent of the *Imperial Poems for the Mountain Resort* by them together, the best-known work was *Imperial Poems for Farming and Weaving* [poems by Xuanye (Emperor Kangxi) and paintings by Jiao Bingzhen, a court painter in the Qing Dynasty] published in the 35th year under the reign of Emperor Kangxi.

⑦ The table is based on information from *Bishu shanzhuang qishier jing* (Beijing: Dizhi chubanshe, 1993). *Qinding Rehe zhi* also contains several illustrations of the seventy-two sights in the Mountain Resort.

⑧ In his article "Translating the Landscape: Genre, Style, and Pictorial Technology in the Thirty-Six Views of the Mountain Estate for Escaping the Heat" (in Richard Strassberg, *Thirty-Six Views: The Kangxi Emperor's Mountain Estate in Poetry and Prints* [Cambridge, MA: Harvard University Press, 2016], introduction), the American scholar Stephan H. Whiteman investigates the new artistic approach in Shen Yu's paintings of the Mountain Resort that reflects the state governance philosophy of the Kangxi emperor through comparison with Ming-dynasty painting albums of private gardens.

⑨ In total there are thirty-eight sights in Jingji shanzhuang: the eight sights inside the mountain resort, the eight outer sights, the six new sights, and the sixteen additional sights.

⑩ Harry F. Mallgrave, *Modern Architectural Theory: A Historical Survey, 1673-1968*, Cambridge: Cambridge University Press, 2009. p. 54.

⑪ This division is not recorded in Qing-dynasty literature; in his 1956 article *Chengde Bishu shanzhuang*, Lu Sheng divided the Mountain Resort into two parts according to their function—a temporary palace zone and a scenic garden zone; in his 1960 article *Bishu shanzhuang de yuanlin yishu*, Zhou Weiquan further divided the garden zone into three subparts—the flatland, lake, and mountain districts; *Chengde gujianzhu*, jointly compiled by the Department of Architecture at Tianjin University and the Chengde Administration of Cultural Heritage (Wenwuju) in 1982, suggests that "the whole garden area of the Mountain Resort can be divided into four parts, each with different characteristics"—namely the palace, lake, flatland, and mountain districts.

⑫ Namely, Ruyi Lake, Cheng Lake, Upper Lake, Lower Lake, Jing Lake, Yin Lake, Banyue Lake, and Inner Lake (or West Lake).

⑬ In the northern part of the lake district are located the scenic areas of Shuixinxie, Qingshu shanguan, Wenyuan shizilin, Jiedetang, Huifang zongchun Temple, Jinshan (Island), and Xiangyuan yiqin; in the central part are located the scenic areas of Yuese jiangsheng, Cailingdu, Ruyi (Island), and Yanyulou; in the western part are located the scenic areas of Fangyuanju, Fangzhu linliu, Changhong yinlin, Linfangshu, and Zhiyuji.

⑭ In his article *Leng Mei jiqi Bishu shanzhuang tu* (Palace Museum Journal 1 [1979]), Yang Boda suggest that the painting was created before emperor Kangxi 's sixtieth birthday in the fifty-second reign year; *Chengde gujianzhu* gives the forth-seventh or forty-eighth reign year of emperor Kangxi as the date of creation; other scholars have confirmed that it was drawn before the expansion of the Mountain Resort in the first year of emperor Qianlong's reign.

⑮ Ding Haibin and Yu Xin, "Xiadu de sheli jiqi zai Yuan, Qing liangdai bianjiang zhili zhong de teshu zuoyong," *Journal of Minzu University of China* (philosophy and social science edition) 2 (2015), pp. 79-83.

⑯ The *nianban* was a system adopted by the Qing court to consolidate its political alliance with Mongolian tribes, i.e. Mongolian princes and nobles should come to the capital and pay tribute to the Qing court and offer respect to the emperor during the New Year Festival.

⑰ Gao Yong, *Qingchao tianhua de fangzhi he yingxiang* (Prevention, treatment, and impact of smallpox in the Qing dynasty), master's thesis, Inner Mongolia University, 2005. pp. 28-30.

⑱ See the poem "Quting" written by emperor Qianlong in 1781 and included in *Rehe zhi* (section Wenjinge) edited by He Shen and Liang Guozhi and published in the forty-sixth year of emperor Qianlong's reign.

⑲ See the poem "Quting" written by emperor Qianlong in 1781 and included in *Rehe zhi* (section Wenjinge) edited by He Shen and Liang Guozhi and published in the forty-sixth year of emperor Qianlong's reign.

⑳ Liu Yu, Yang Jing, and Wang Yingni, "The Chinese Golden Pavilion at the Chicago World's Fair and Chinese-style Architecture at Early World Expositions," *World Architecture* 8 (2013), pp. 116-119.

图版

Darwings

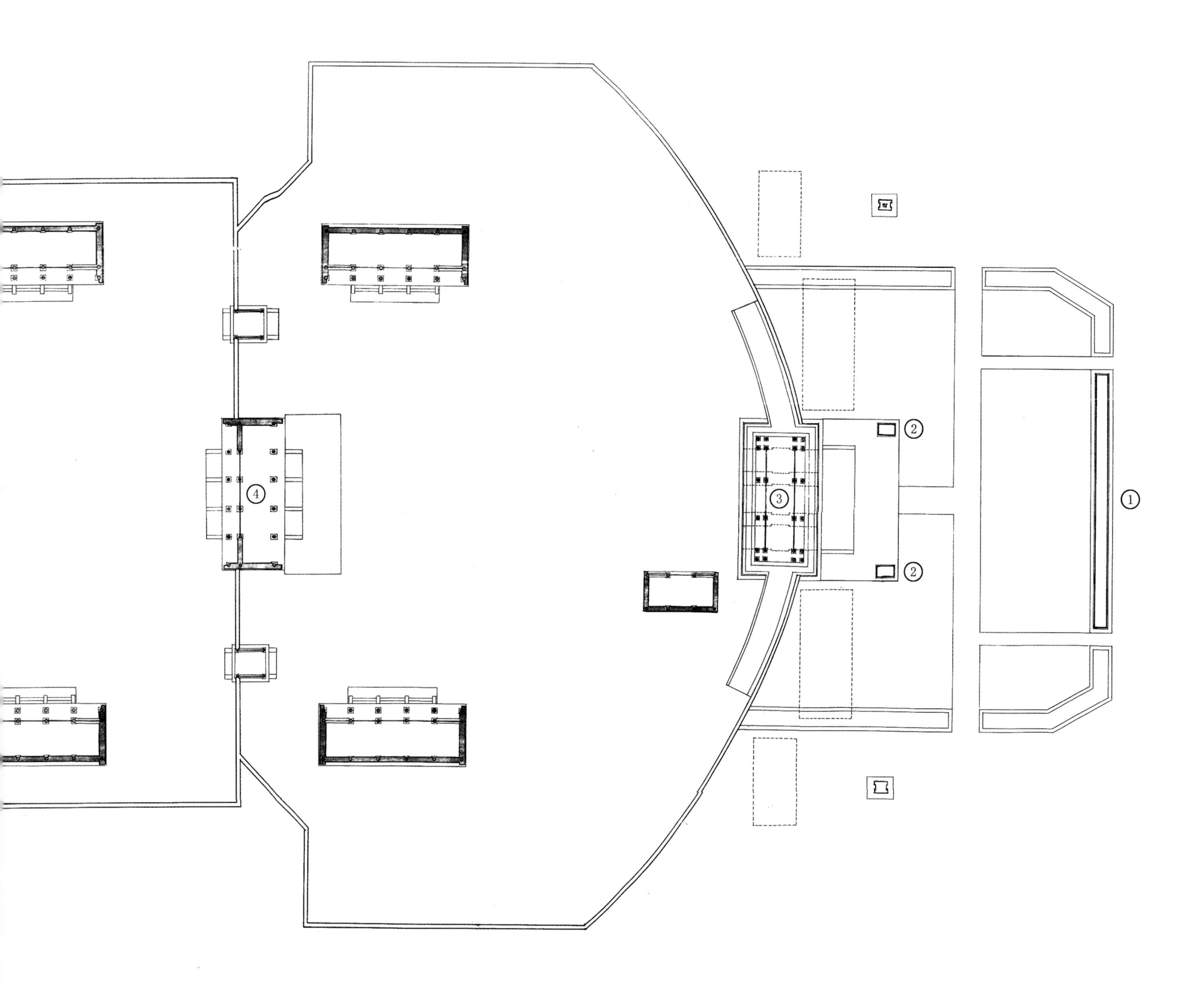

1 照壁　Zhaobi (Screen Wall)
2 石狮　Stone Lions
3 丽正门　Lizheng Gate
4 午门　Meridian Gate
5 铜狮　Bronze Lions
6 宫门　Palace Gate
7 乐亭　Leting (Performance Pavilion)
8 配殿　Peidian (Side Halls)
9 澹泊敬诚殿　Danbo jingcheng (Hall)
10 依清旷殿　Yiqingkuang (Hall)
11 十九间殿　Shijiujian (Hall)

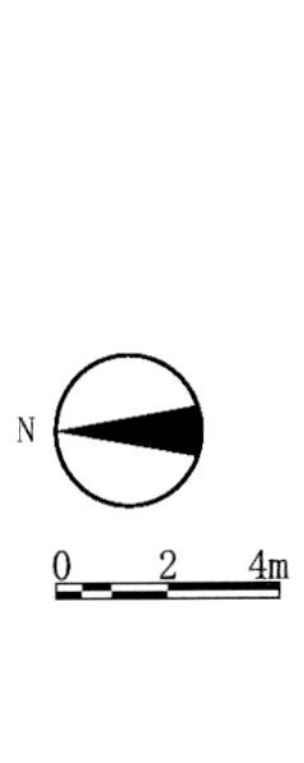

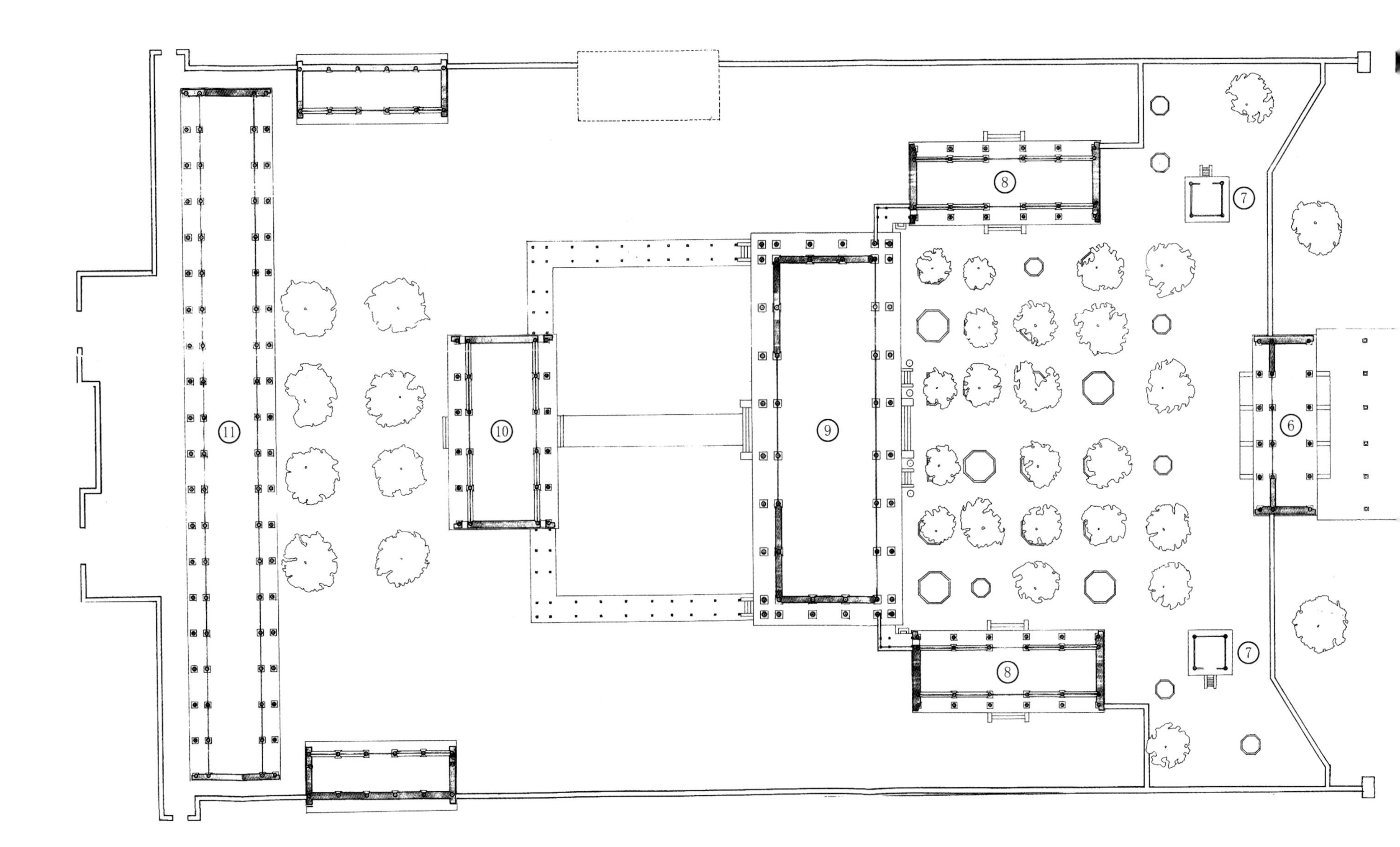

澹泊敬诚殿建筑群总平面图

Site plan of buildings around Danbo jingcheng Hall

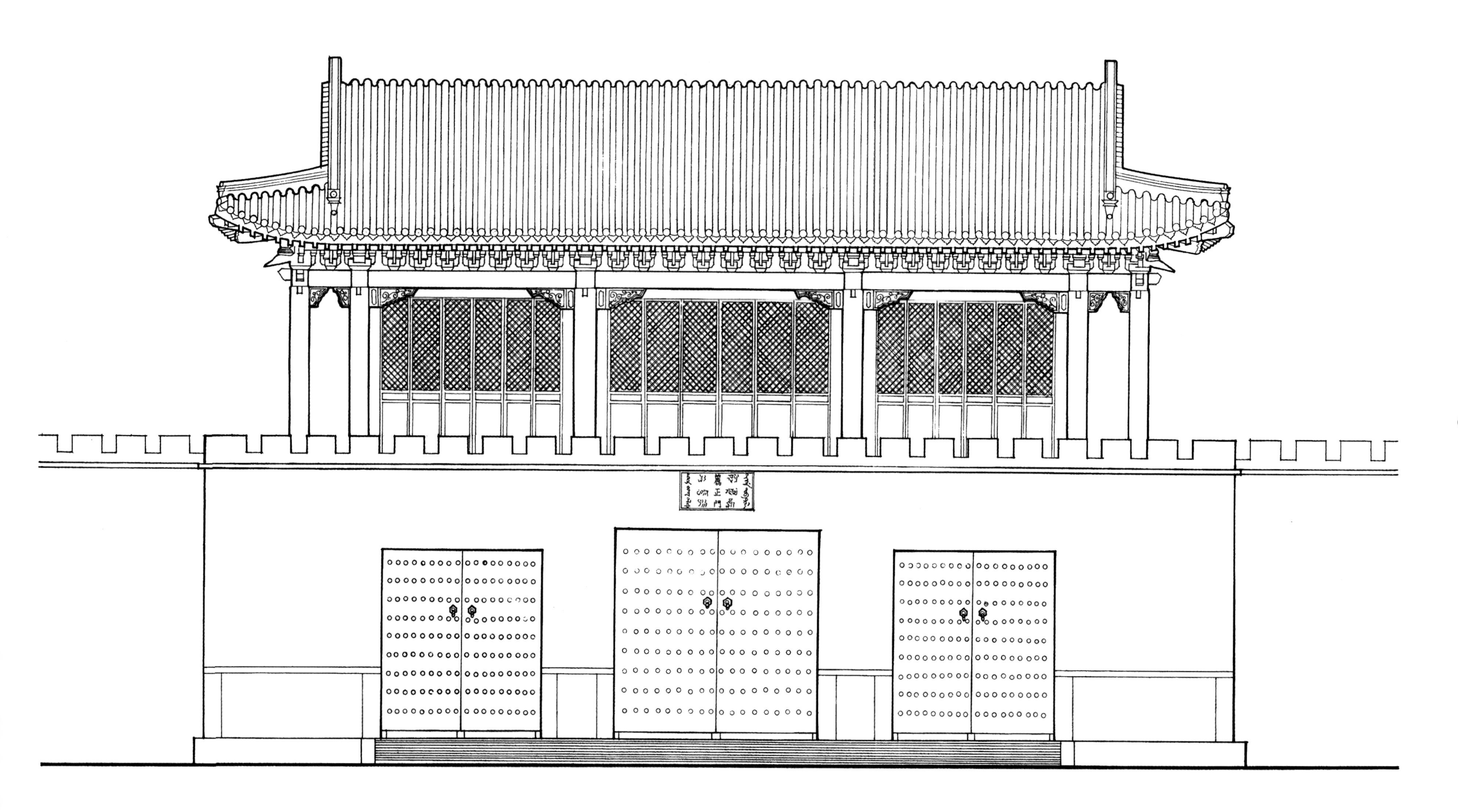

丽正门正立面图

Front elevation of Lizheng Gate

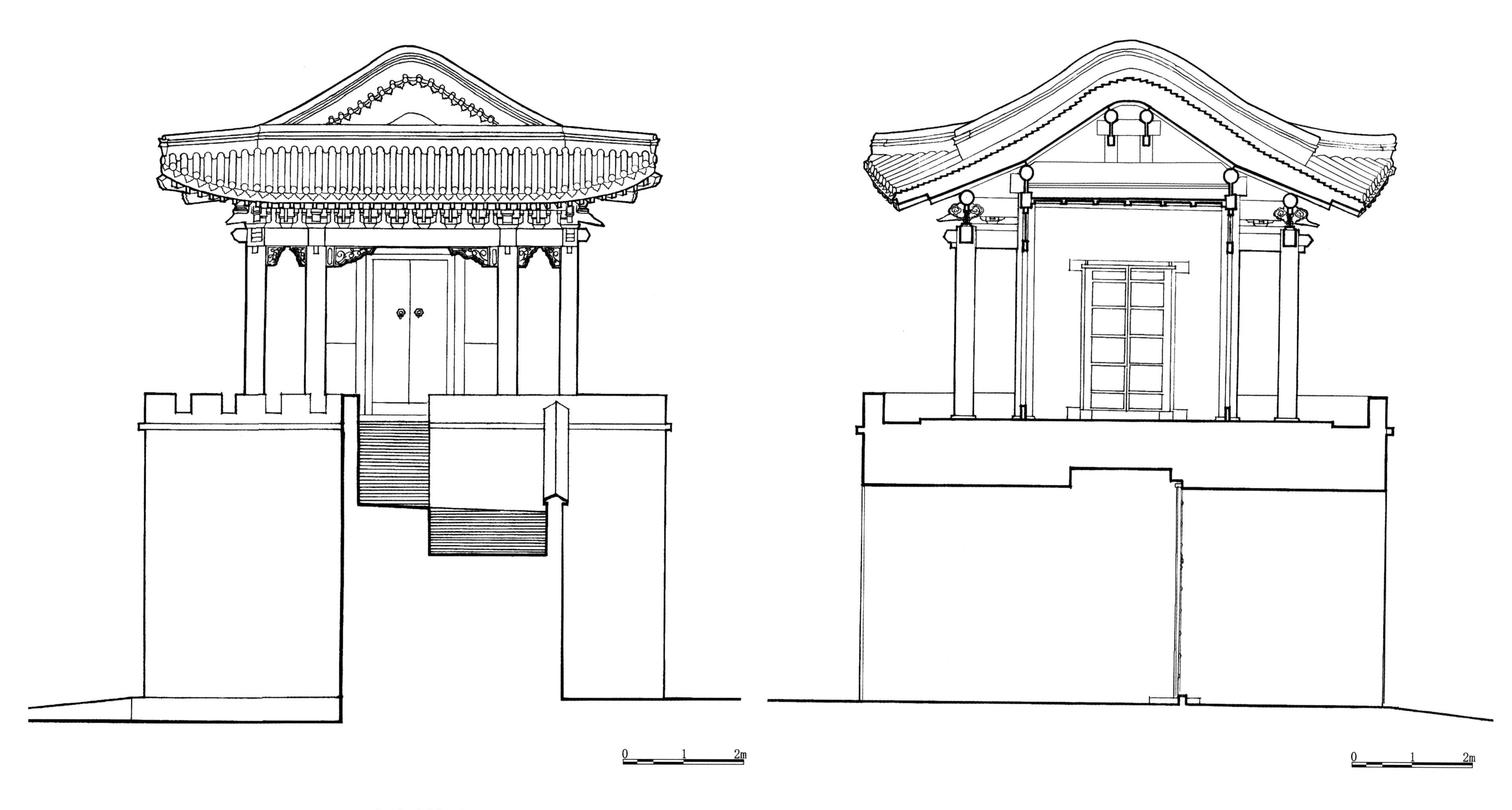

丽正门侧立面图
Side elevation of Lizheng Gate

丽正门横剖面图
Cross-section of Lizheng Gate

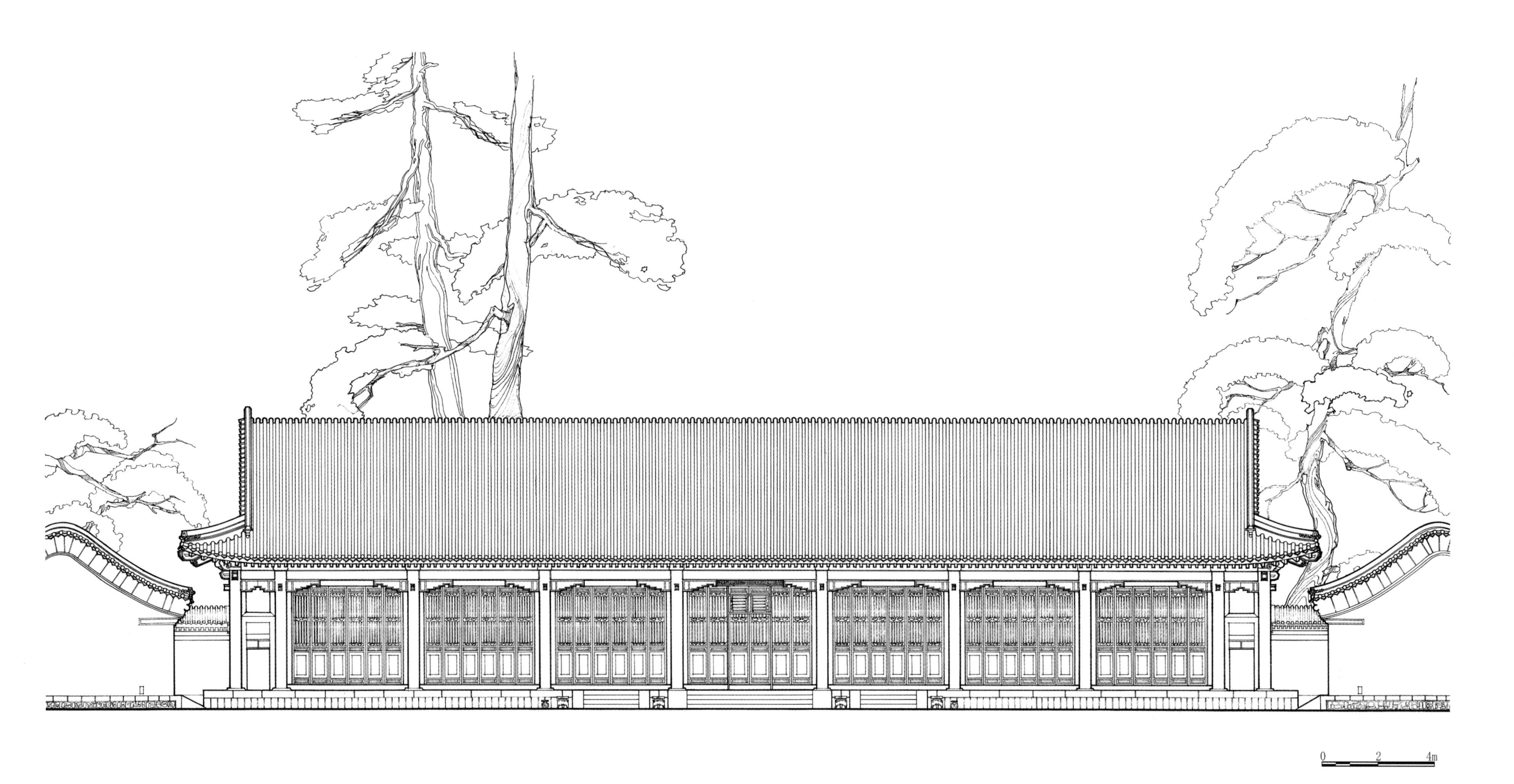

澹泊敬诚殿正立面图
Front elevation of Danbo jingcheng Hall

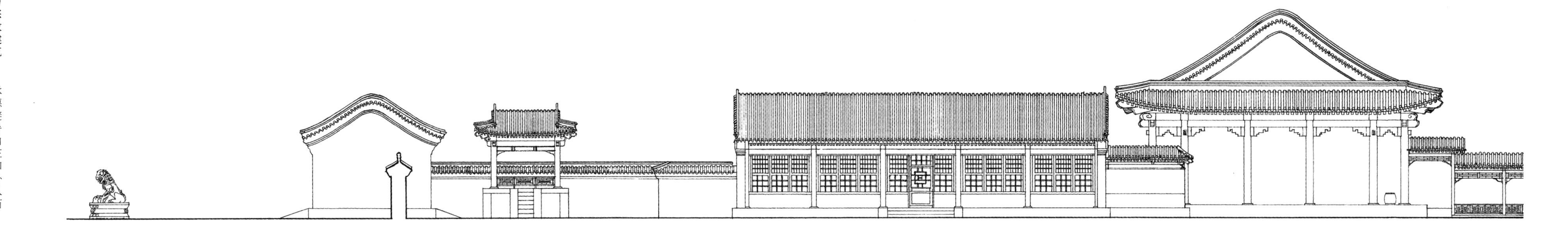

澹泊敬诚殿组群侧立面图
Side elevation of building cluster around Danbo jingcheng Hall

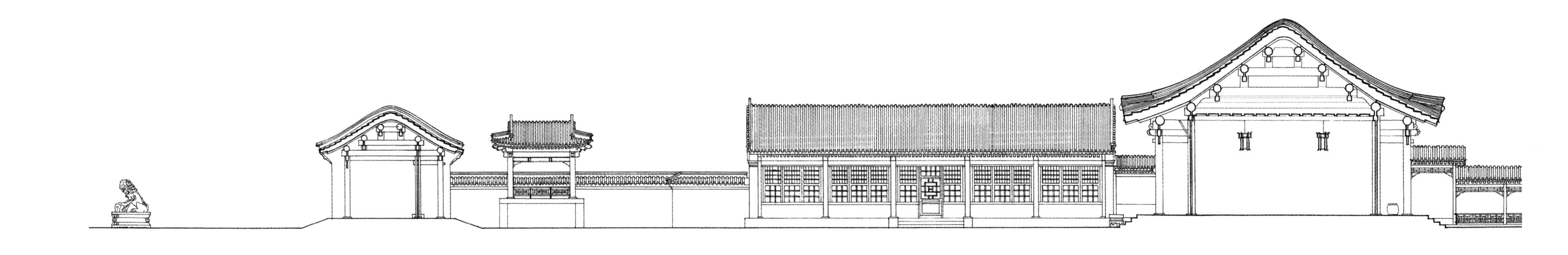

澹泊敬诚殿组群横剖面图
Cross-section of building cluster around Danbo jingcheng Hall

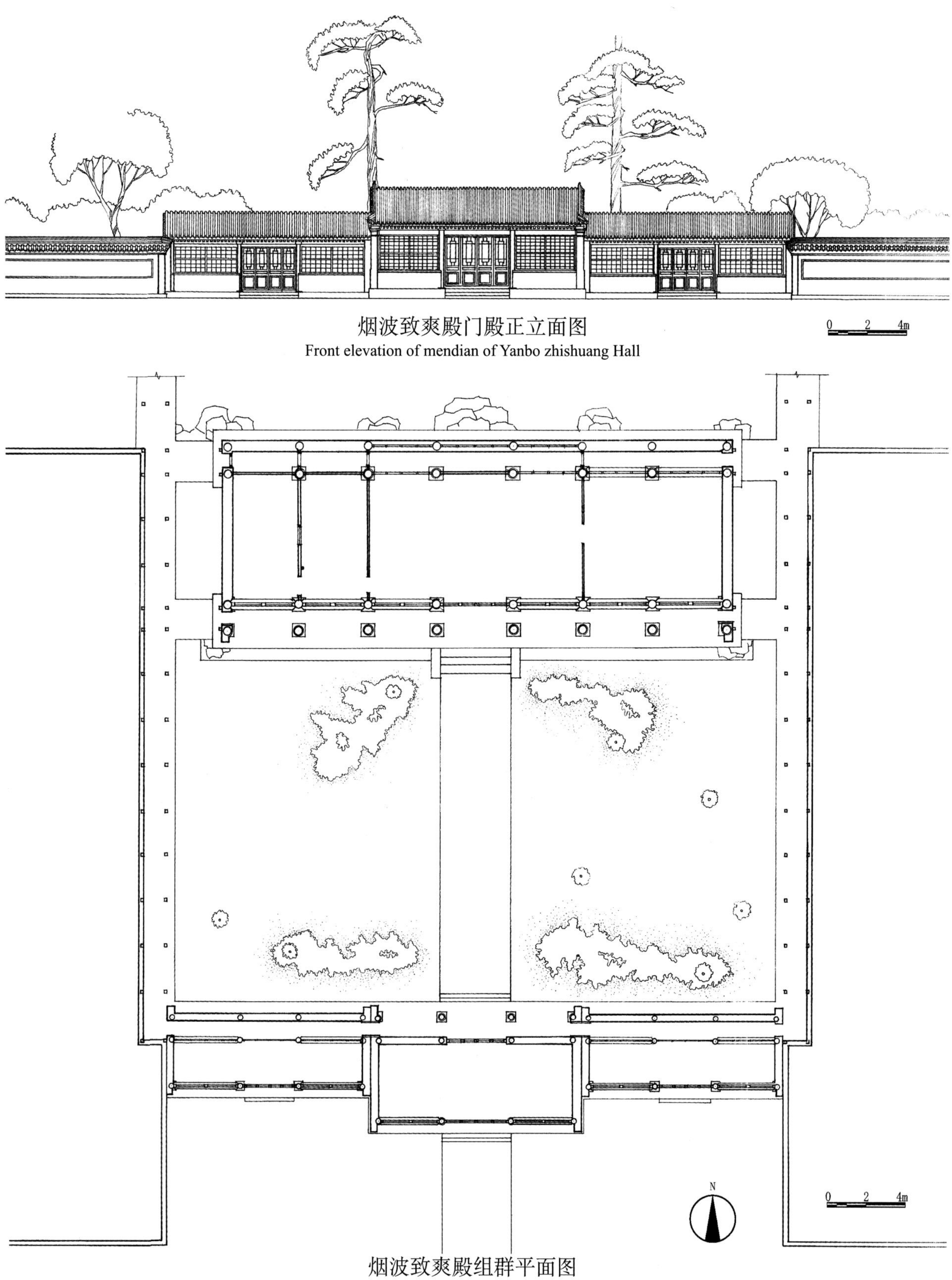

烟波致爽殿门殿正立面图

Front elevation of mendian of Yanbo zhishuang Hall

烟波致爽殿组群平面图

Plan of building cluster around Yanbo zhishuang Hall

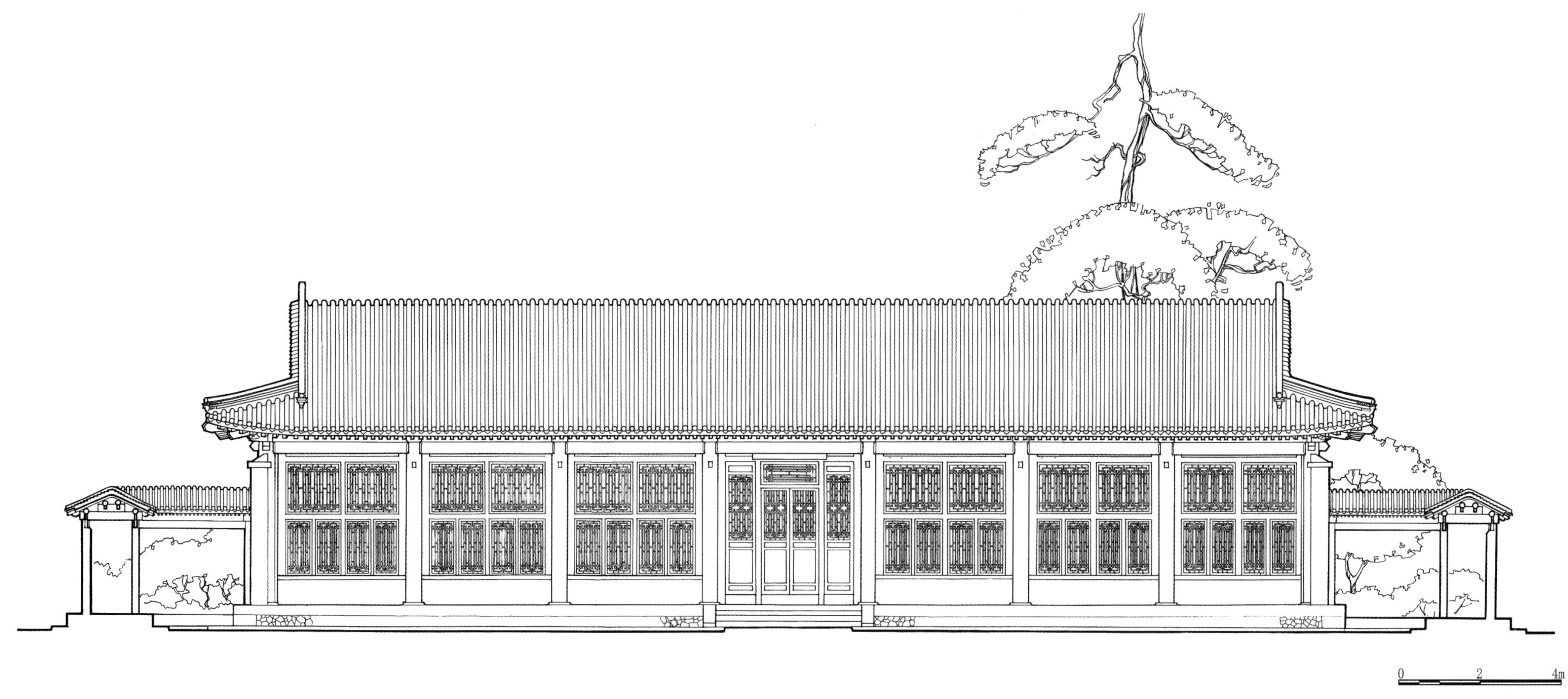

烟波致爽殿正立面图
Front elevation of Yanbo zhishuang Hall

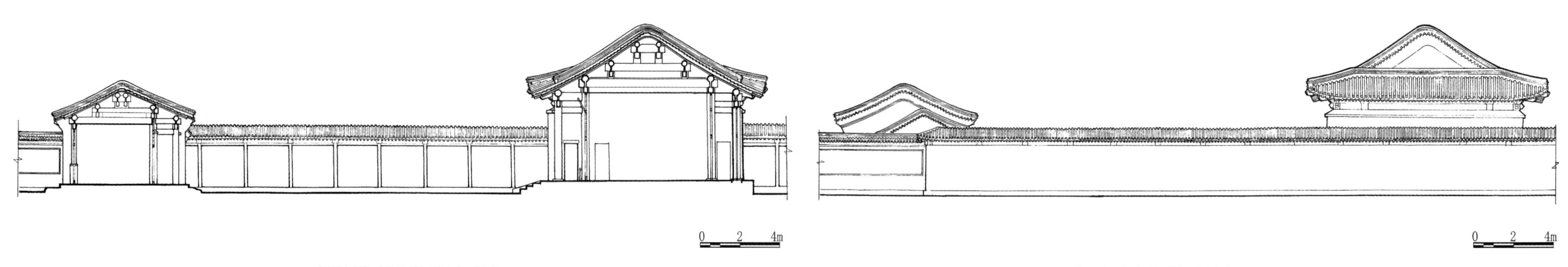

烟波致爽殿组群剖面图
Section of building cluster around Yanbo zhishuang Hall

烟波致爽殿组群侧立面图
Side elevation of building cluster around Yanbo zhishuang Hall

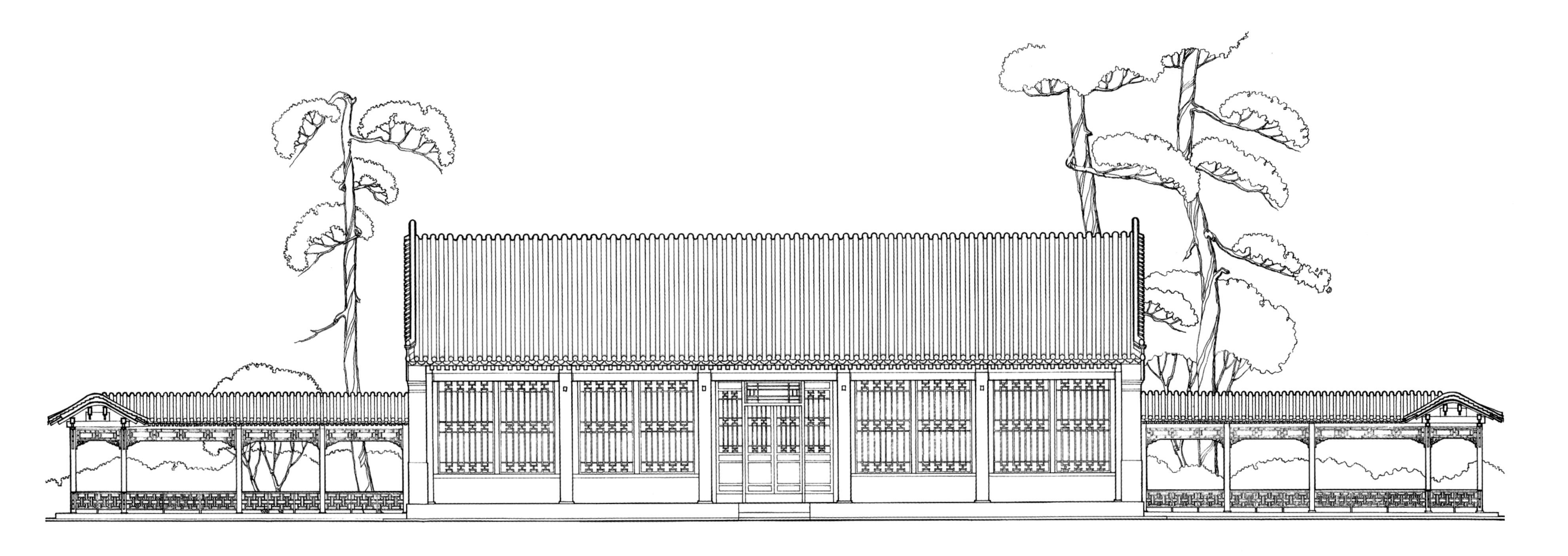

依清旷殿正立面图
Front elevation of Yiqingkuang Hall

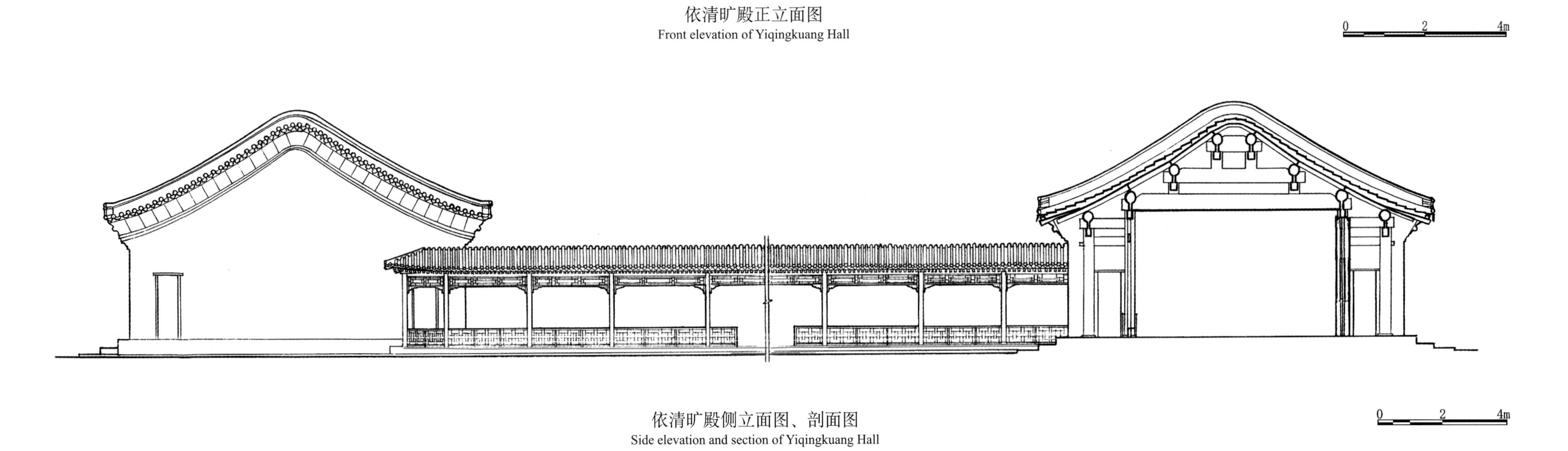

依清旷殿侧立面图、剖面图
Side elevation and section of Yiqingkuang Hall

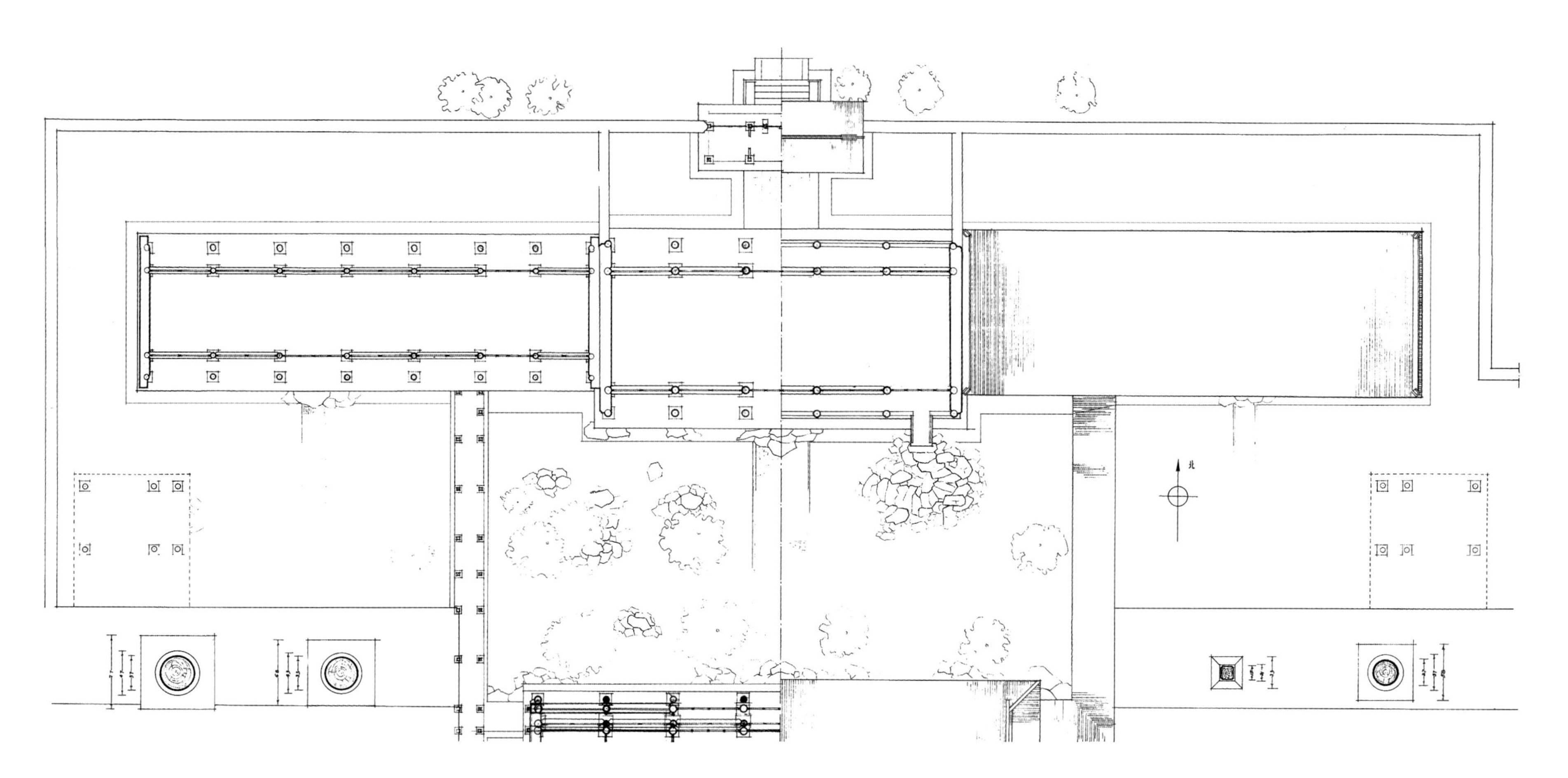

云山胜地楼组群总平面图
Site plan of building cluster around Yunshan shengdi Building

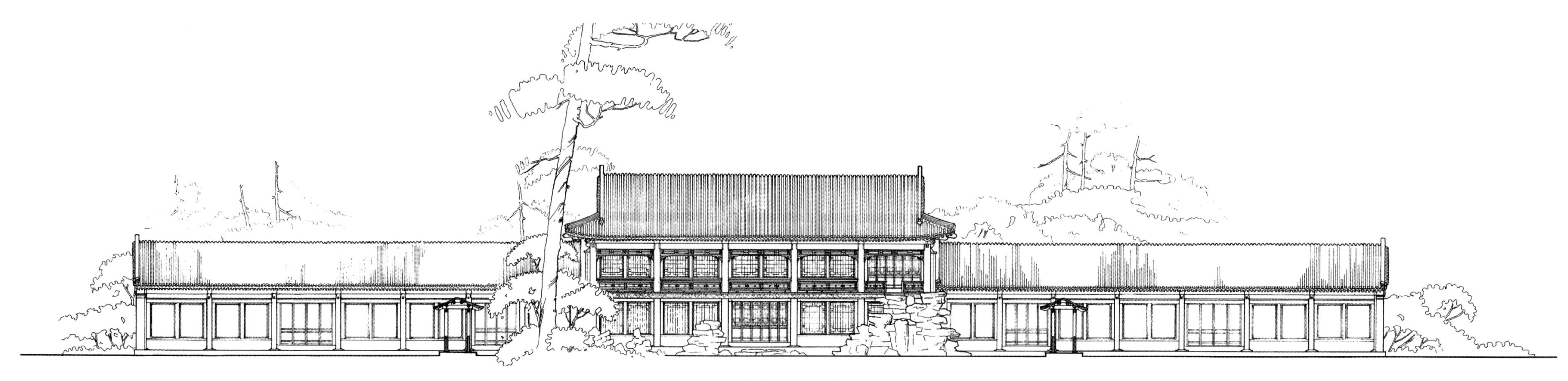

云山胜地楼组群总立面图
Site elevation of building cluster around Yunshan shengdi Building

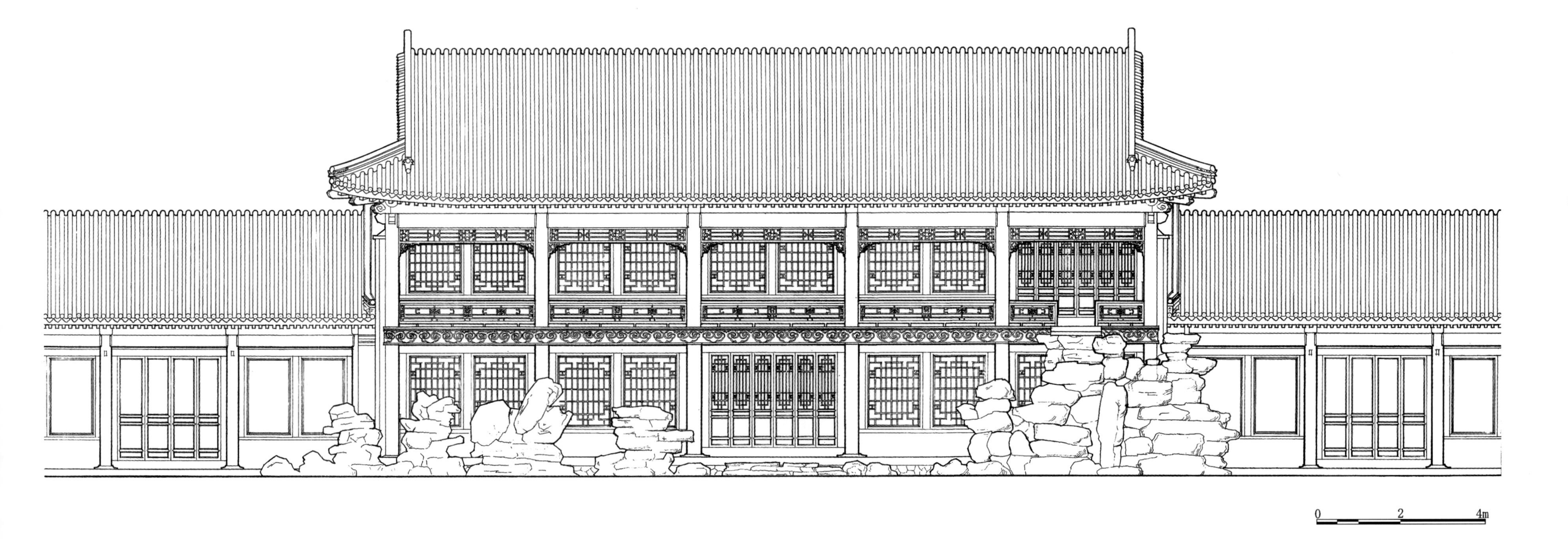

云山胜地楼南立面图
South elevation of Yunshan shengdi Building

云山胜地楼组群东立面图
East elevation of building cluster around Yunshan shengdi Building

云山胜地楼组群剖面图
Section of building cluster around Yunshan shengdi Building

云山胜地楼垂花门立面图
Elevation of chuihuamen of Yunshan shengdi Building

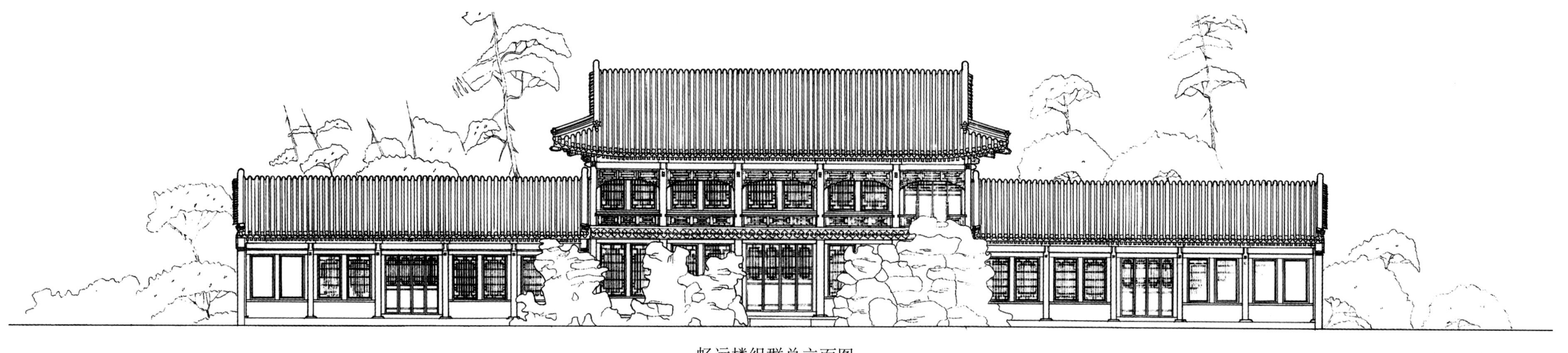

畅远楼组群总立面图
Site elevation of building cluster around Changyuan Building

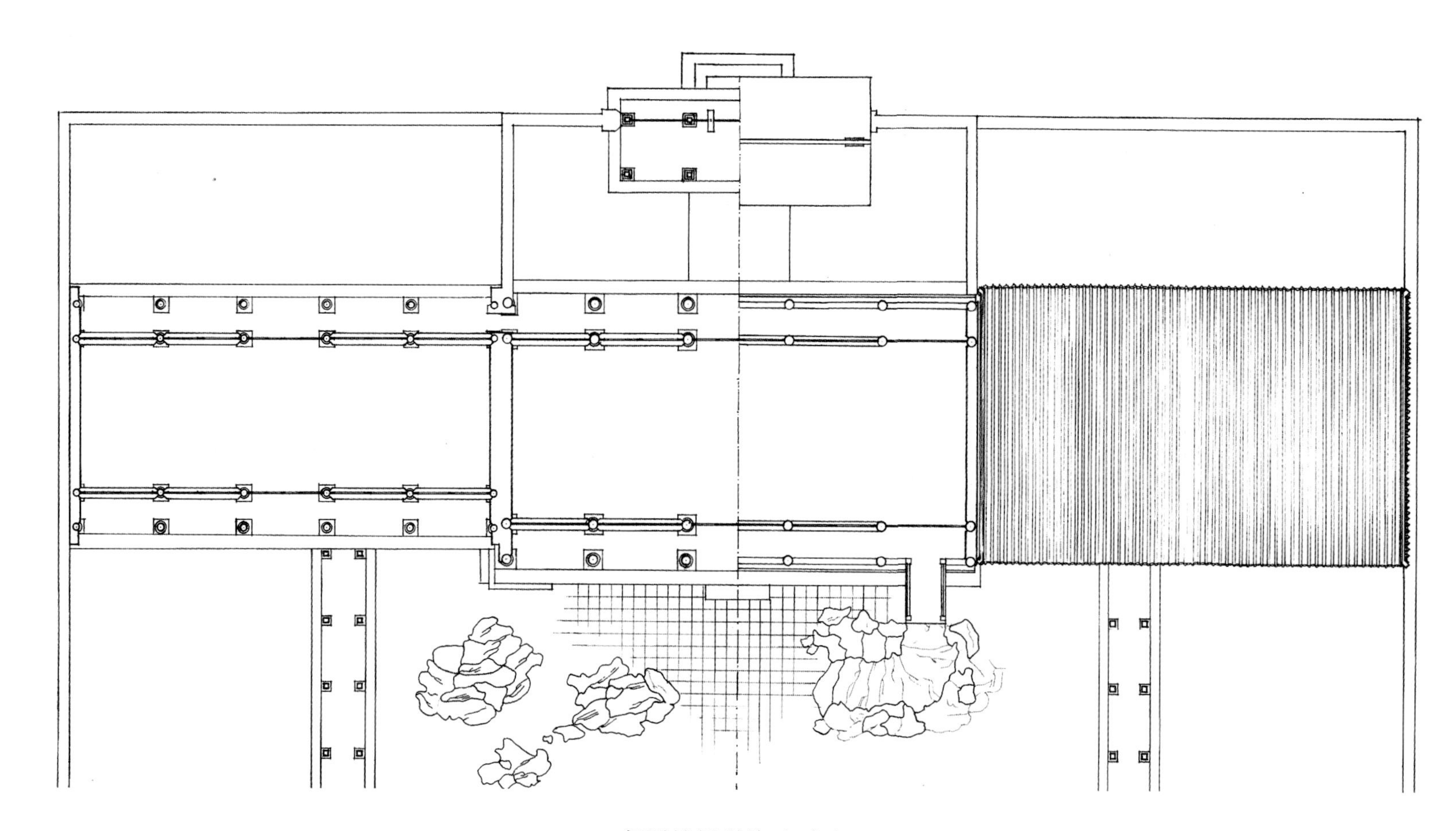

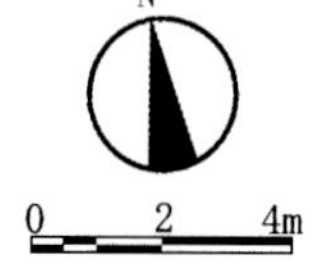

畅远楼组群总平面图
Site plan of building cluster around Changyuan Building

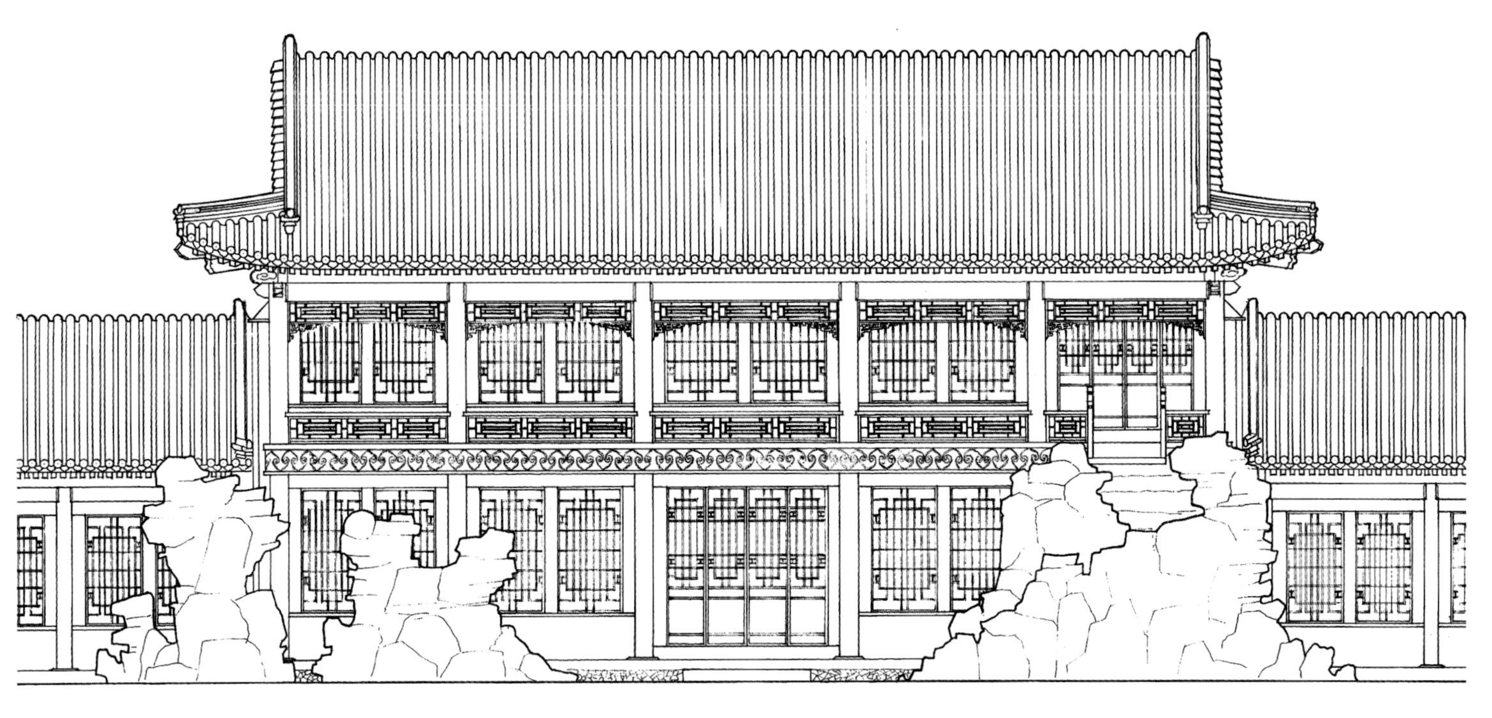

畅远楼正立面图
Front elevation of Changyuan Building

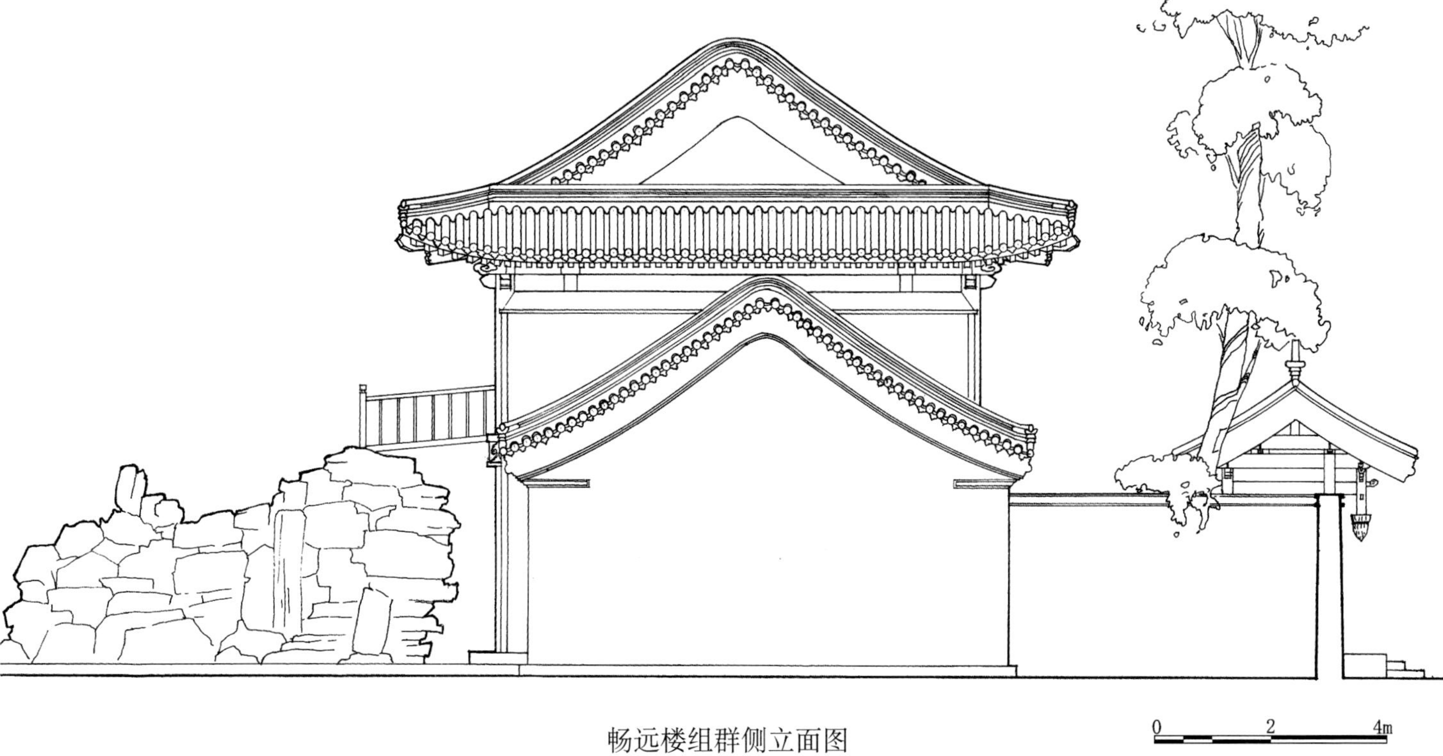

畅远楼组群侧立面图
Side elevation of building cluster around Changyuan Building

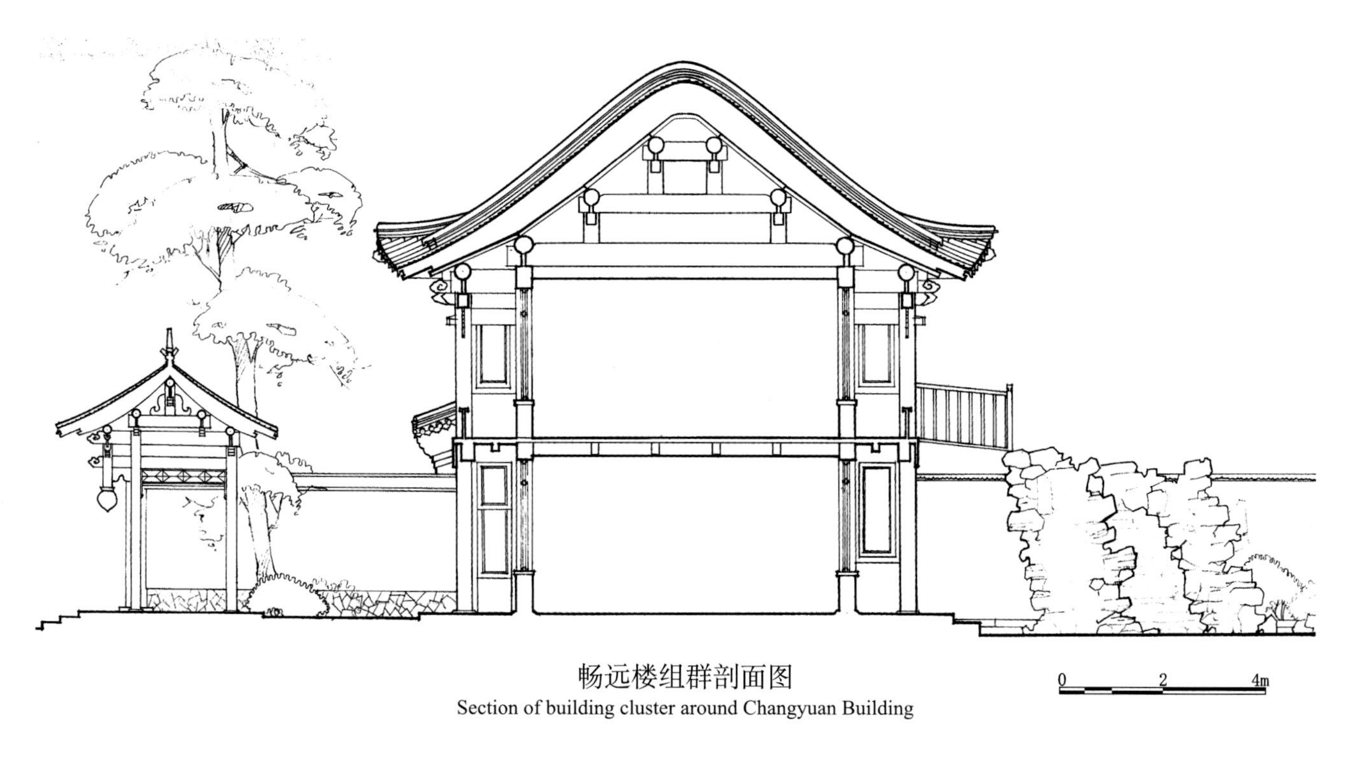

畅远楼组群剖面图

Section of building cluster around Changyuan Building

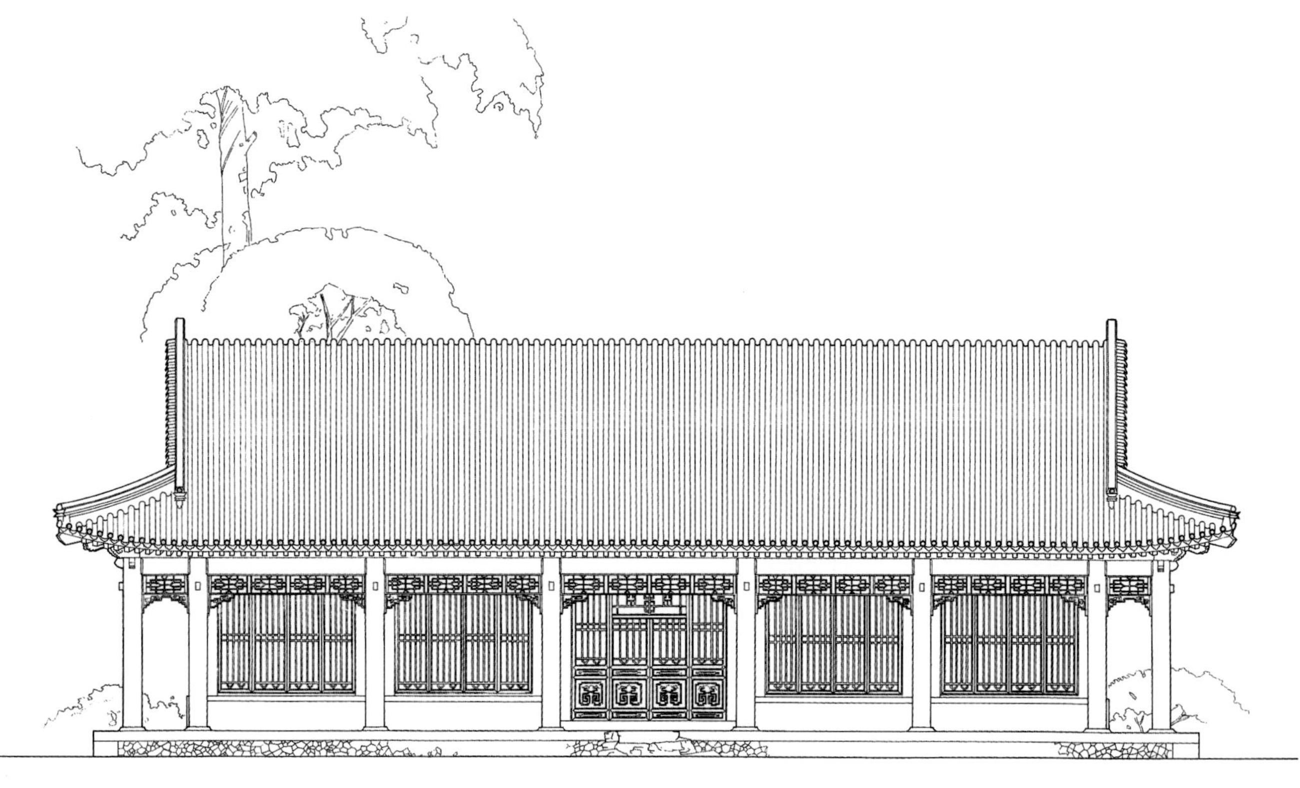

万壑松风纪恩堂正立面图

Front elevation of Ji'entang of Wanhe songfeng

万壑松风建筑群总立面图

Site elevation of buildings around Wanhe songfeng

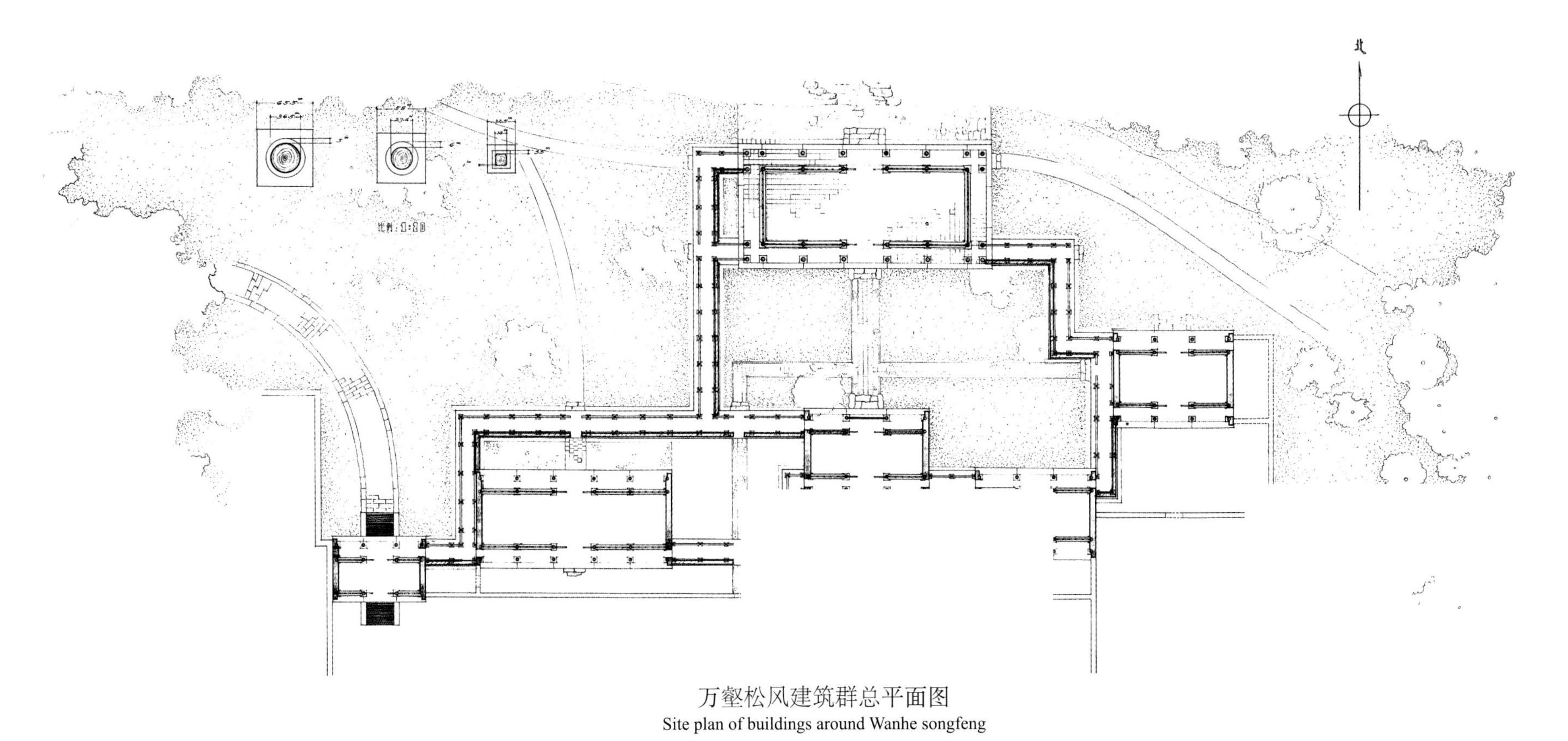

万壑松风建筑群总平面图

Site plan of buildings around Wanhe songfeng

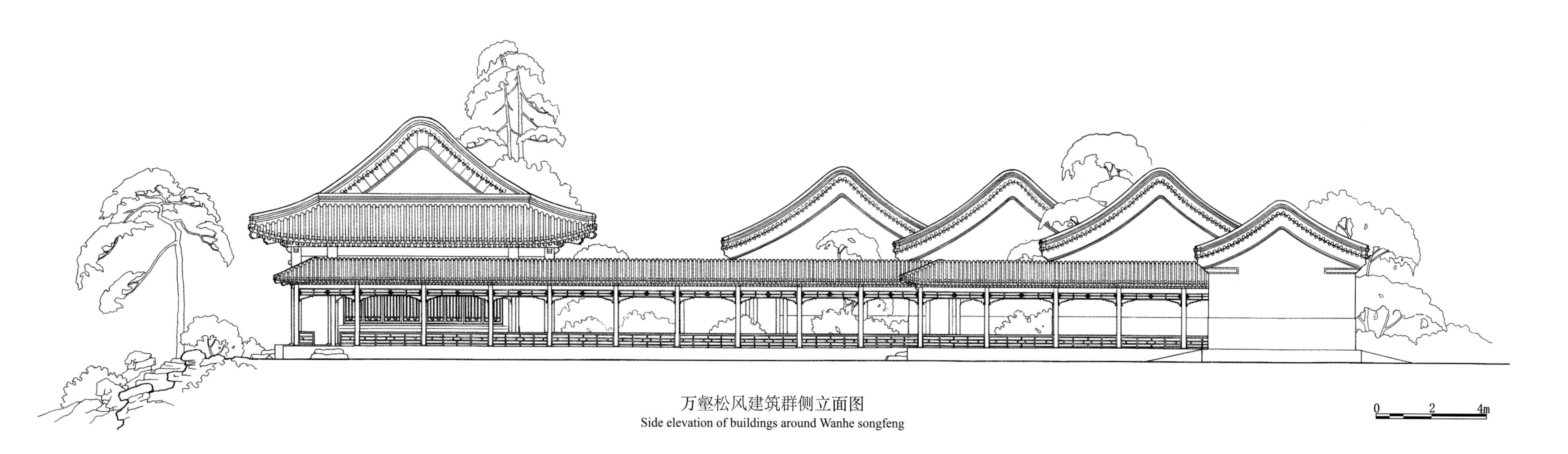

万壑松风建筑群侧立面图
Side elevation of buildings around Wanhe songfeng

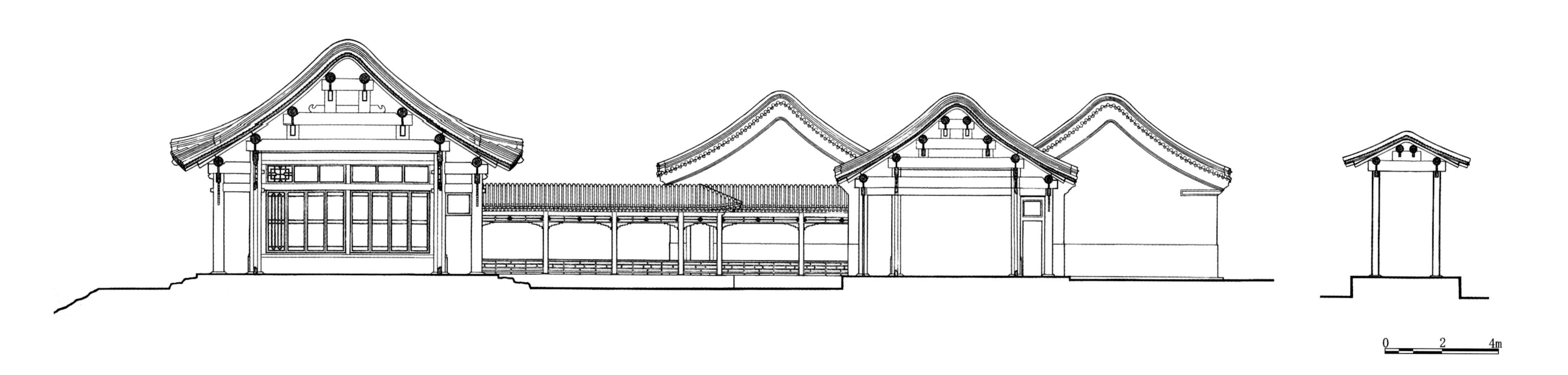

万壑松风建筑群总剖面图
Site section of buildings around Wanhe songfeng

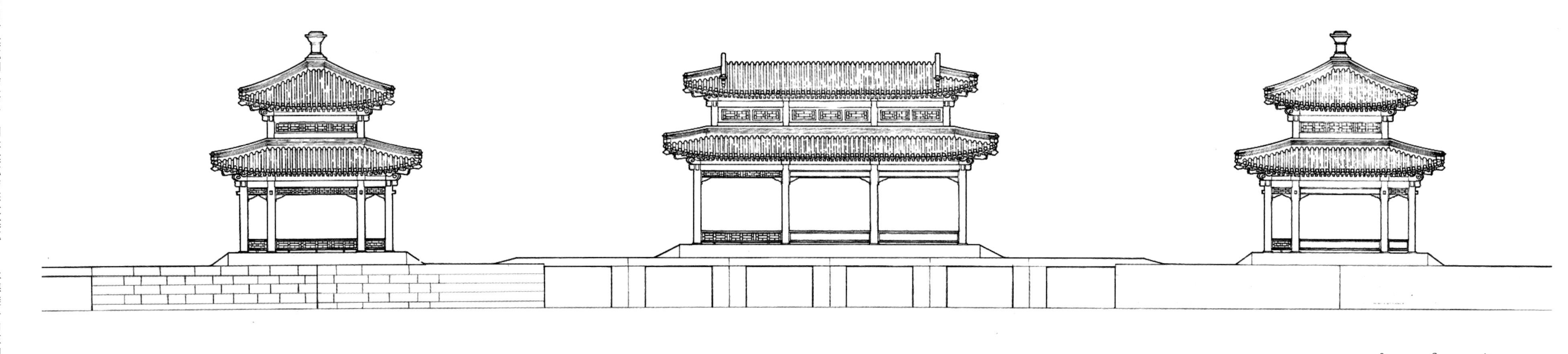

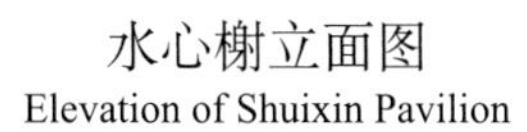
水心榭立面图
Elevation of Shuixin Pavilion

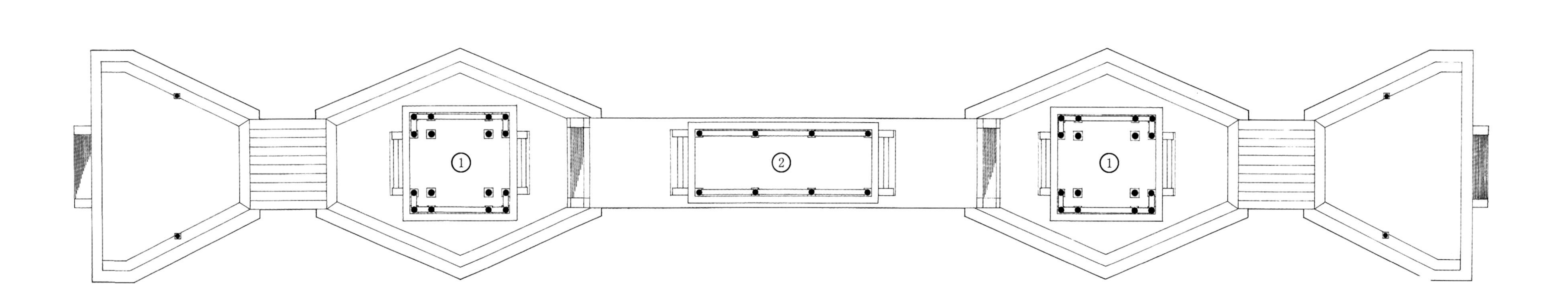

1 方亭　　2 长亭

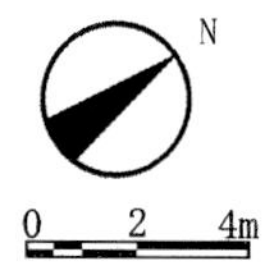

水心榭平面图
Plan of Shuixin Pavilion

水心榭长亭侧立面图
Side elevation of changting of Shuixin Pavilion

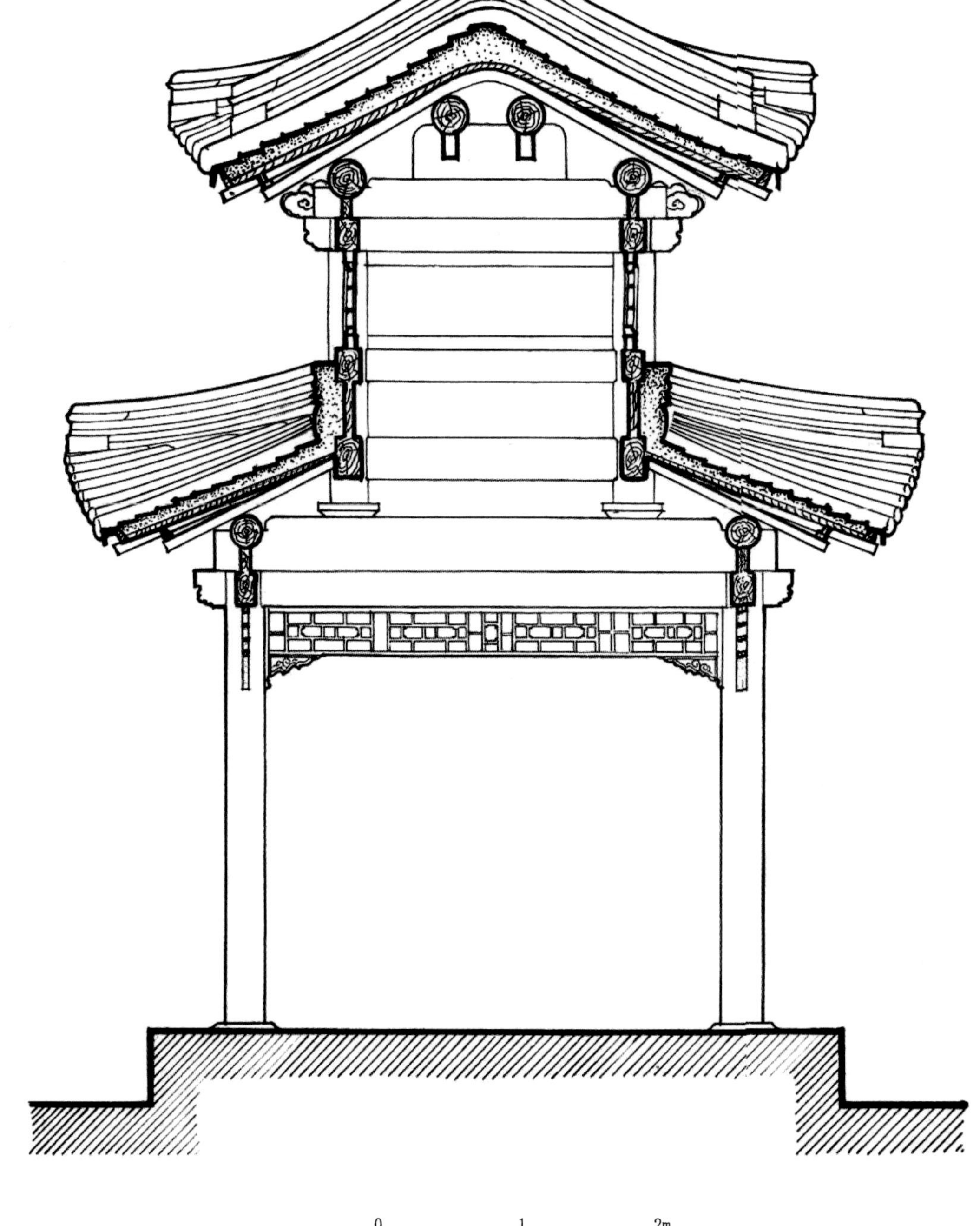

水心榭长亭横剖面图
Cross-section of changting of Shuixin Pavilion

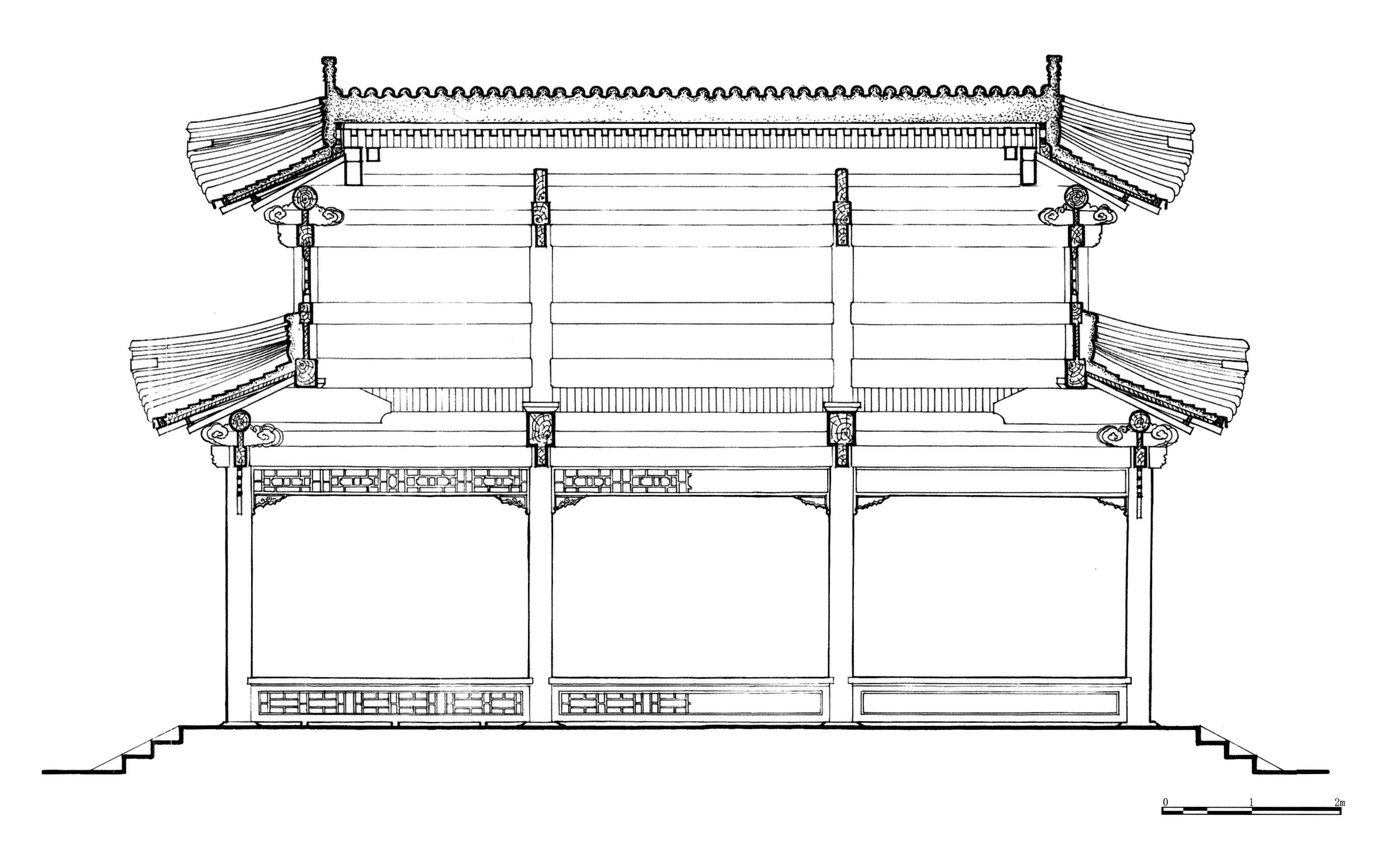

水心榭长亭纵剖面图

Longitudinal section of changting of Shuixin Pavilion

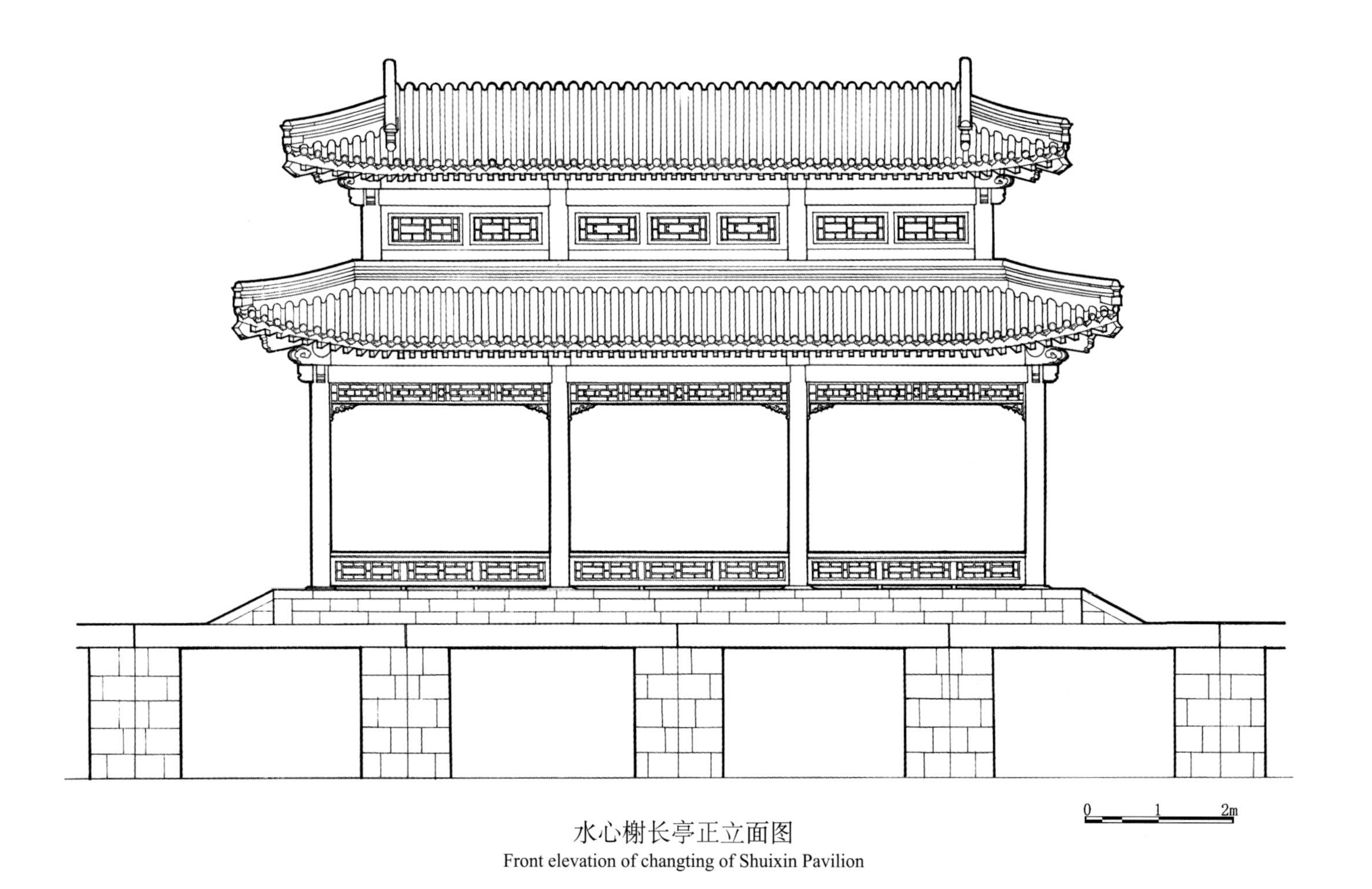

水心榭长亭正立面图
Front elevation of changting of Shuixin Pavilion

水心榭长亭平面图、梁架仰视图
Plan and plan of framework as seen from below of changting of Shuixin Pavilion

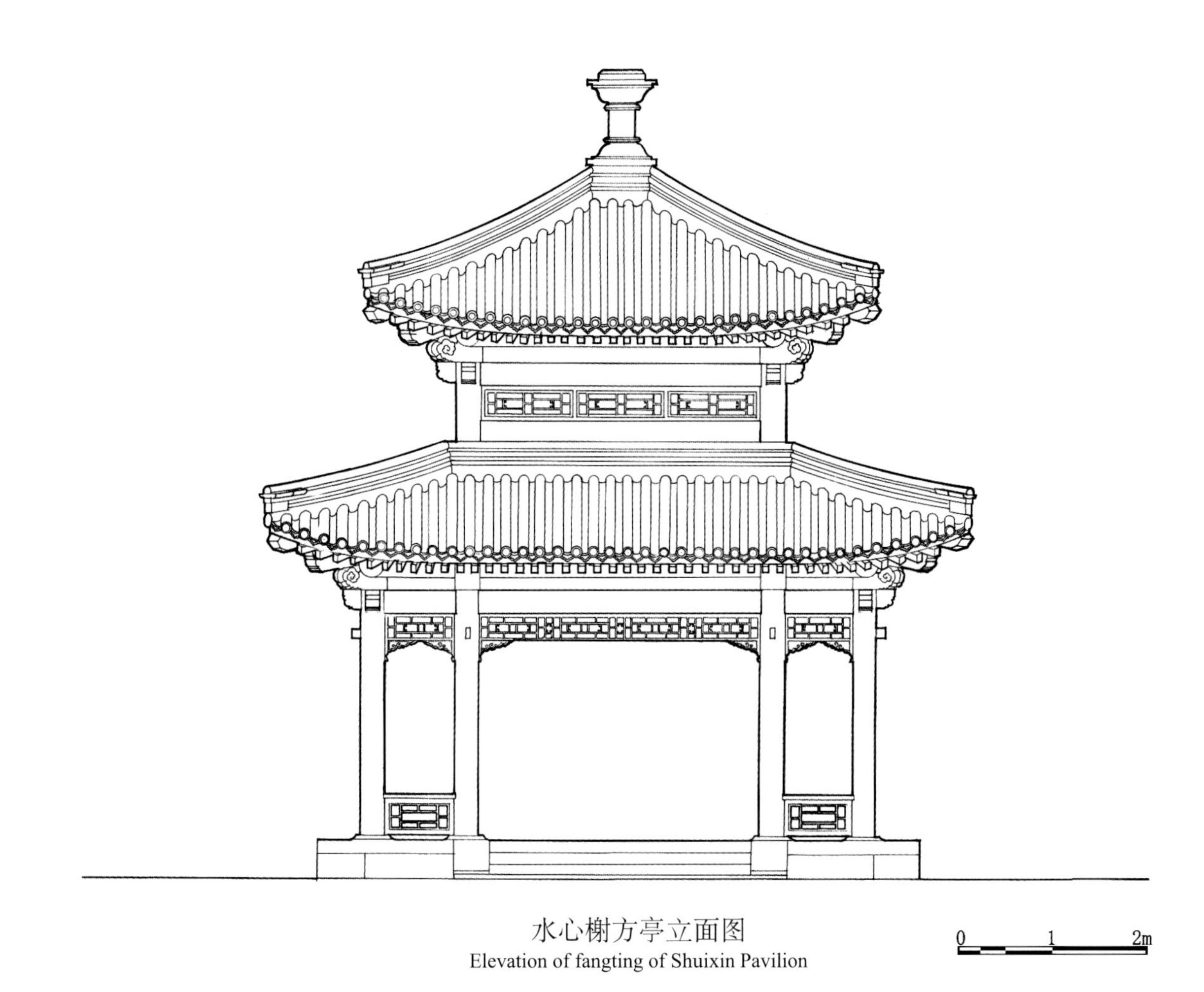

水心榭方亭立面图
Elevation of fangting of Shuixin Pavilion

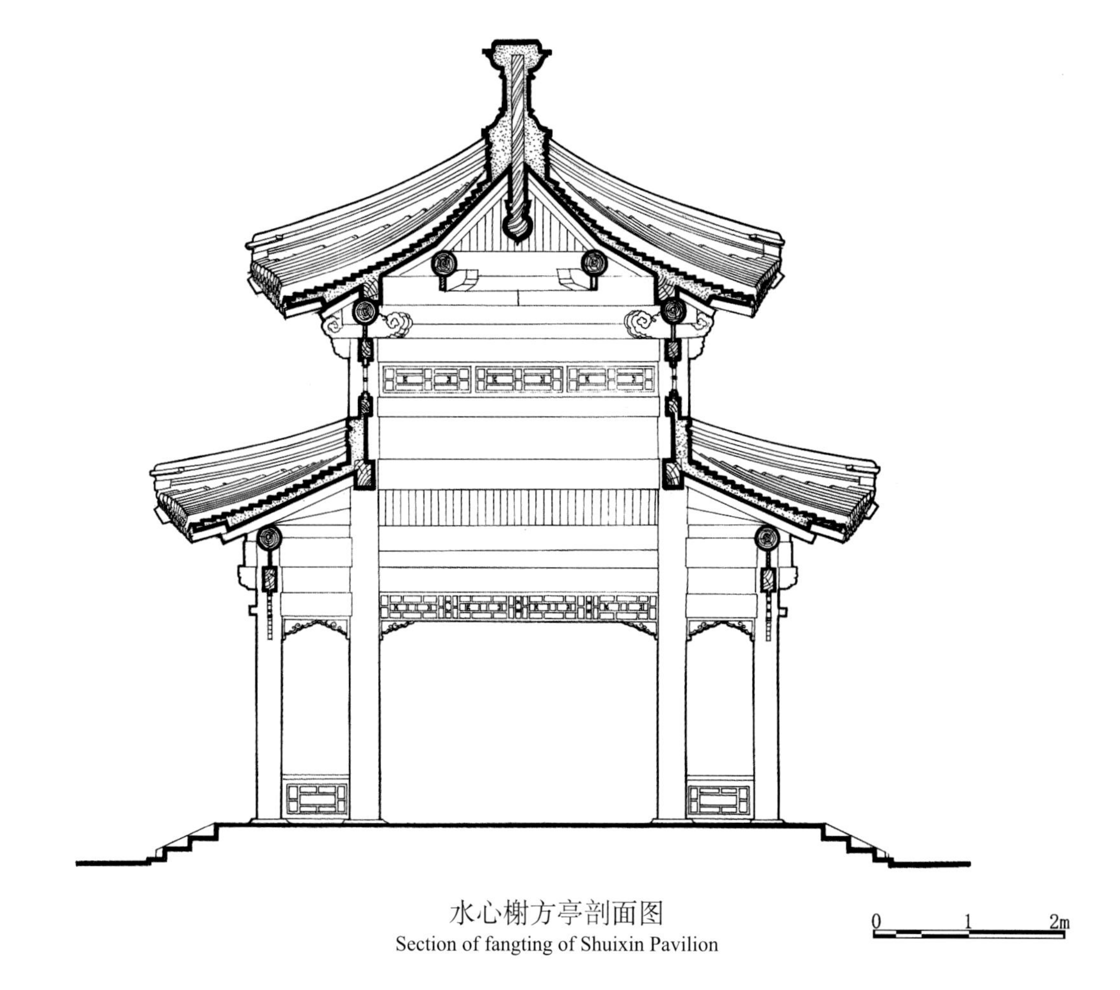

水心榭方亭剖面图
Section of fangting of Shuixin Pavilion

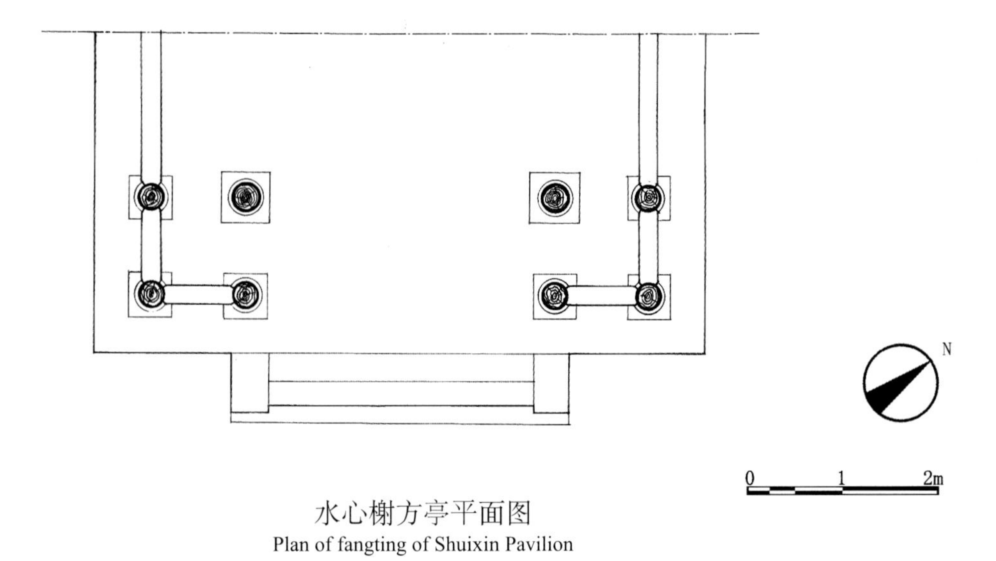

水心榭方亭平面图
Plan of fangting of Shuixin Pavilion

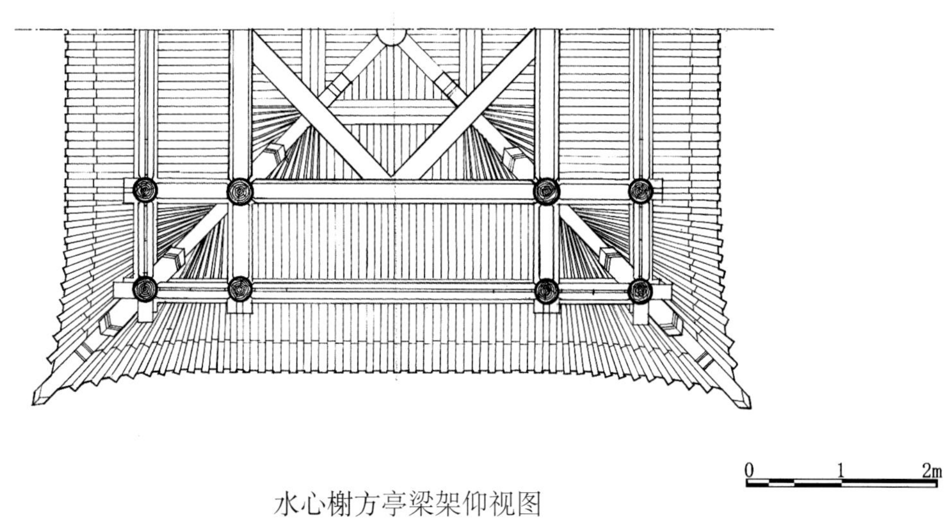

水心榭方亭梁架仰视图
Plan of framework as seen from below of fangting of Shuixin Pavilion

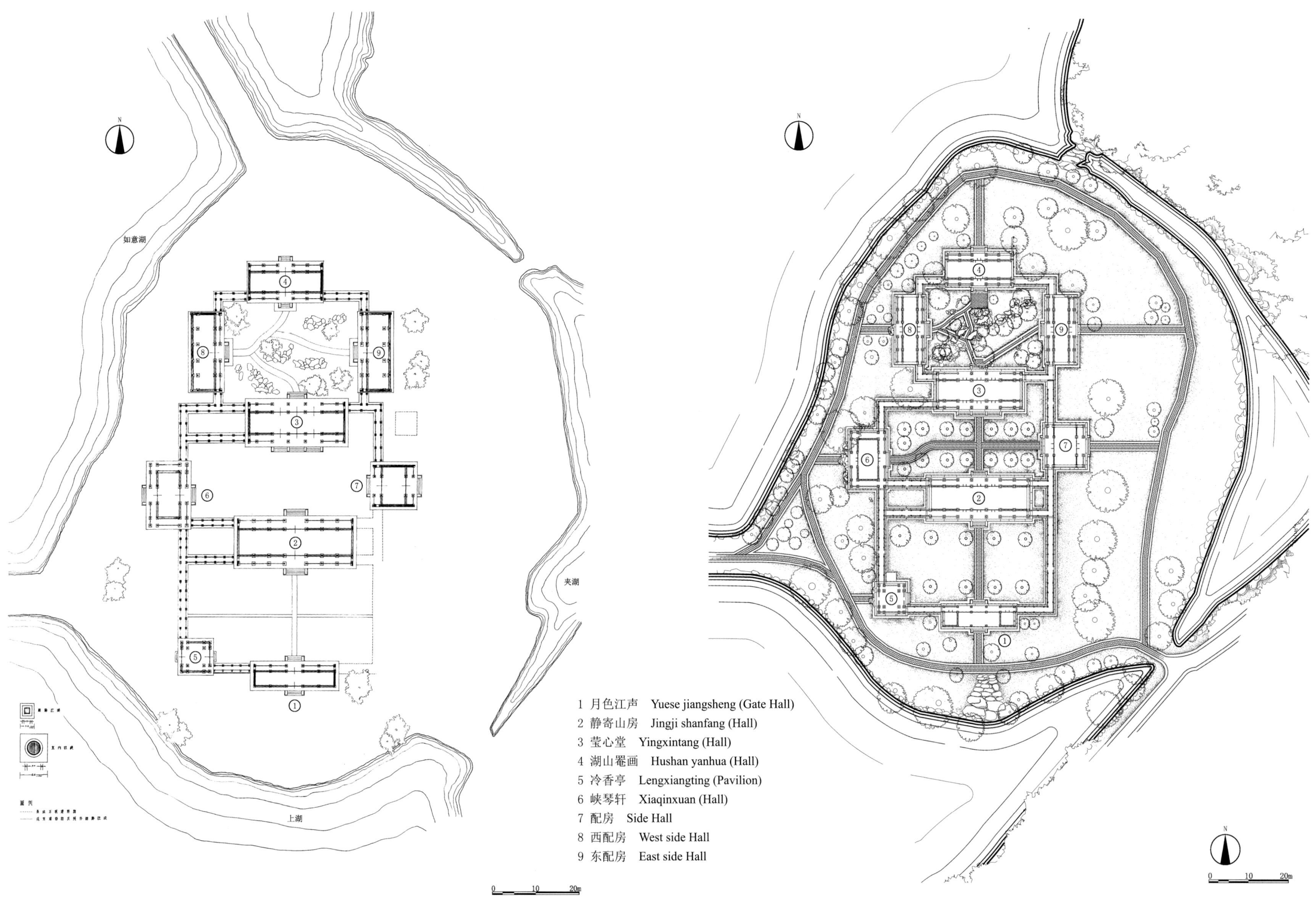

月色江声建筑群总平面图
Site plan of buildings around Yuese jiangsheng

月色江声建筑群总平面图（2011 年测绘）
Site plan of buildings around Yuese jiangsheng (2011 survey)

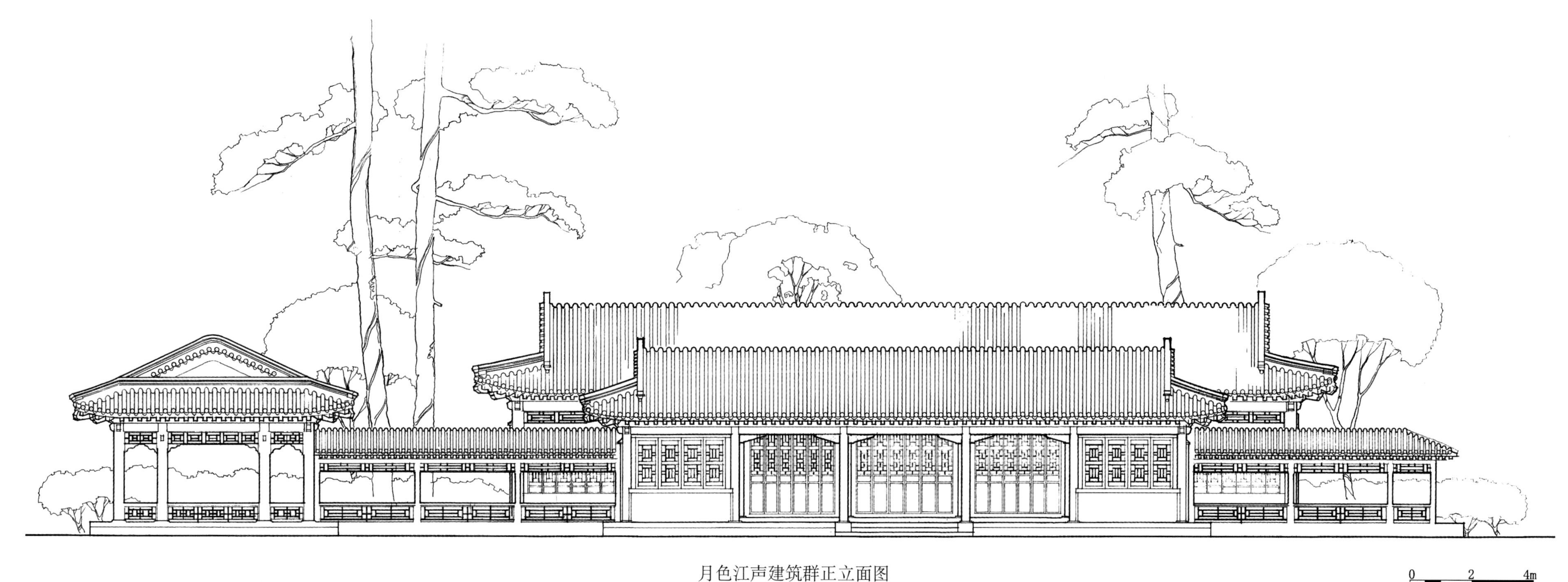

月色江声建筑群正立面图

Yuese jiangsheng buildings around Front elevation of

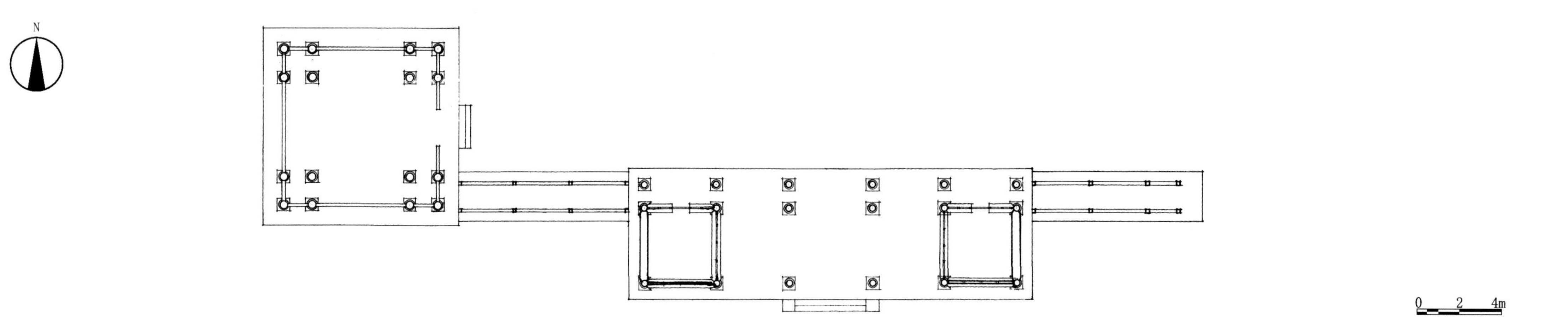

月色江声门殿、冷香亭平面图

Plan of mendian and Lengxiangting of Yuese jiangsheng

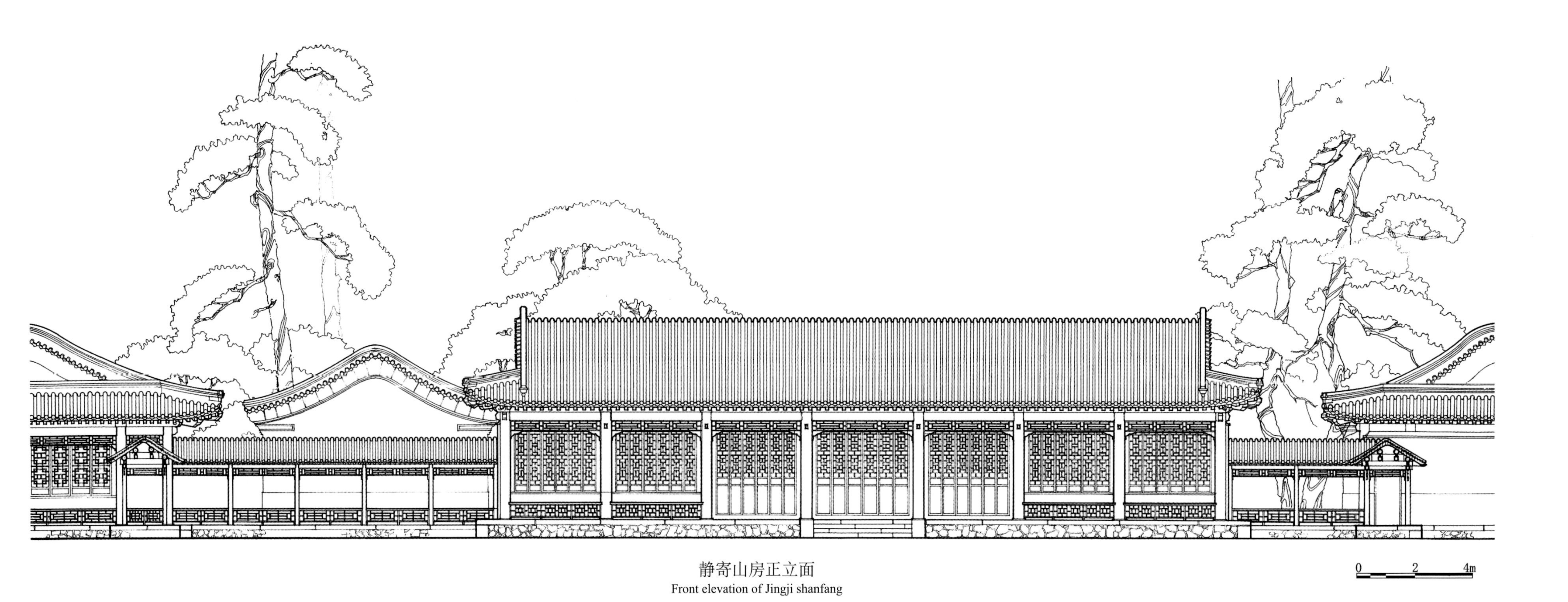

静寄山房正立面
Front elevation of Jingji shanfang

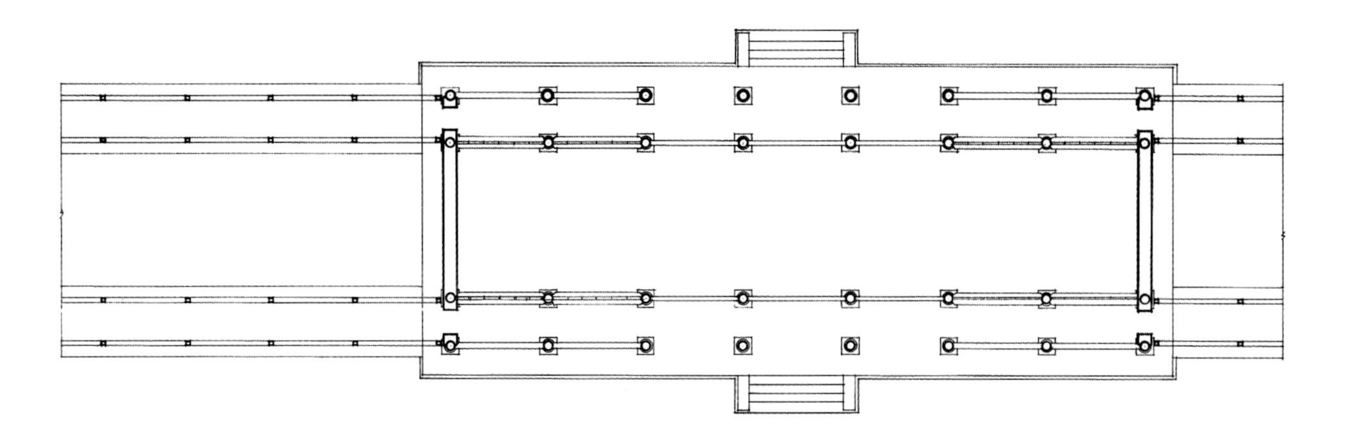

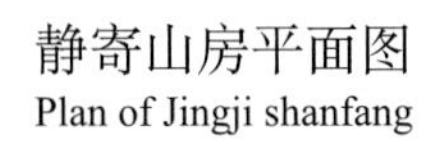

静寄山房平面图
Plan of Jingji shanfang

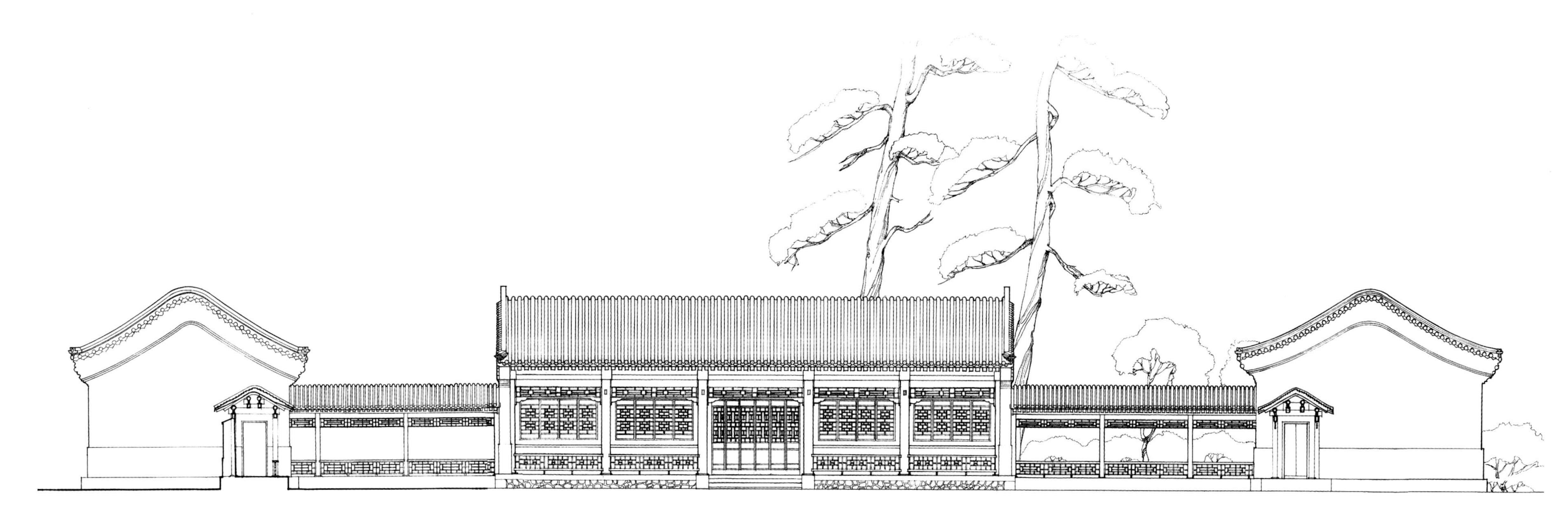

湖山罨画正立面图
Front elevation of Hushan yanhua

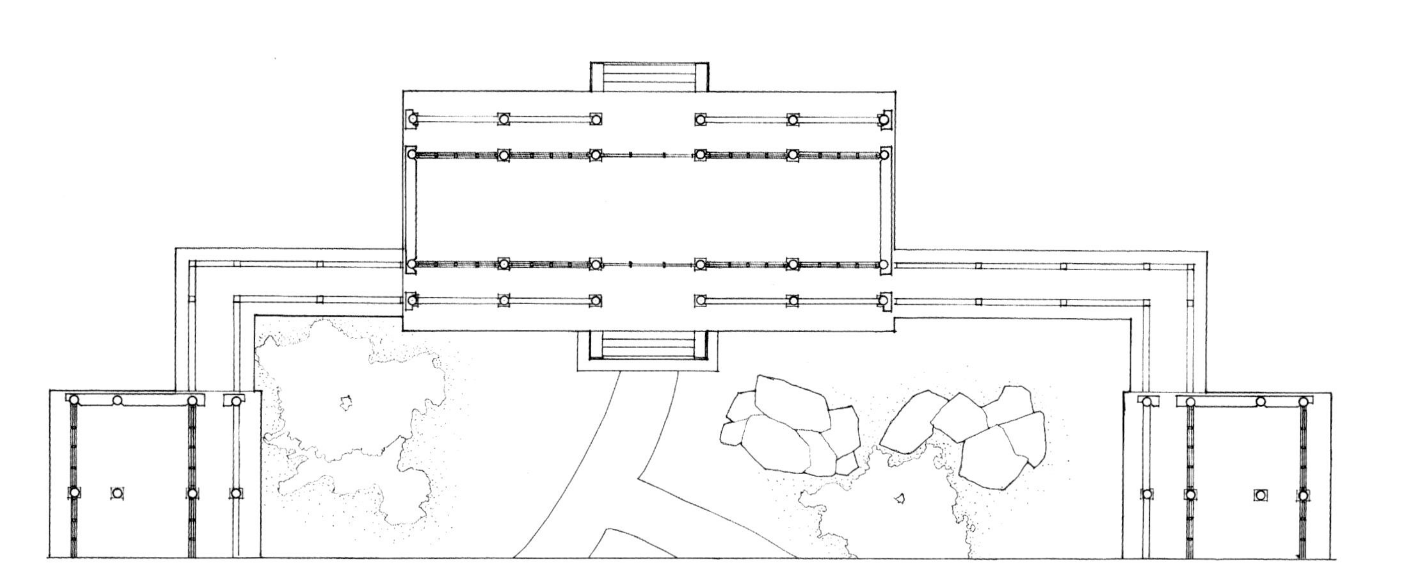

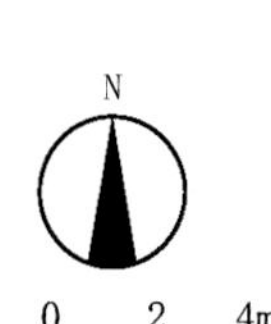

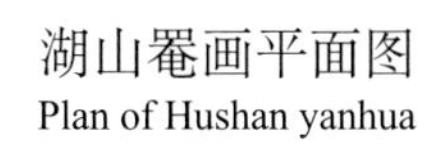

湖山罨画平面图
Plan of Hushan yanhua

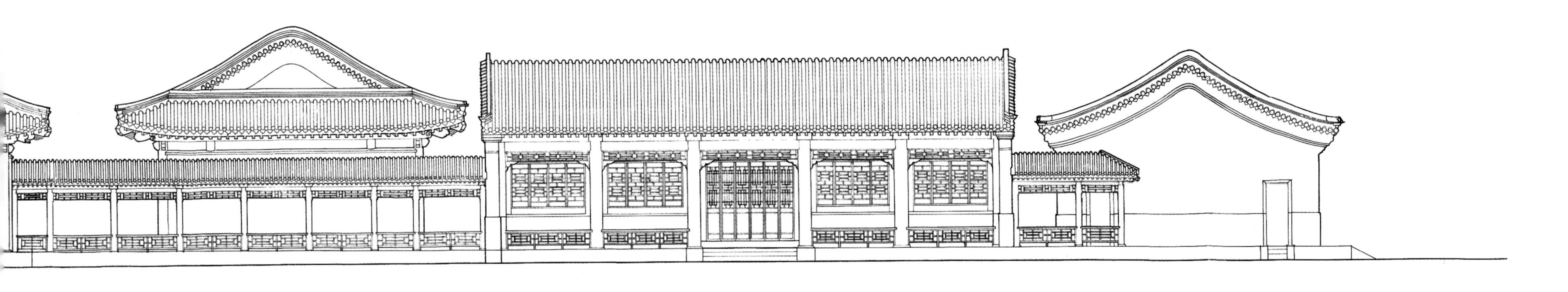
0 2 4m

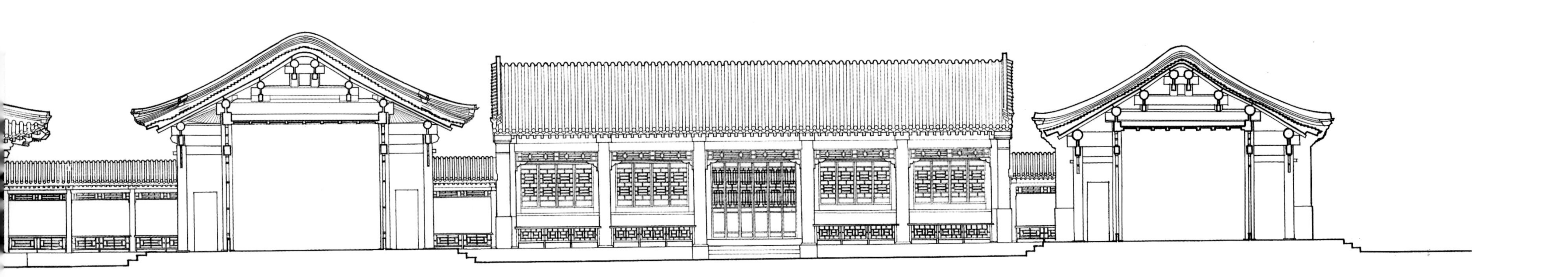
0 2 4m

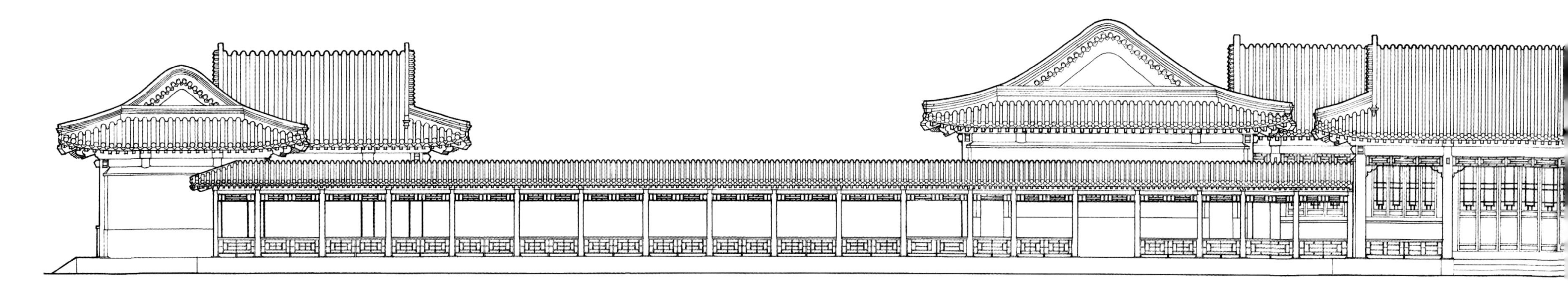

月色江声建筑群东立面图
East elevation of buildings around Yuese jiangsheng

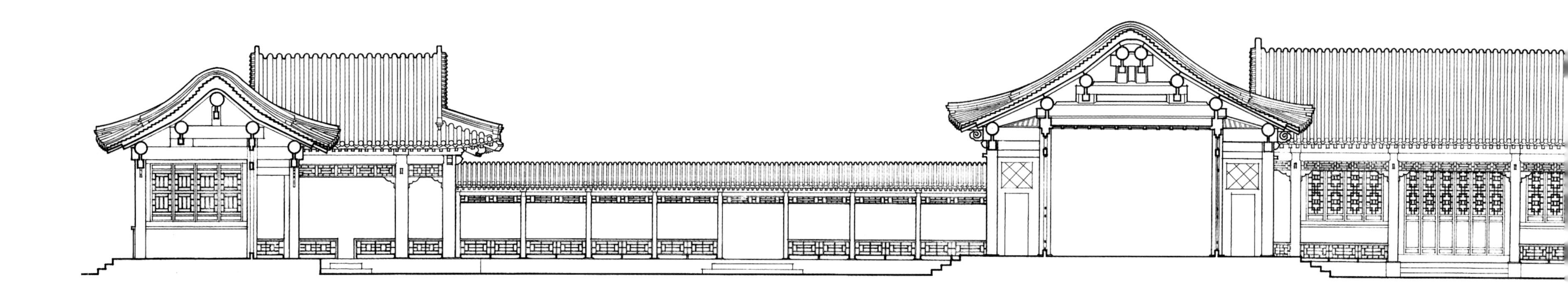

月色江声建筑群总剖面图
Site section of buildings around Yuese jiangsheng

1 无暑清凉 Wushu qingliang (Gate Hall)
2 延薰山馆 Yanxun shanguan (Hall)
3 乐寿堂 Leshoutang (Hall)
4 一片云 Yipianyun (Pavilion)
5 群楼 Qunlou (Podium)
6 西配殿 West peidian (side Hall)
7 东配殿 East peidian (side Hall)
8 西配房 West peidian (side House)
9 金莲映日 Jinlian yingri
10 西堆 Xidui Hall
11 观莲所 Guanliansuo (Pavilion)
12 云帆月舫（遗址） Yunfan yuefang (Pavilion Site)
13 前书房（遗址） Qian shufang (Site)
14 随安堂（遗址） Suiantang
15 西岭晨霞（遗址） Xiling chenxia (Site)
16 沧浪屿（遗址） Canglangyu (Site)
17 殿门（遗址） Dianmen (Gate Hall Site)
18 东殿（遗址） East Hall (Site)
19 东堆（遗址） Dongdui Hall (Site)
20 法林寺 Falin Temple

如意洲建筑群总平面图
Site plan of buildings around Ruyizhou

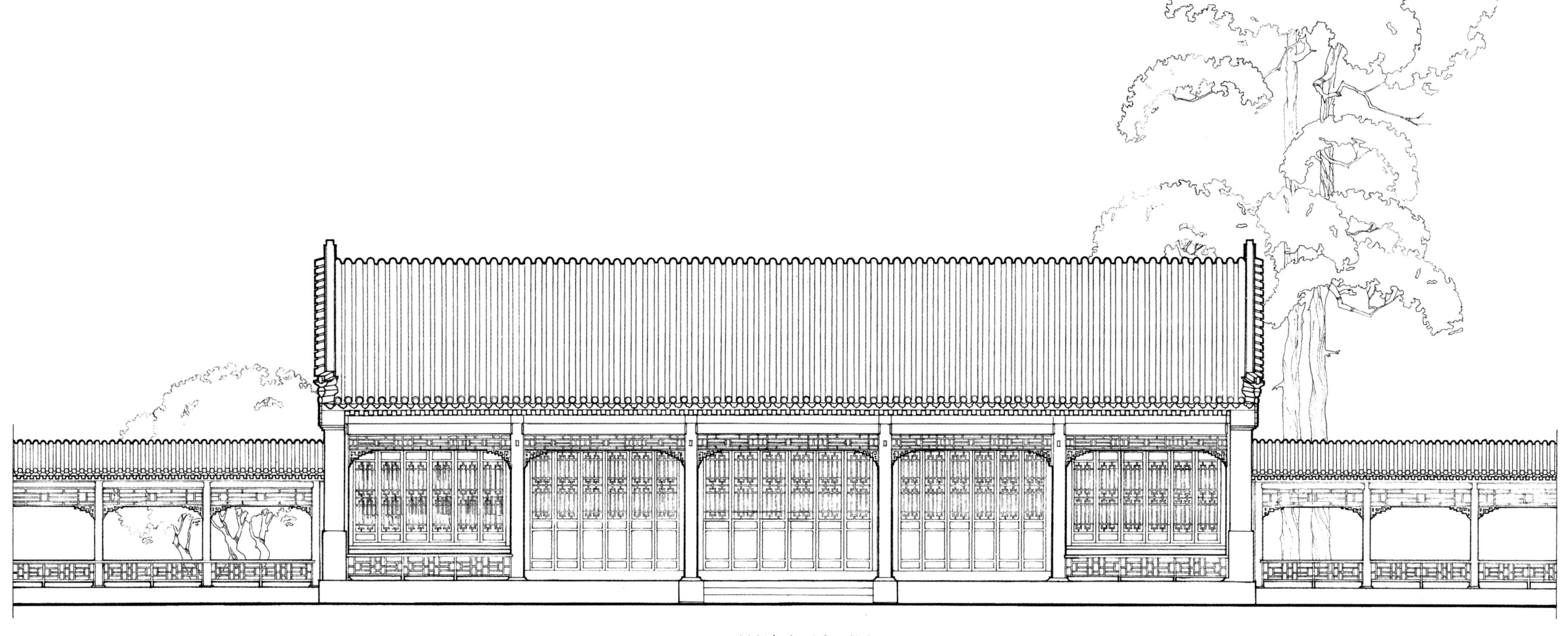

无暑清凉正立面图
Front elevation of Wushu qingliang

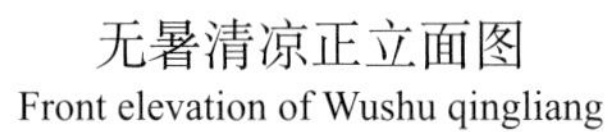

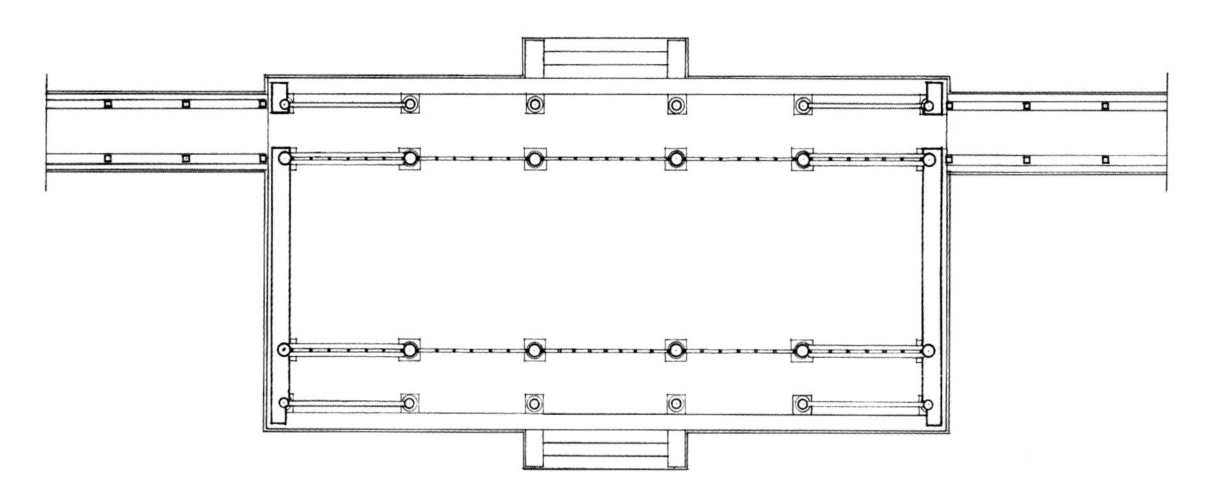

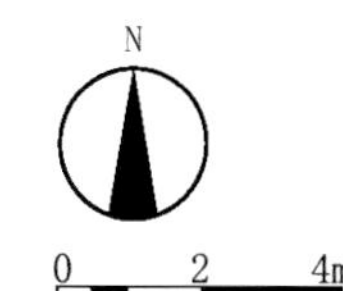

无暑清凉平面图
Plan of Wushu qingliang

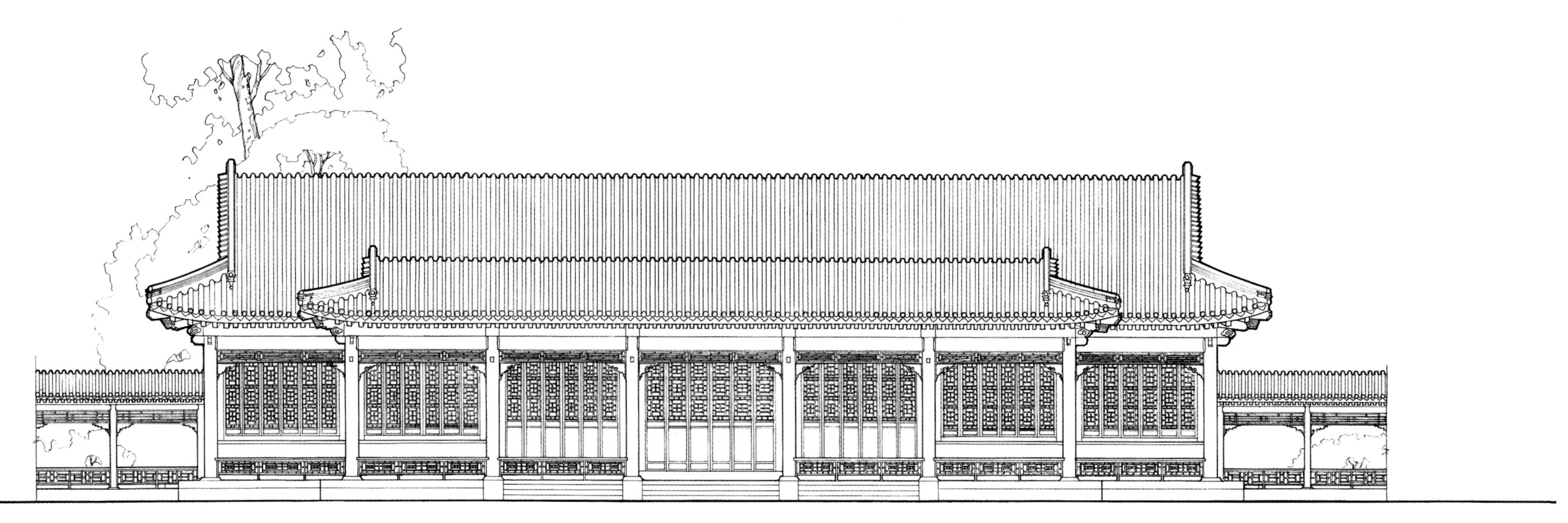

延薰山馆正立面图
Front elevation of Yanxun shanguan

0 2 4m

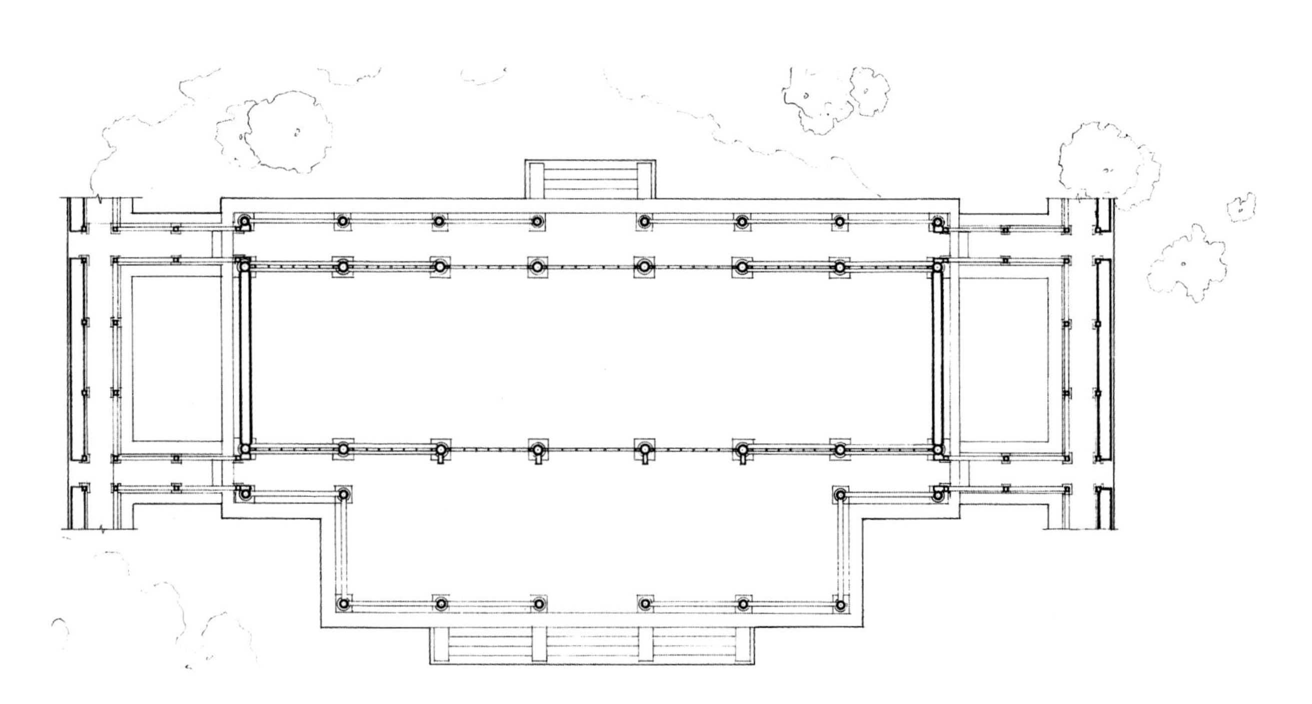

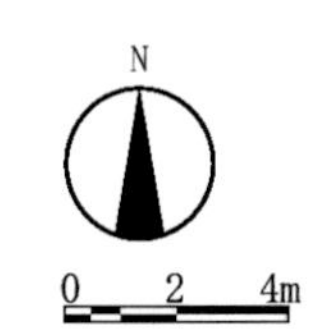

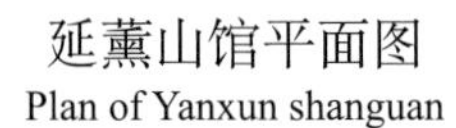
延薰山馆平面图
Plan of Yanxun shanguan

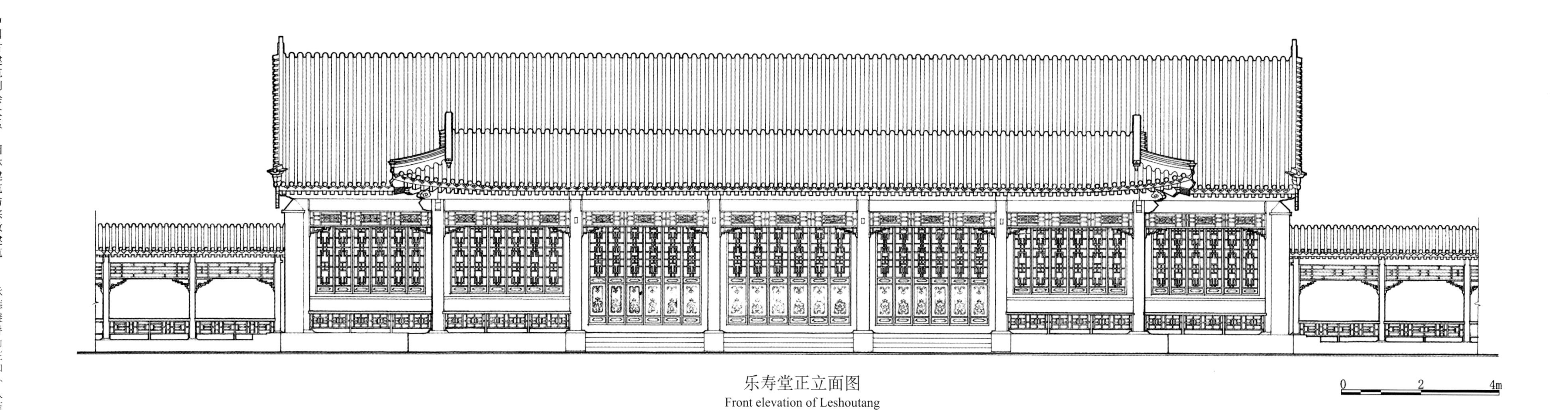

乐寿堂正立面图
Front elevation of Leshoutang

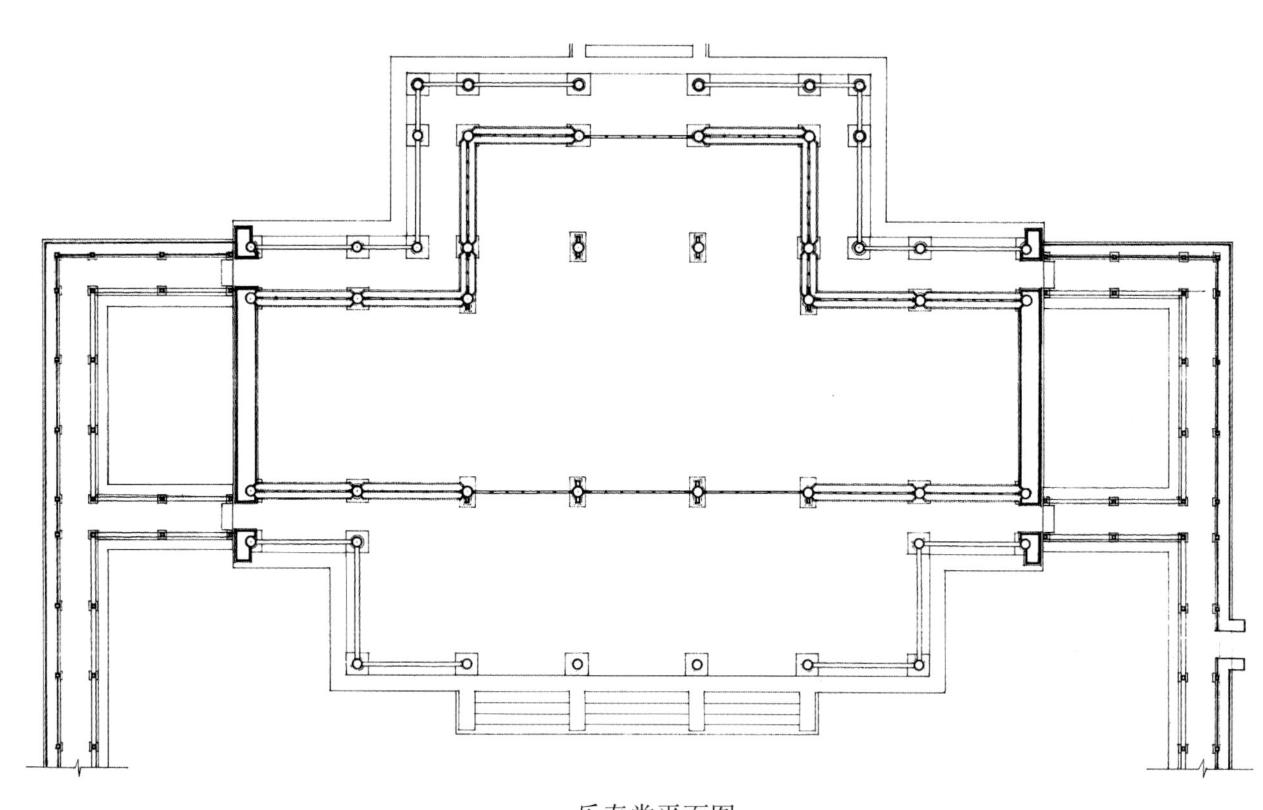

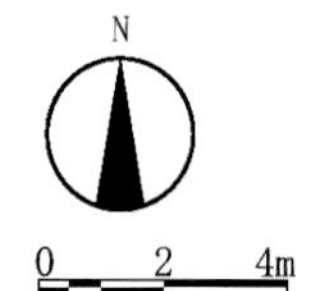

乐寿堂平面图
Plan of Leshoutang

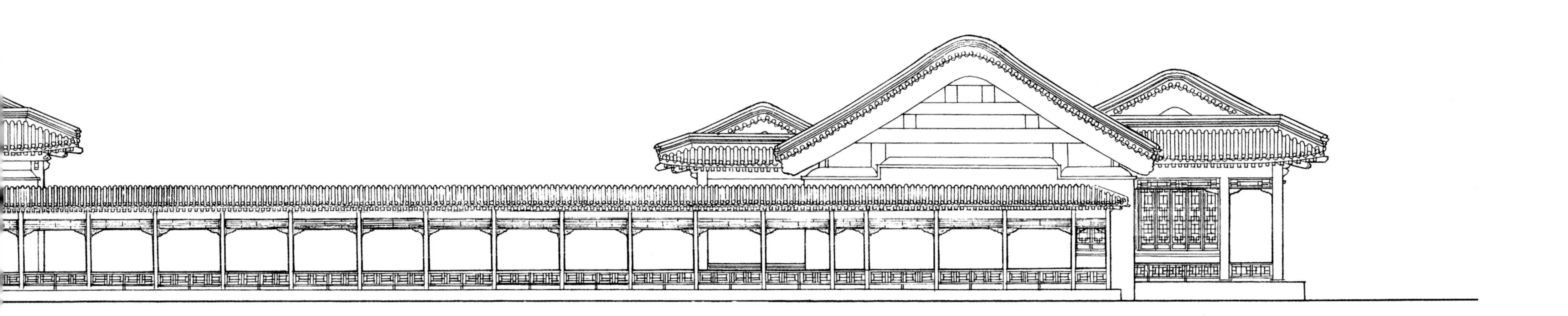
0 2 4m

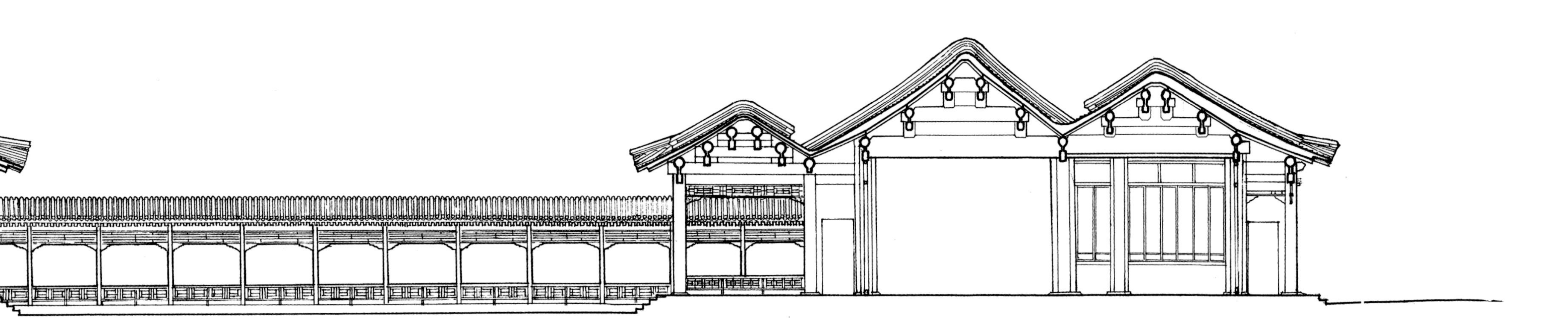
0 2 4m

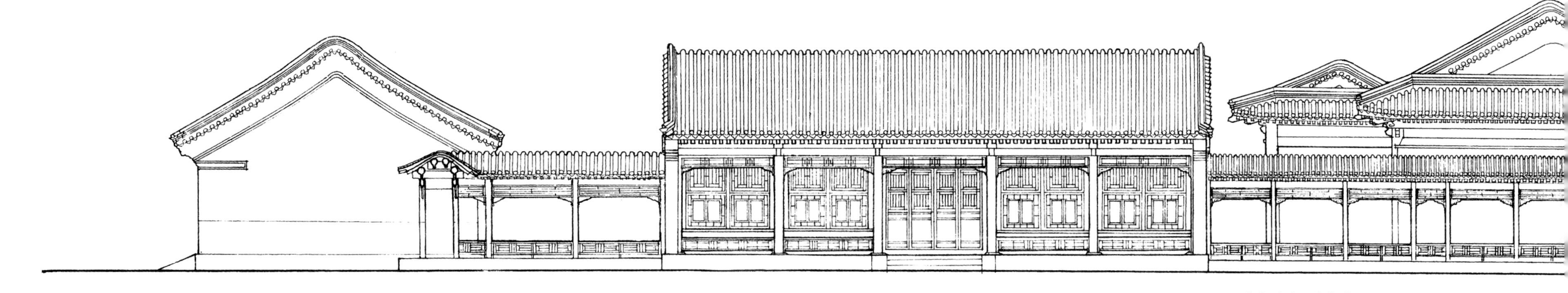

乐寿堂组群东立面图
East elevation of building cluster around Leshoutang

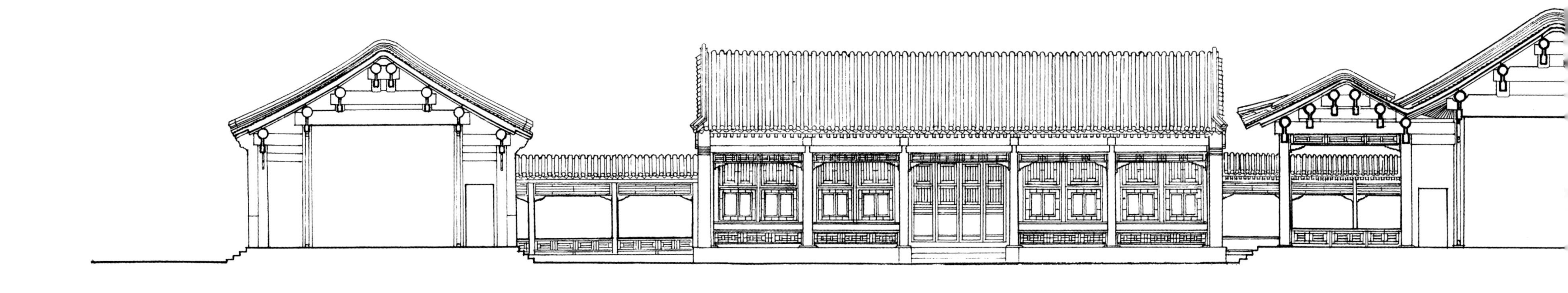

乐寿堂组群纵剖面图
Longitudinal section of building cluster around Leshoutang

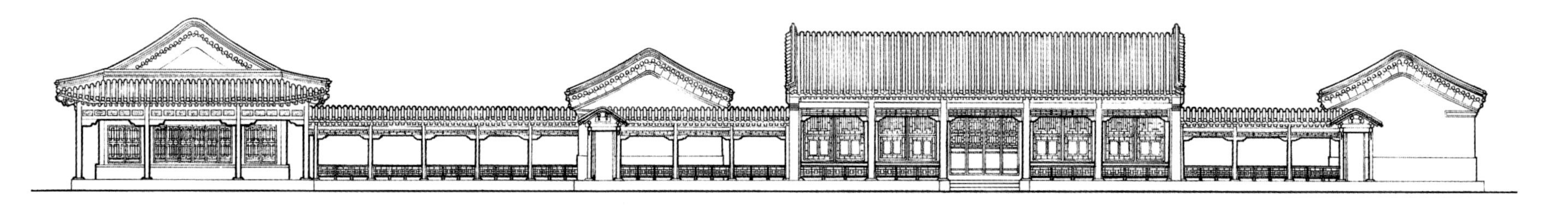

金莲映日组群西立面图
West elevation of building cluster around Jinlian yingri

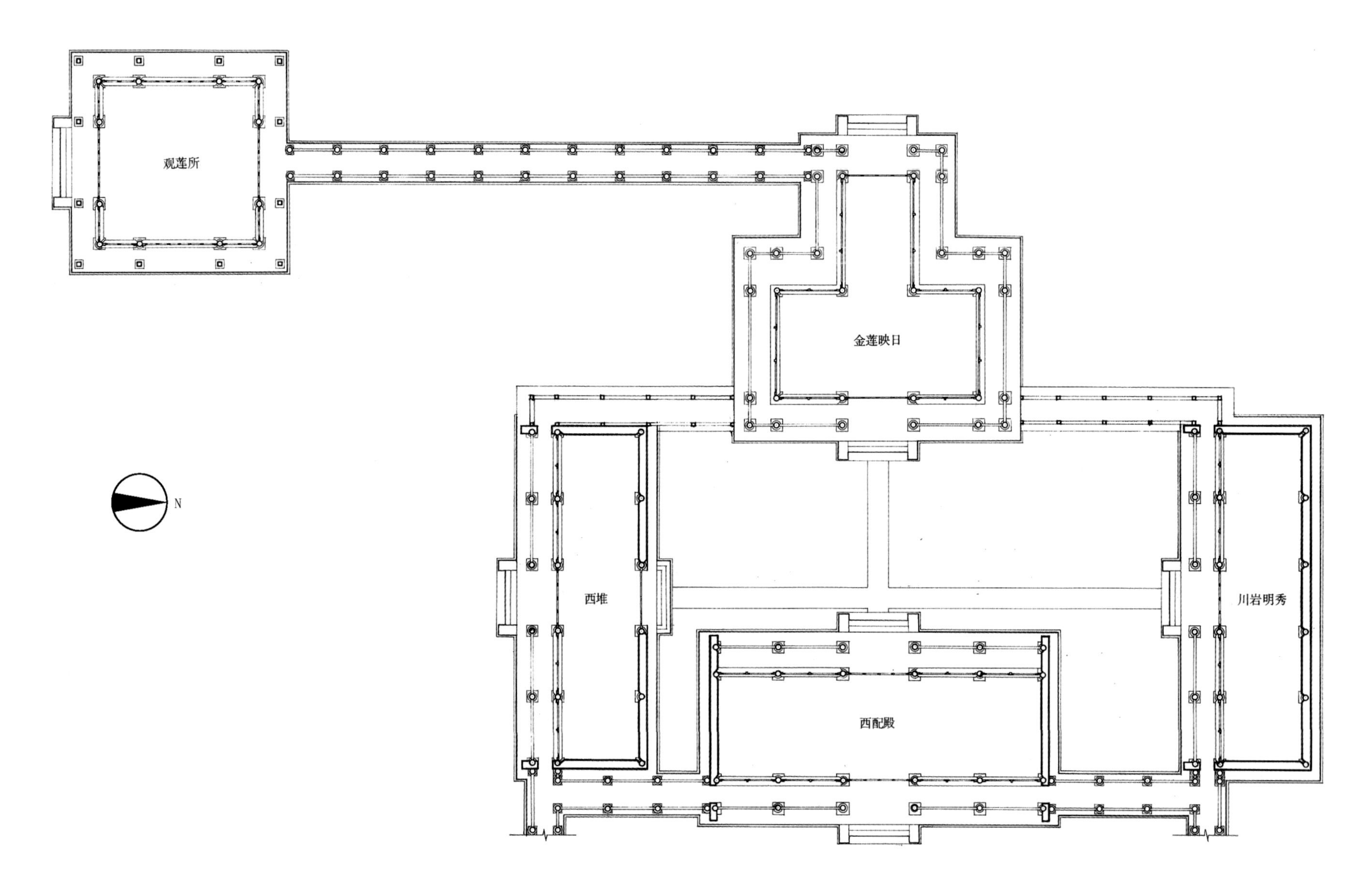

金莲映日组群总平面图
Site plan of building cluster around Jinlian yingri

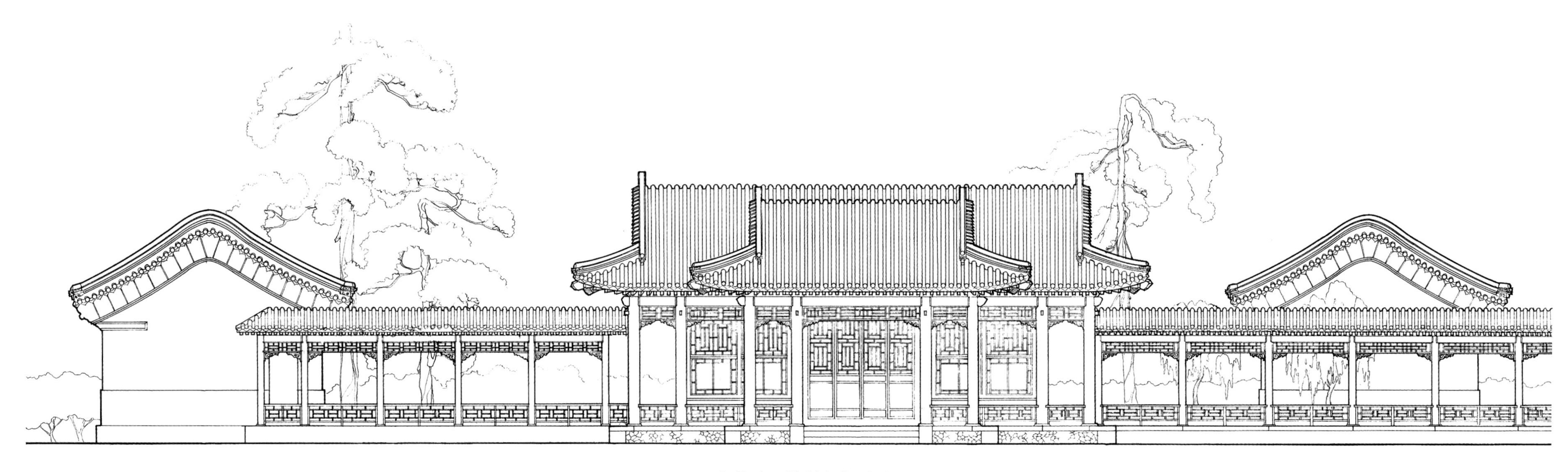

金莲映日组群东立面图
East elevation of building cluster around Jinlian yingri

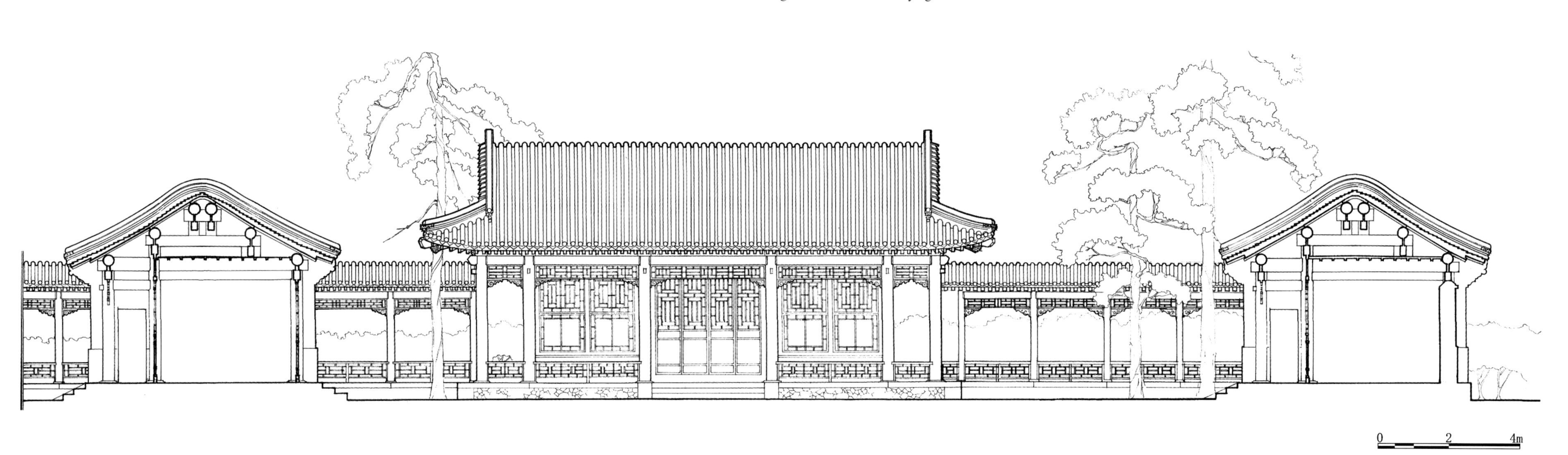

金莲映日组群横剖面图
Cross-section of building cluster around Jinlian yingri

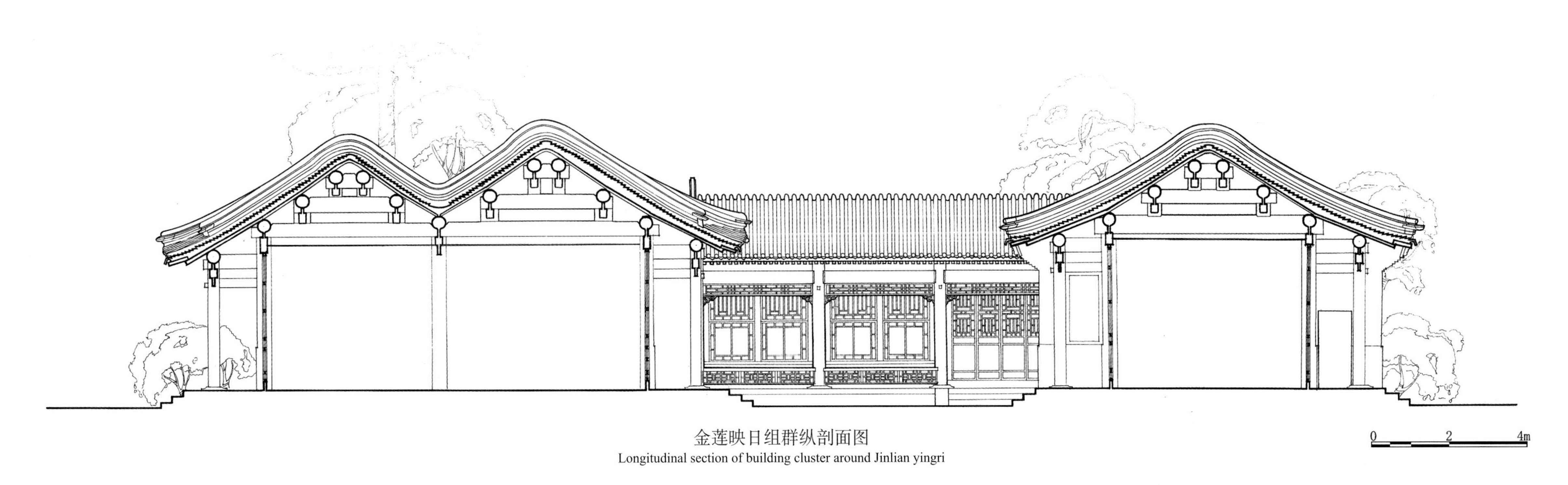

金莲映日组群纵剖面图
Longitudinal section of building cluster around Jinlian yingri

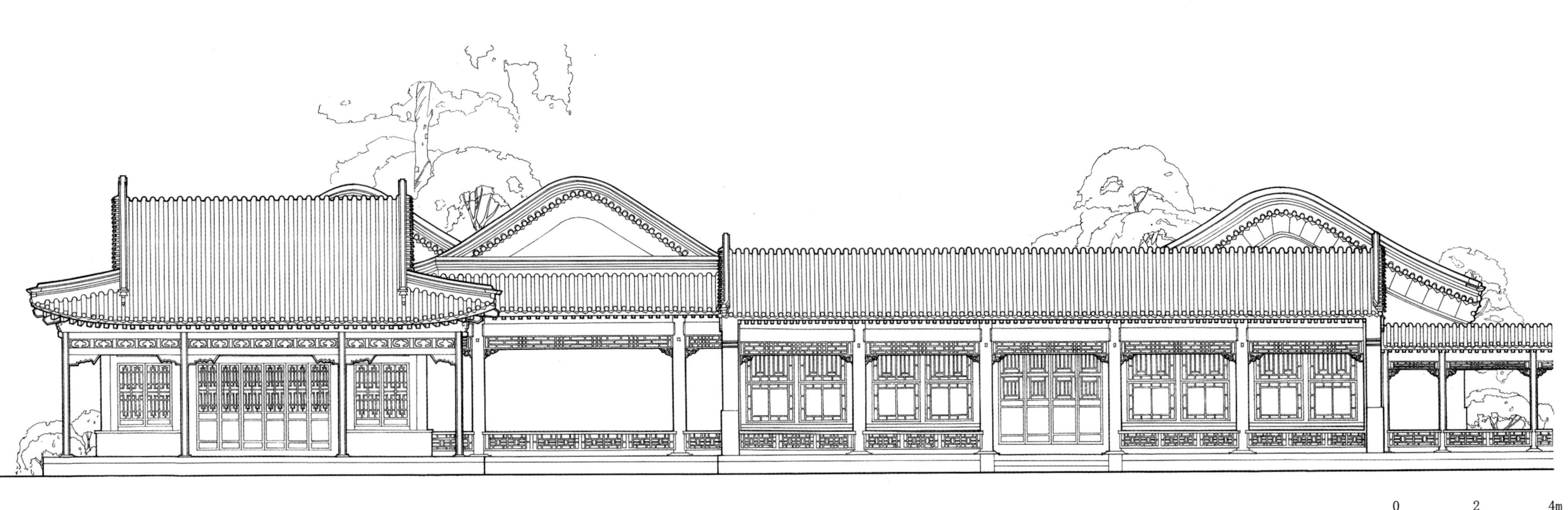

金莲映日组群南立面图
South elevation of building cluster around Jinlian yingri

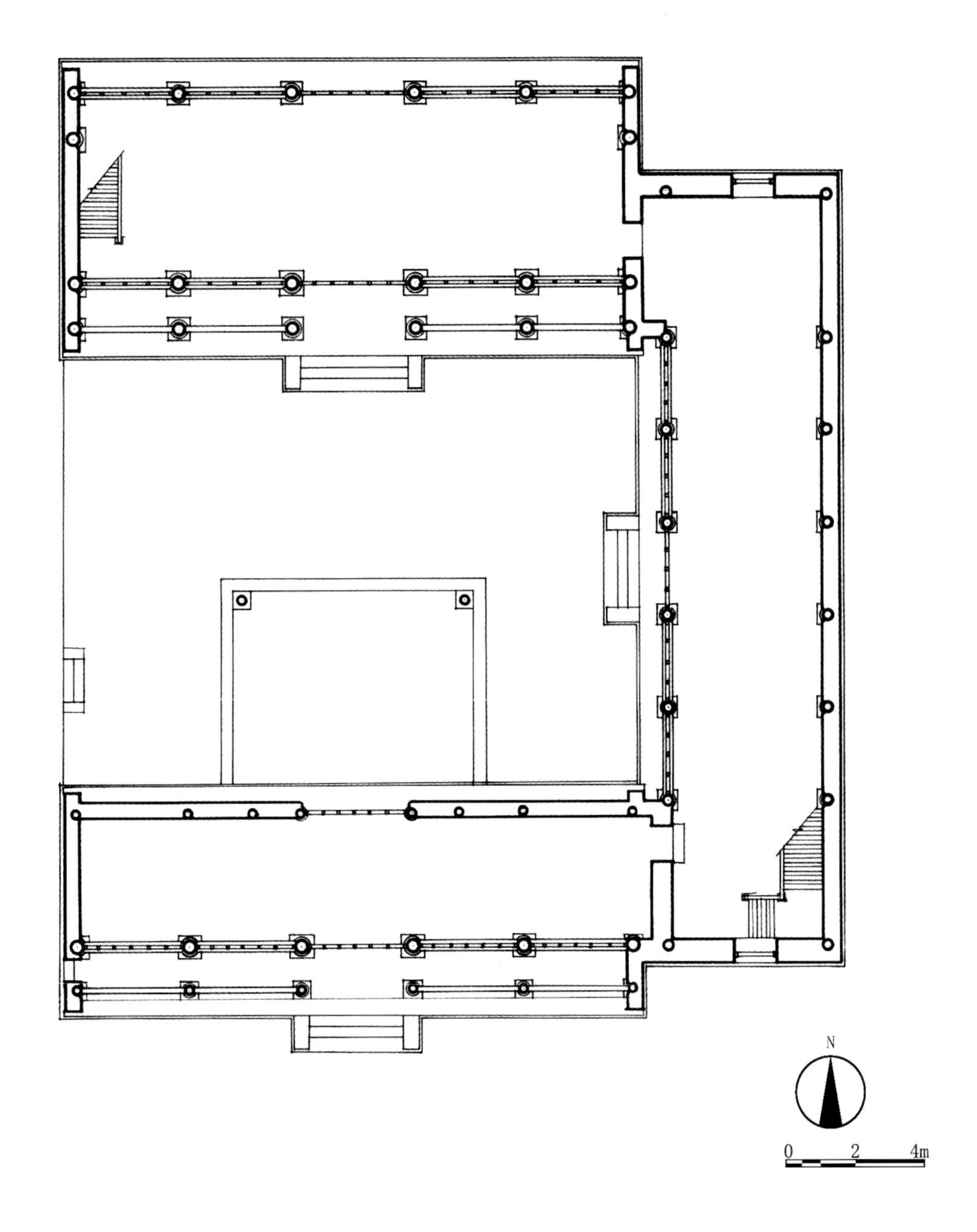

一片云组群一层平面图
Plan of first floor of building cluster around Yipianyun

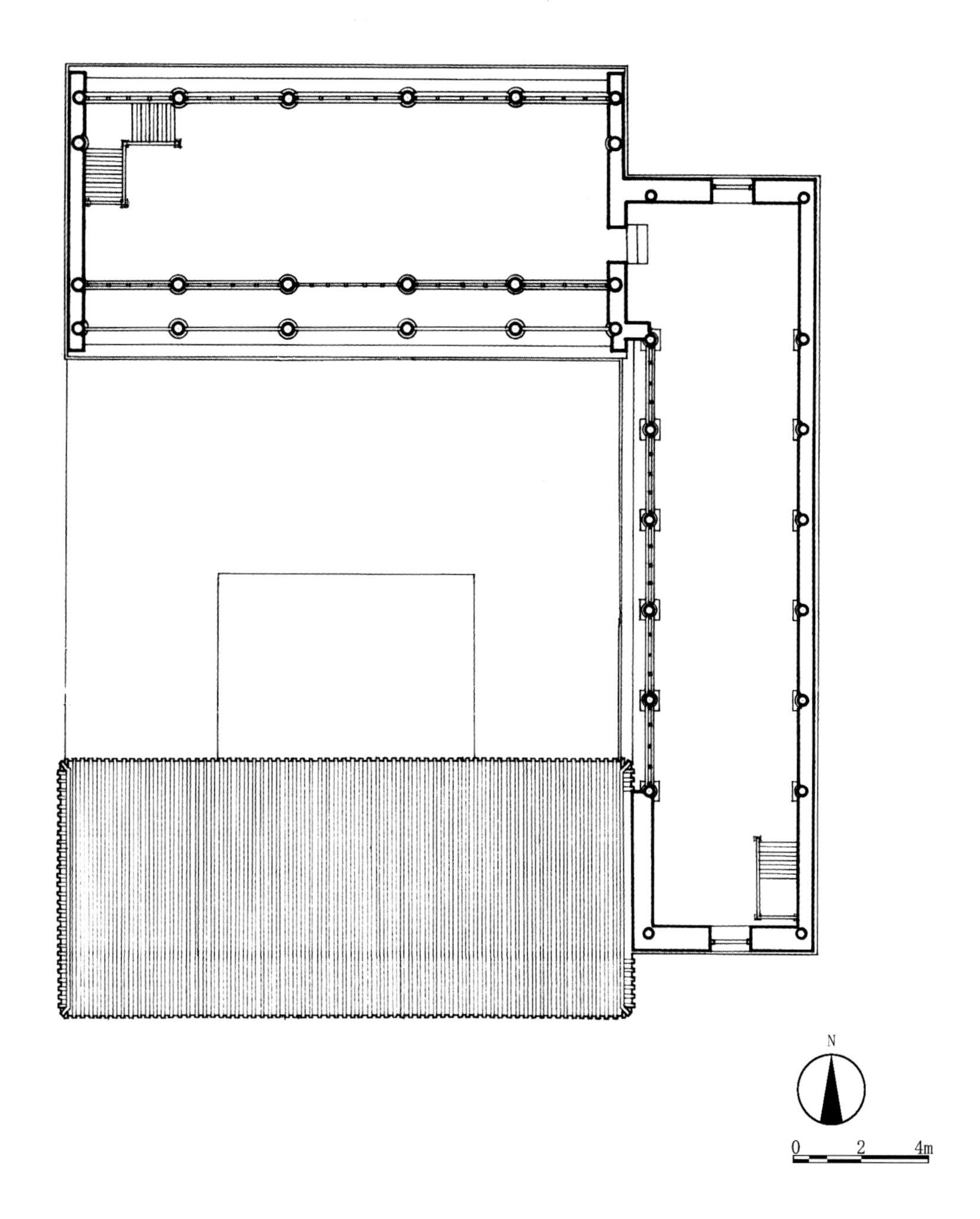

一片云组群二层平面图
Plan of second floor of building cluster around Yipianyun

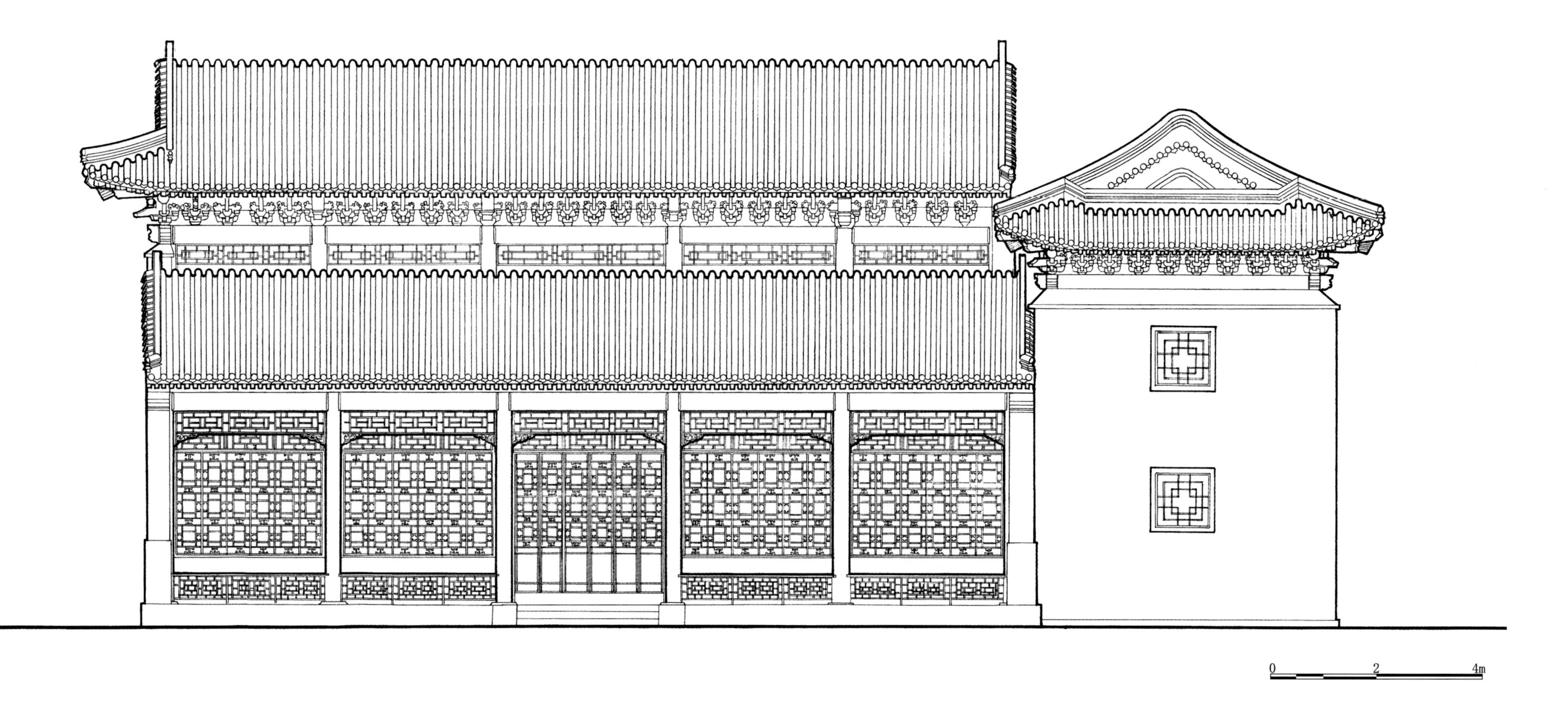

一片云组群南立面图
South elevation of building cluster around Yipianyun

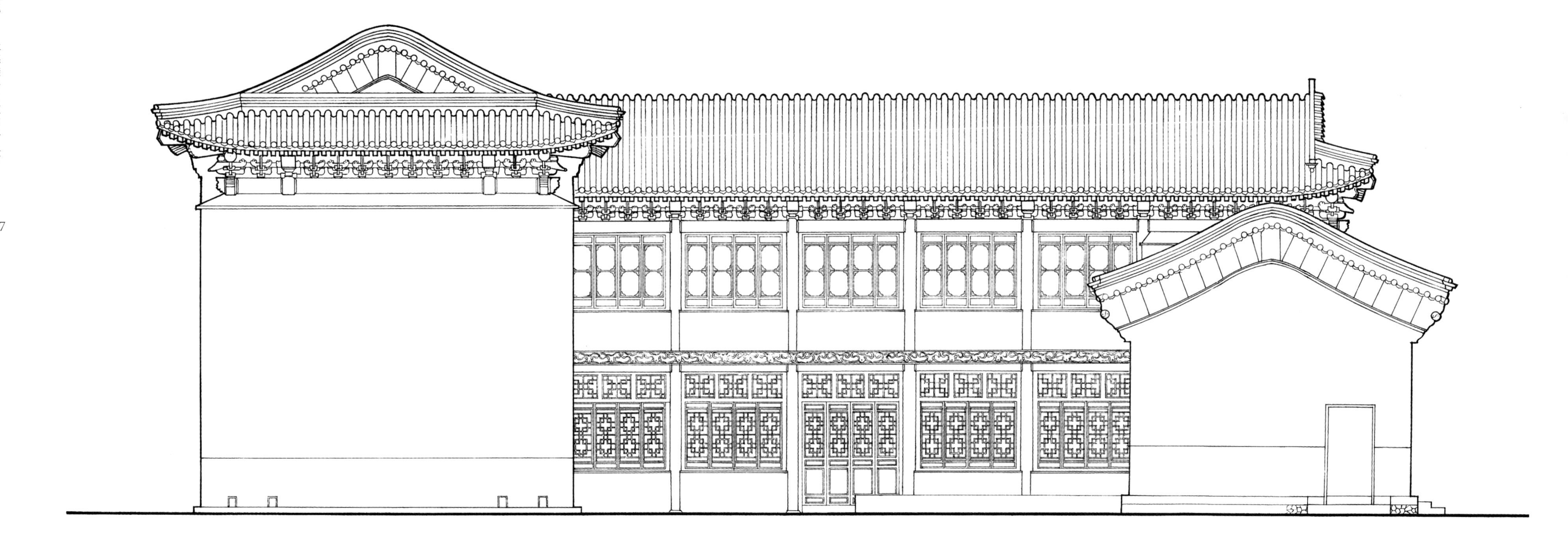

一片云组群西立面图

West elevation of building cluster around Yipianyun

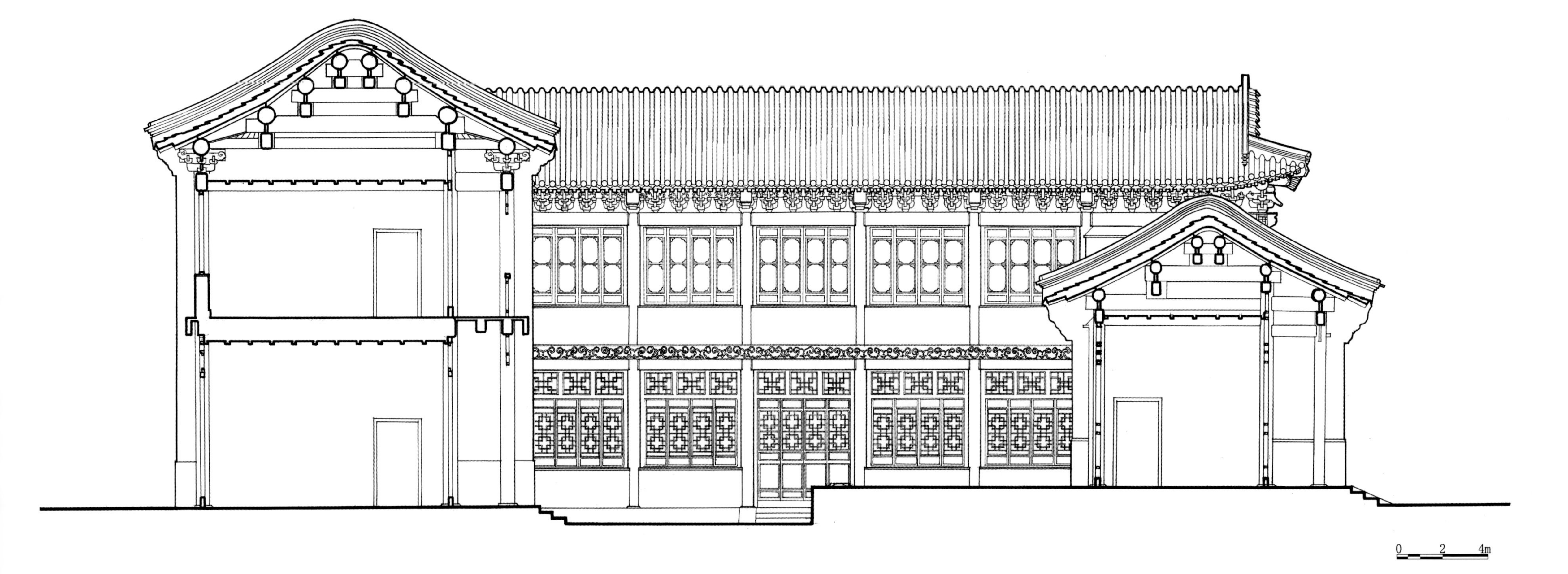

一片云组群戏楼剖面图
Section of building cluster around xilou of Yipianyun

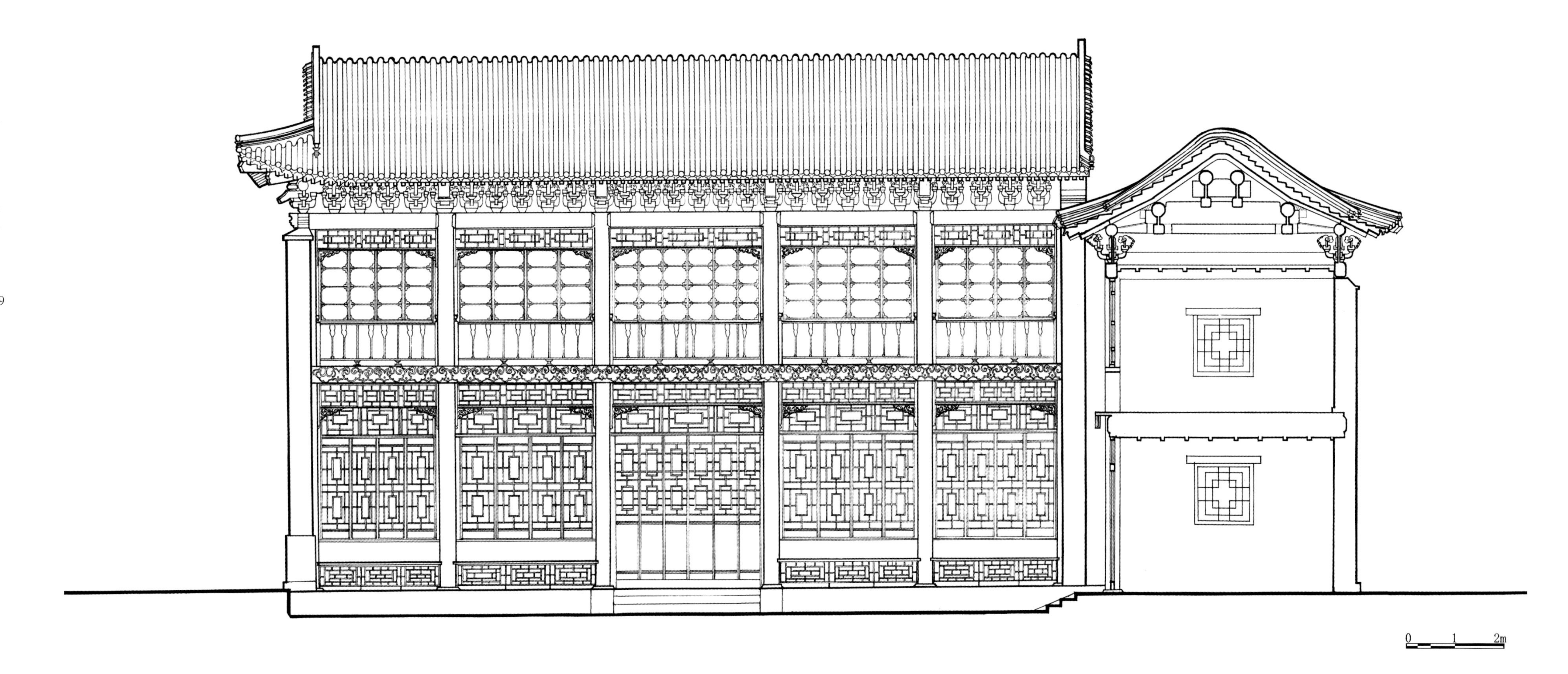

一片云楼南立面图、群楼横剖面图
South elevation and cross-section of qunlou of Yipianyun Building

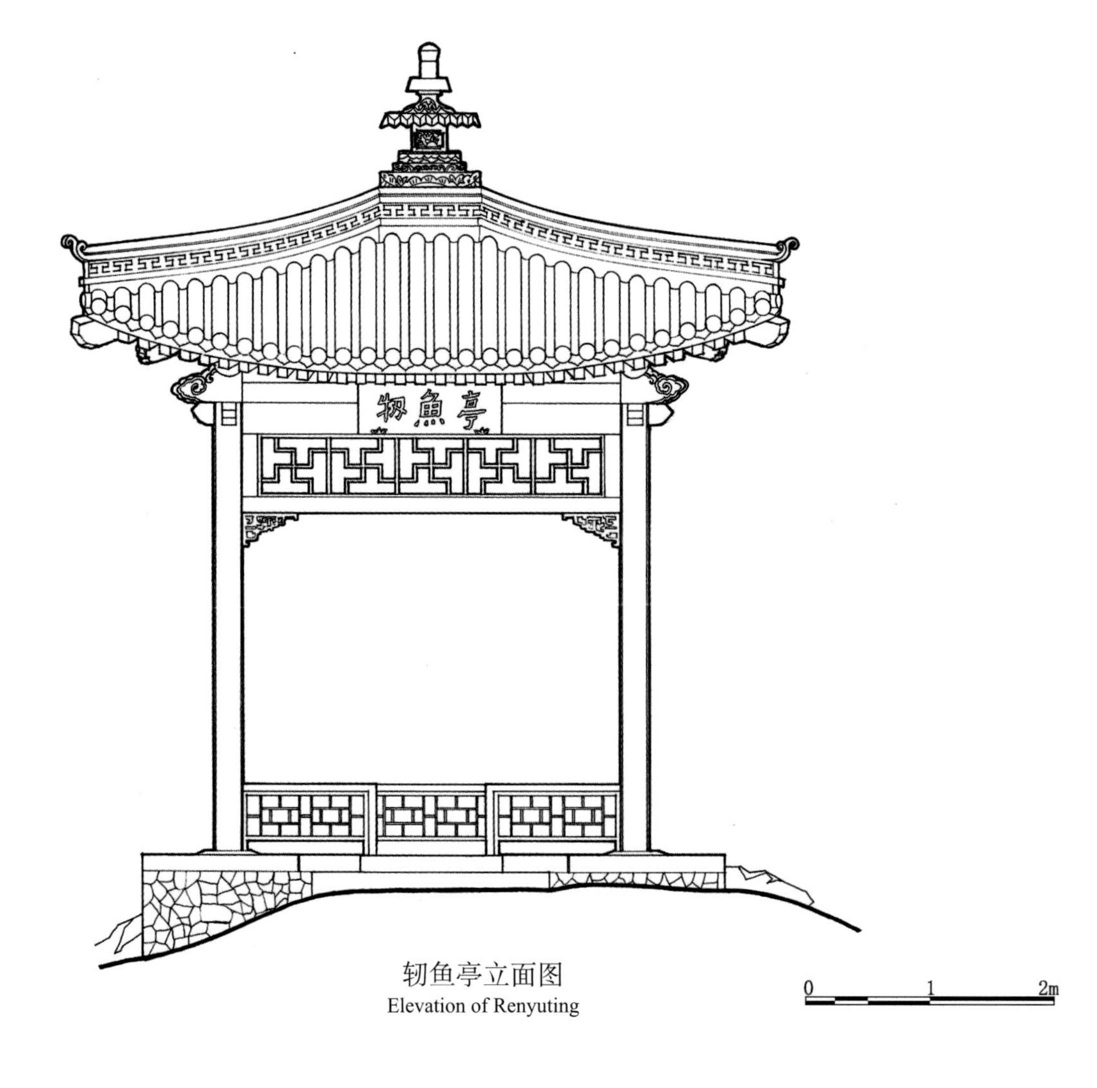

韧鱼亭立面图
Elevation of Renyuting

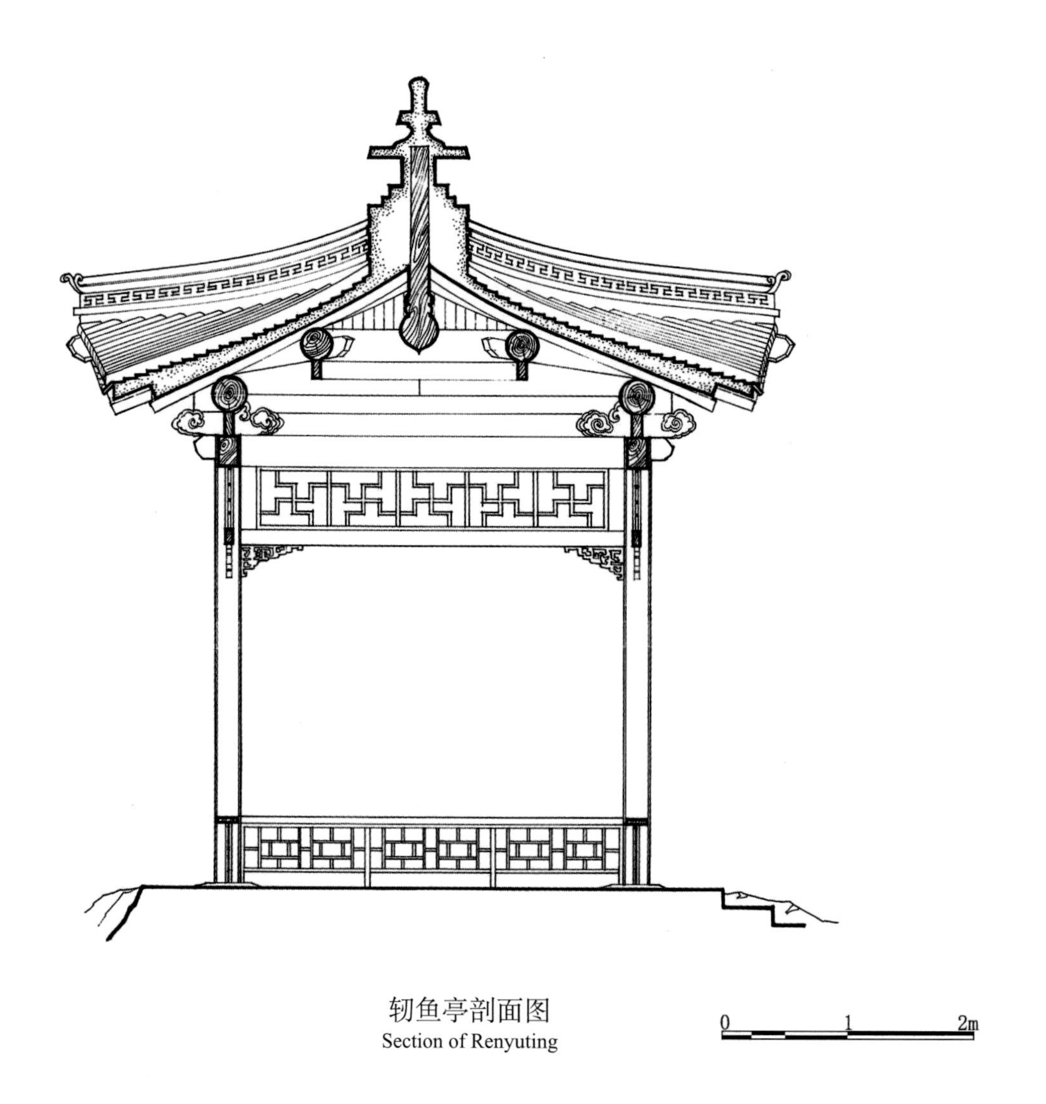

韧鱼亭剖面图
Section of Renyuting

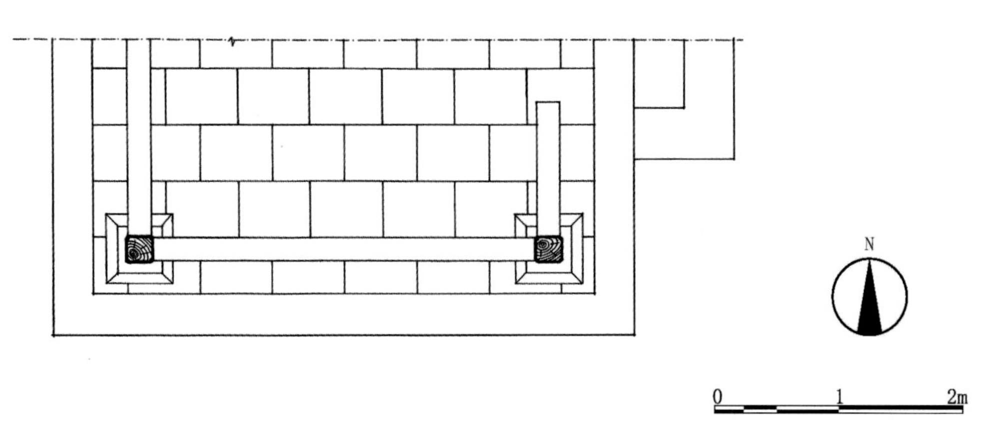

韧鱼亭平面图
Plan of Renyuting

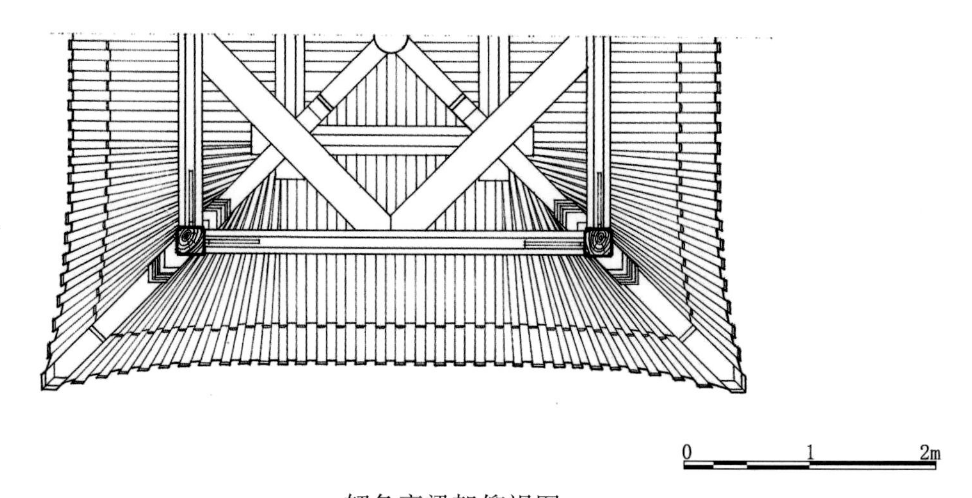

韧鱼亭梁架仰视图
Plan of framework as seen from below of Renyuting

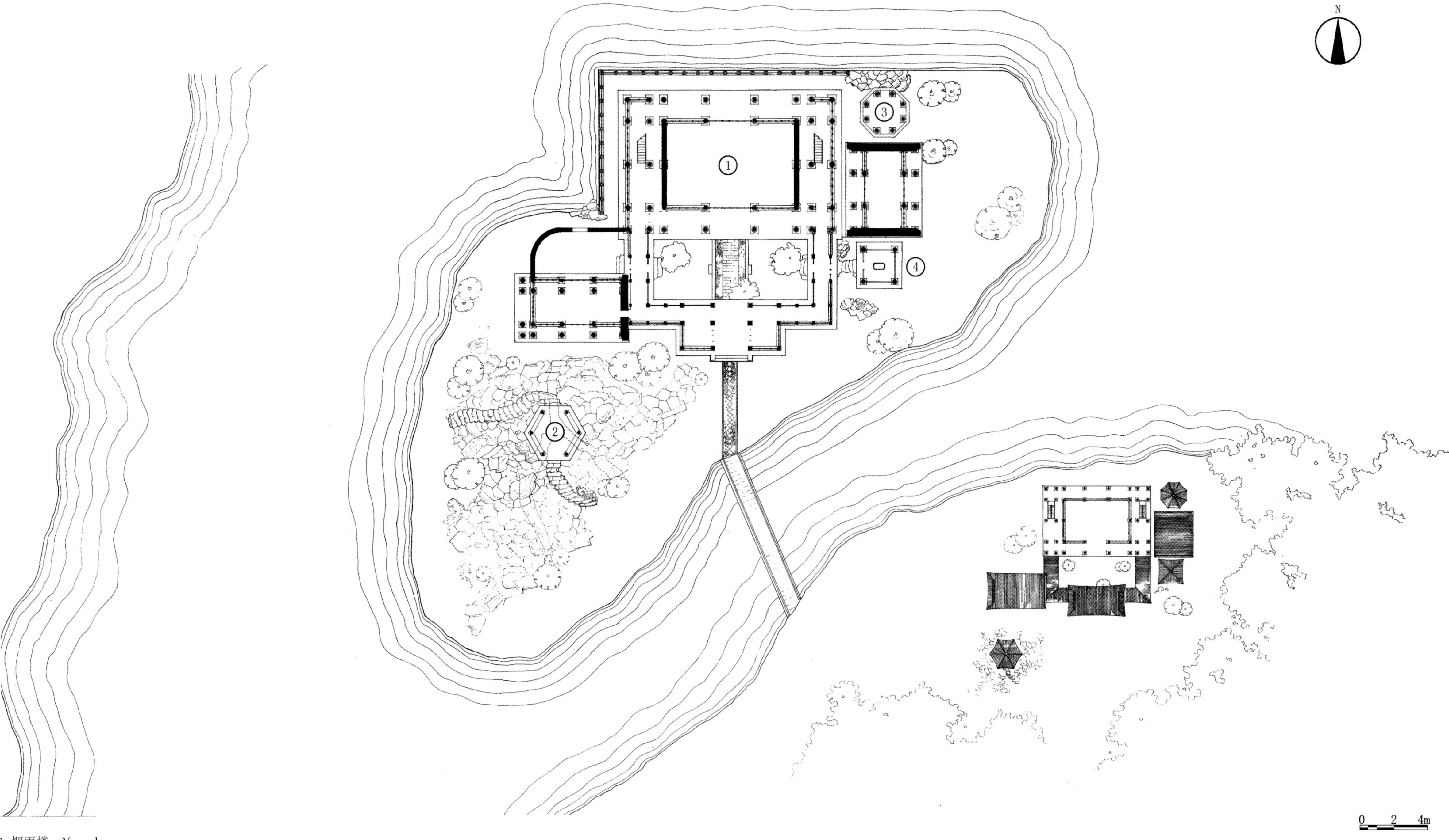

1 烟雨楼 Yanyulou
2 六角亭 Liujiaoting (Hexagonal Pavilion)
3 八角亭 Bajiaoting (Octagonal Pavilion)
4 四角亭 Sijianting (Square Pavilion)

烟雨楼建筑群总平面图
Site plan of buildings around Yanyulou

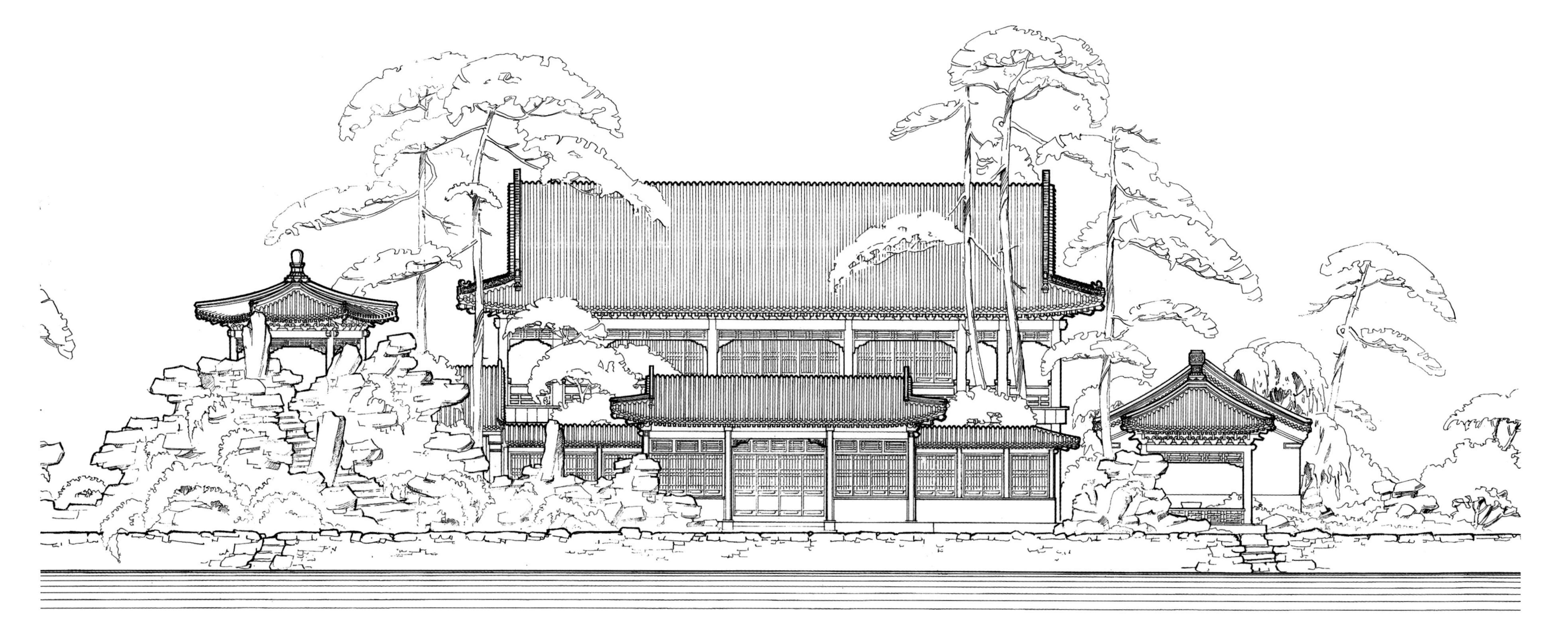

烟雨楼建筑群南立面图
South elevation of buildings around Yanyulou

烟雨楼建筑群北立面图
North elevation of buildings around Yanyulou

烟雨楼建筑群东立面图
East elevation of buildings around Yanyulou

烟雨楼建筑群西立面图
West elevation of buildings around Yanyulou

烟雨楼建筑群总剖面图

Site section of buildings around Yanyulou

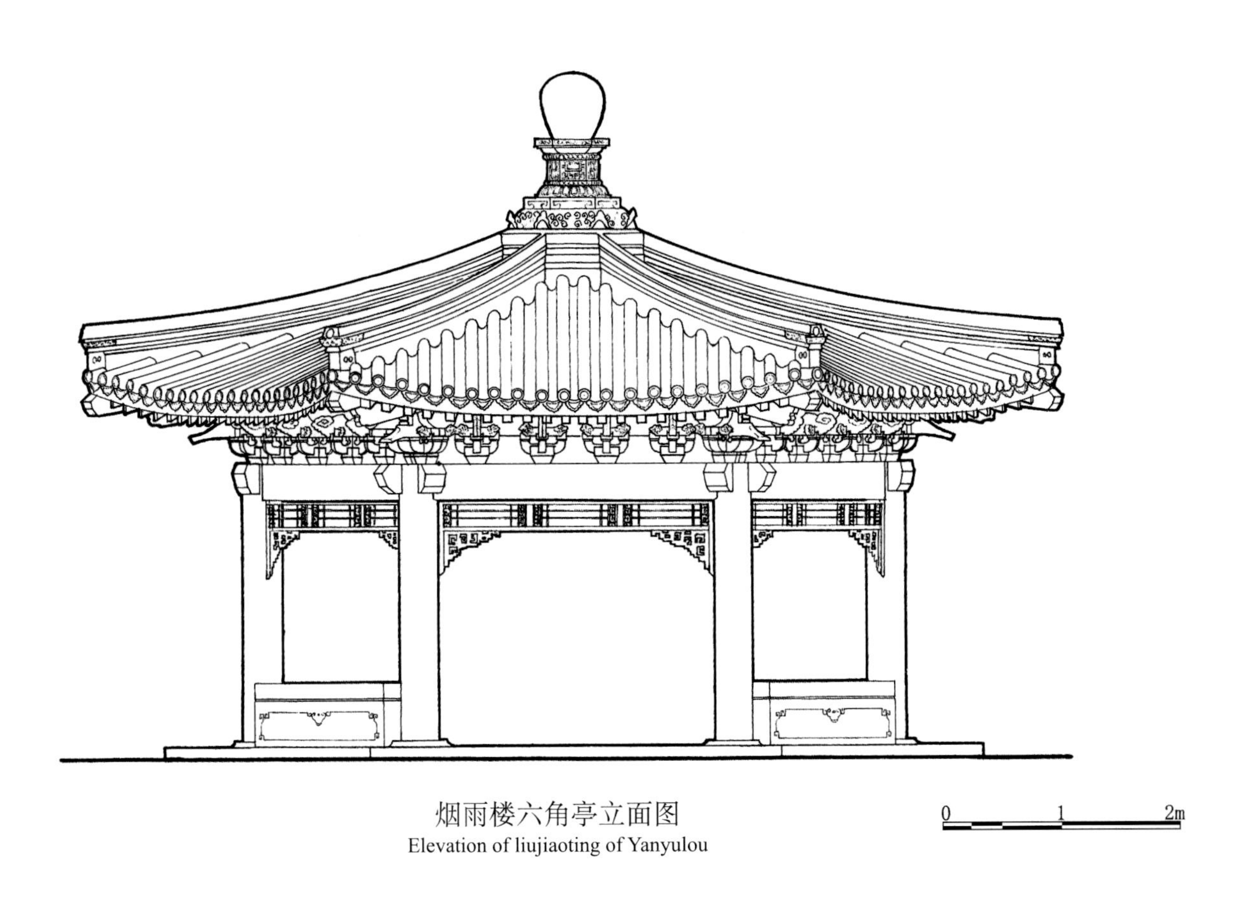

烟雨楼六角亭立面图
Elevation of liujiaoting of Yanyulou

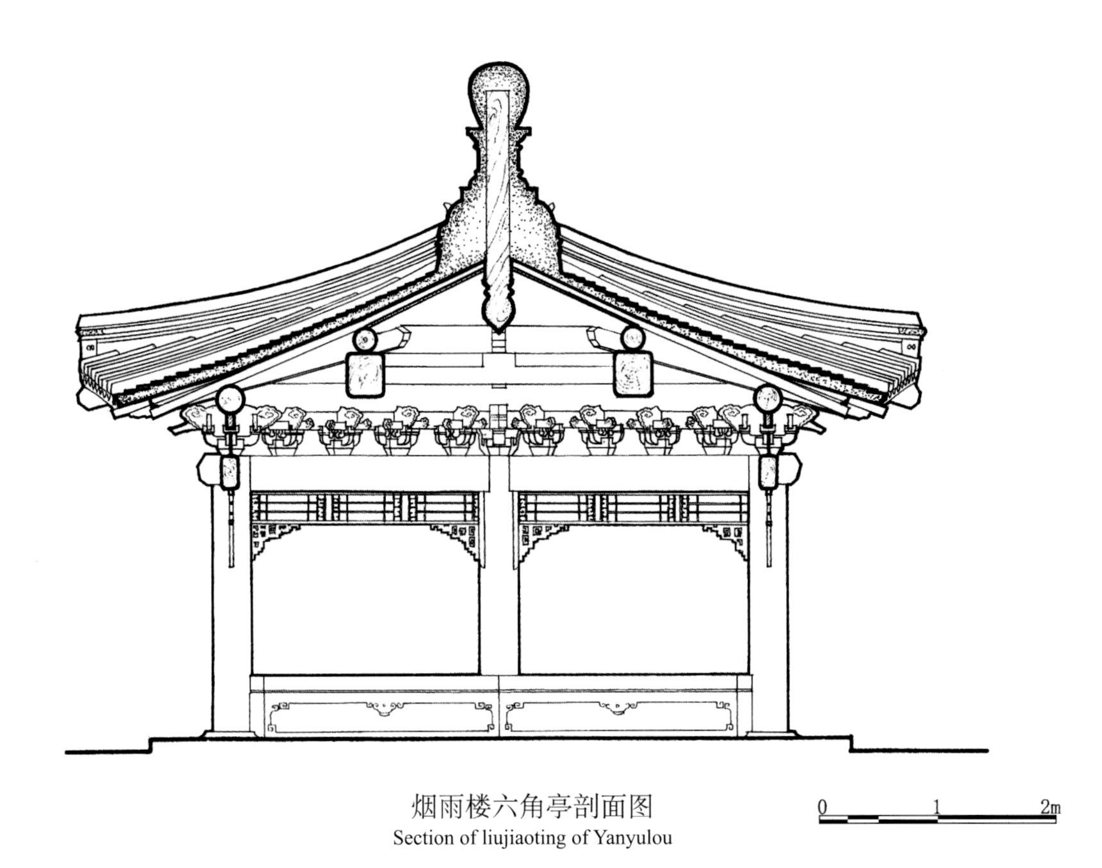

烟雨楼六角亭剖面图
Section of liujiaoting of Yanyulou

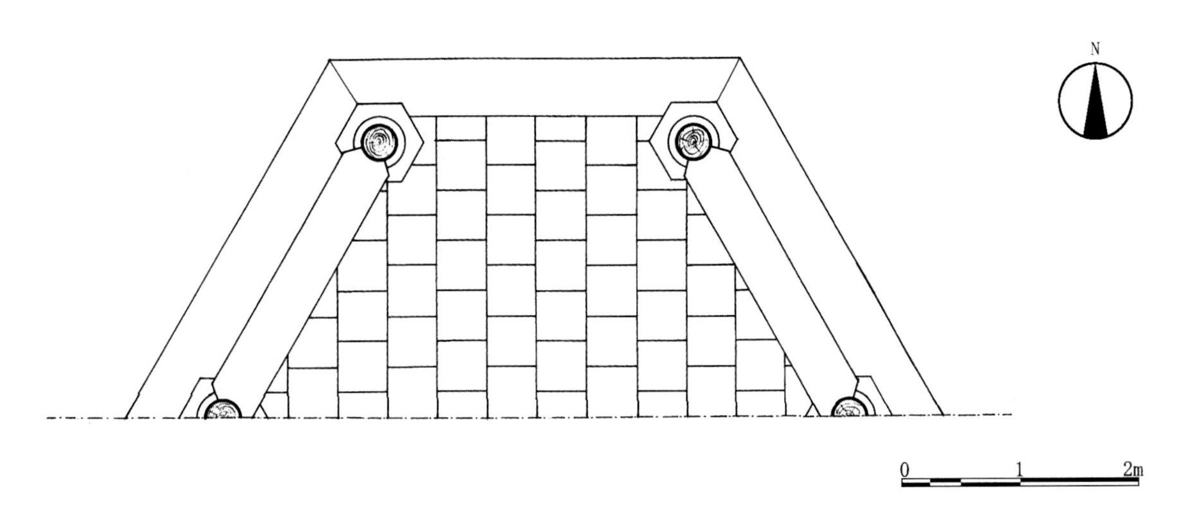

烟雨楼六角亭平面图
Plan of liujiaoting of Yanyulou

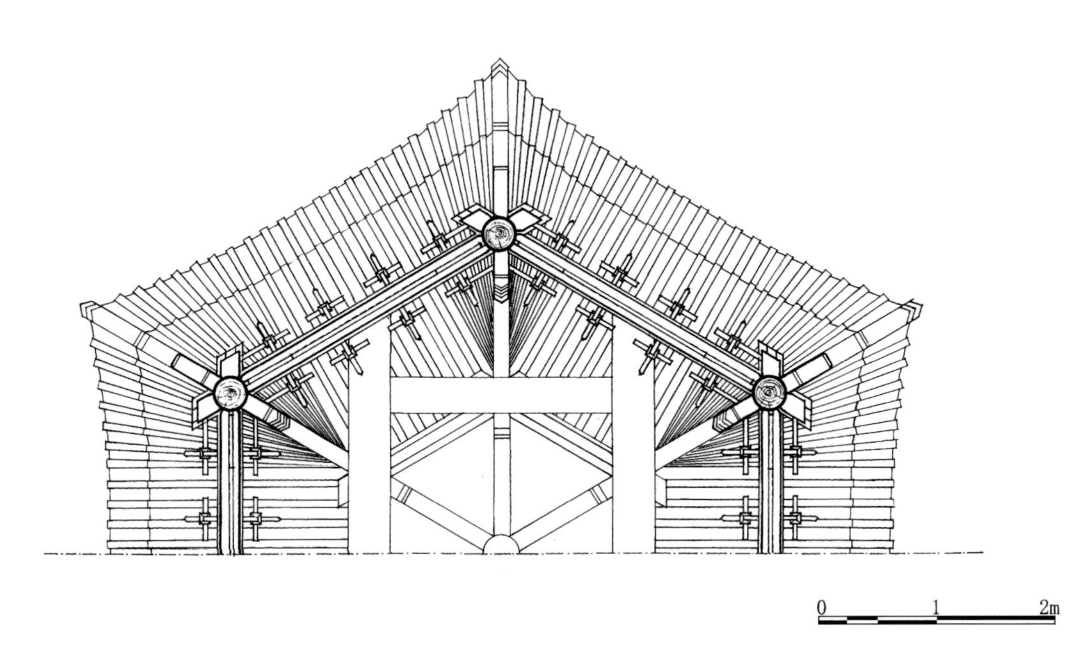

烟雨楼六角亭梁架仰视图
Plan of framework as seen from below of liujiaoting of Yanyulou

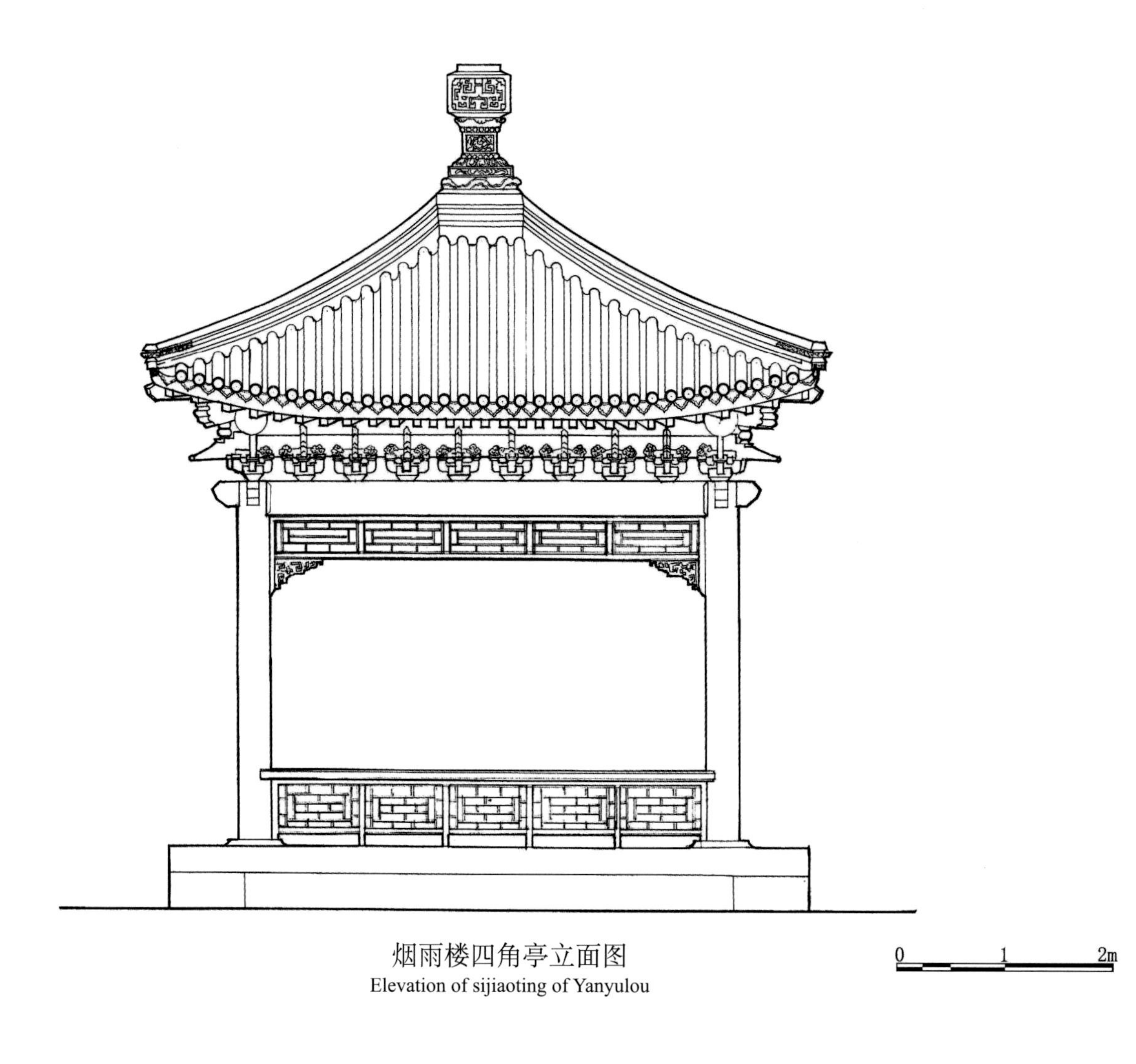

烟雨楼四角亭立面图
Elevation of sijiaoting of Yanyulou

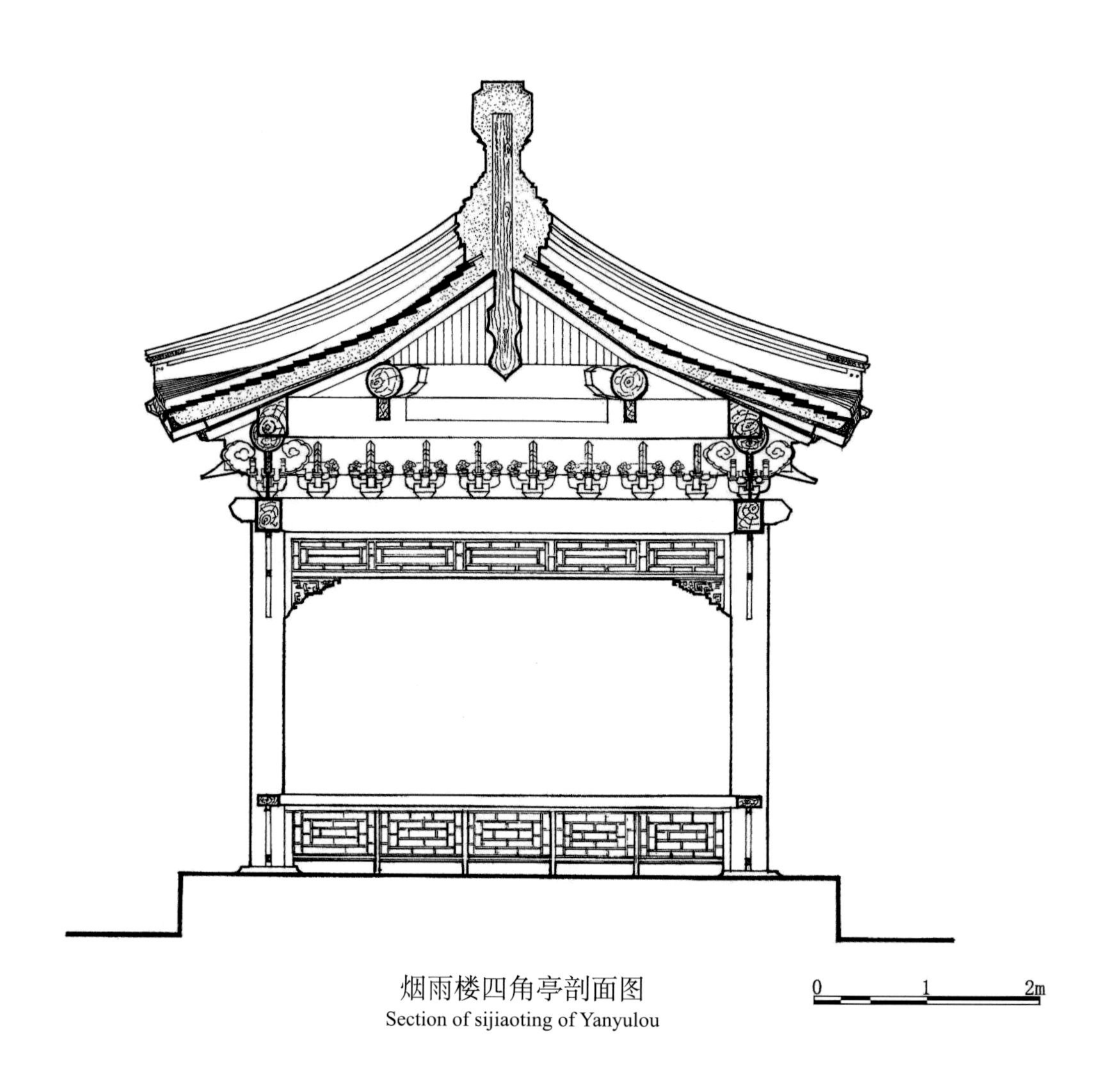

烟雨楼四角亭剖面图
Section of sijiaoting of Yanyulou

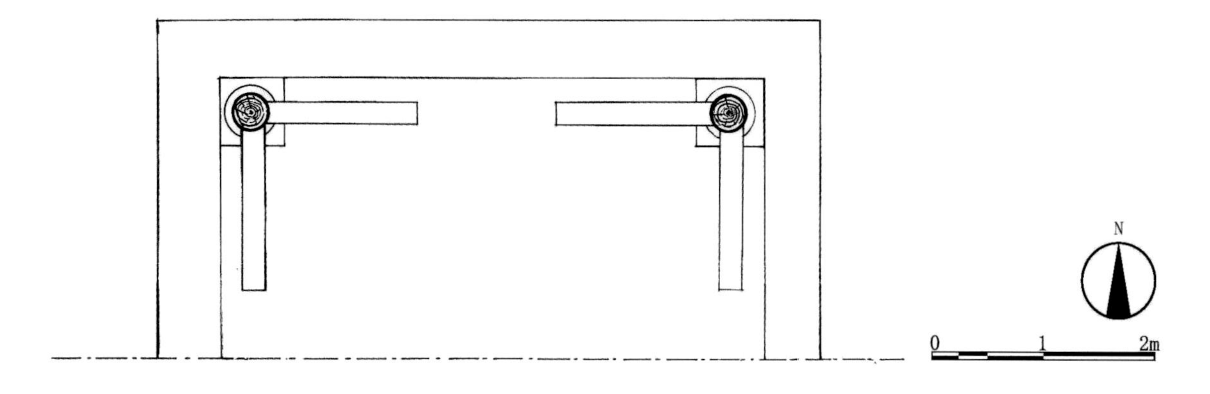

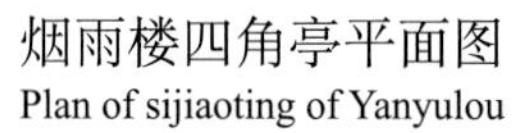

烟雨楼四角亭平面图
Plan of sijiaoting of Yanyulou

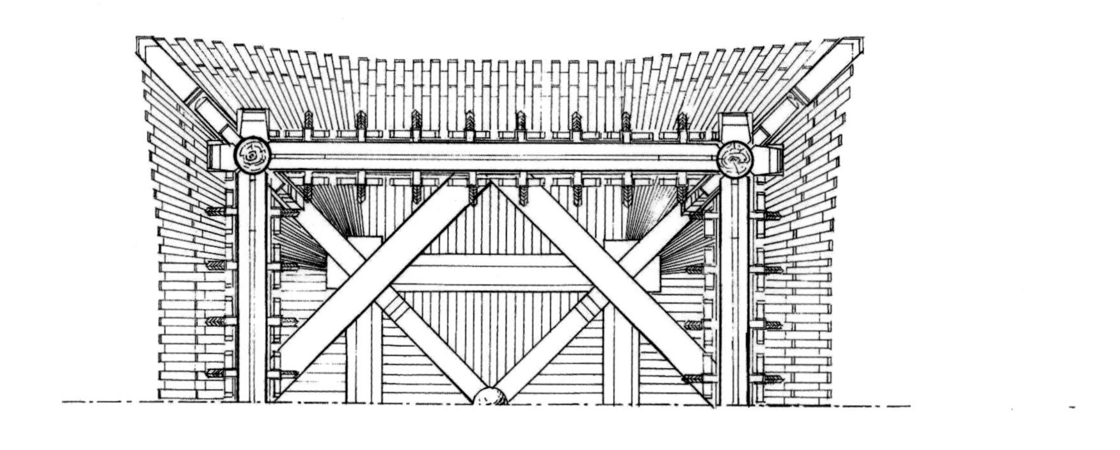

0
1
2m

烟雨楼四角亭梁架仰视图
Plan of framework as seen from below of sijiaoting of Yanyulou

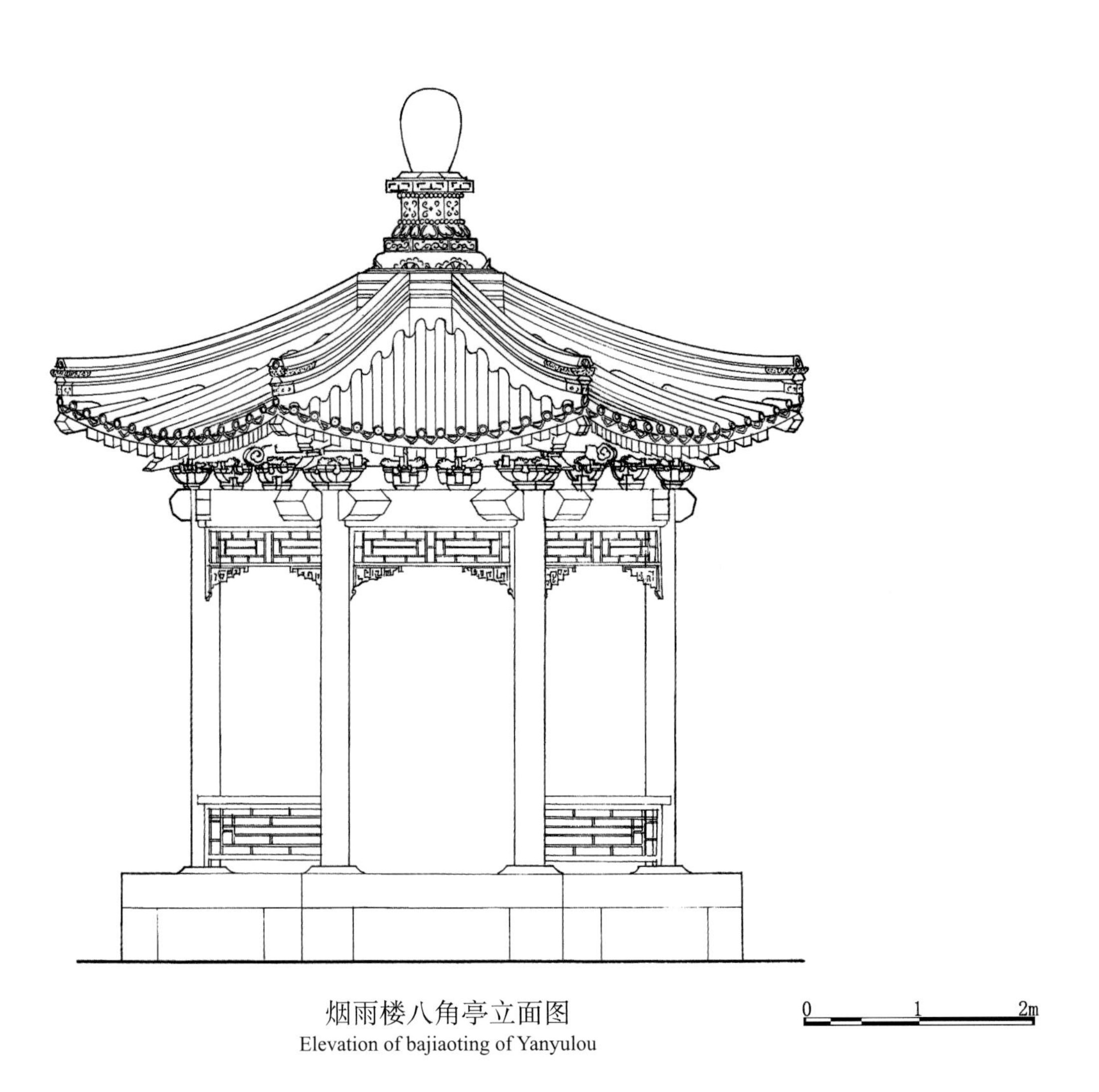

烟雨楼八角亭立面图
Elevation of bajiaoting of Yanyulou

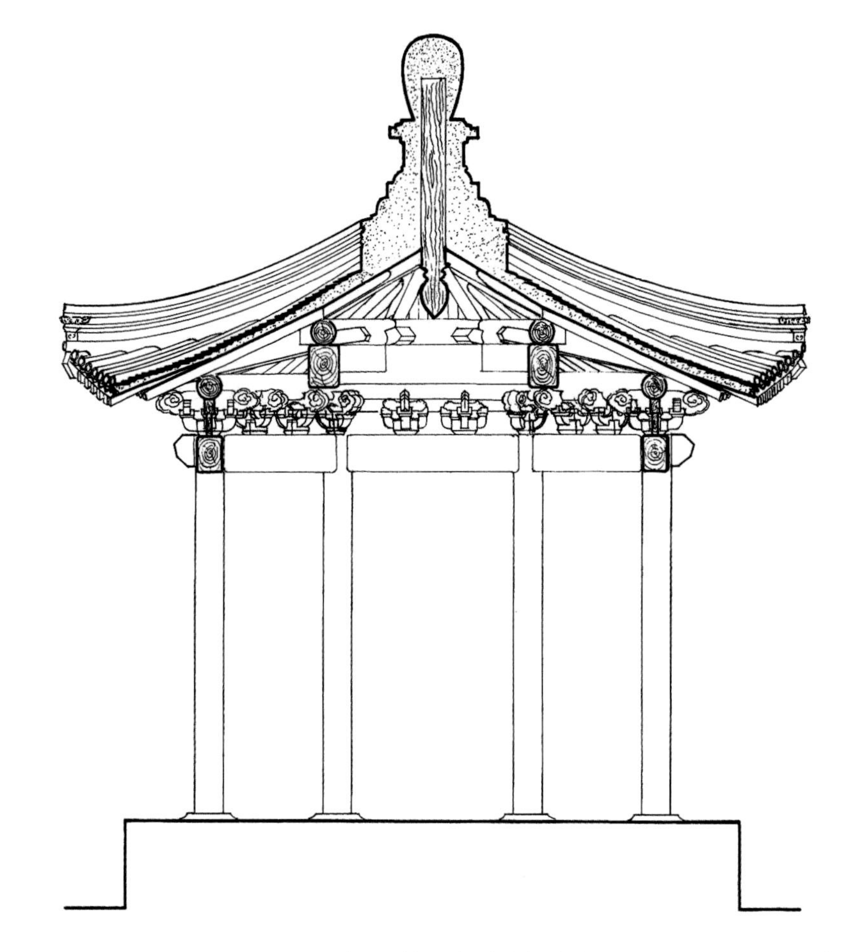

烟雨楼八角亭剖面图
Section of bajiaoting of Yanyulou

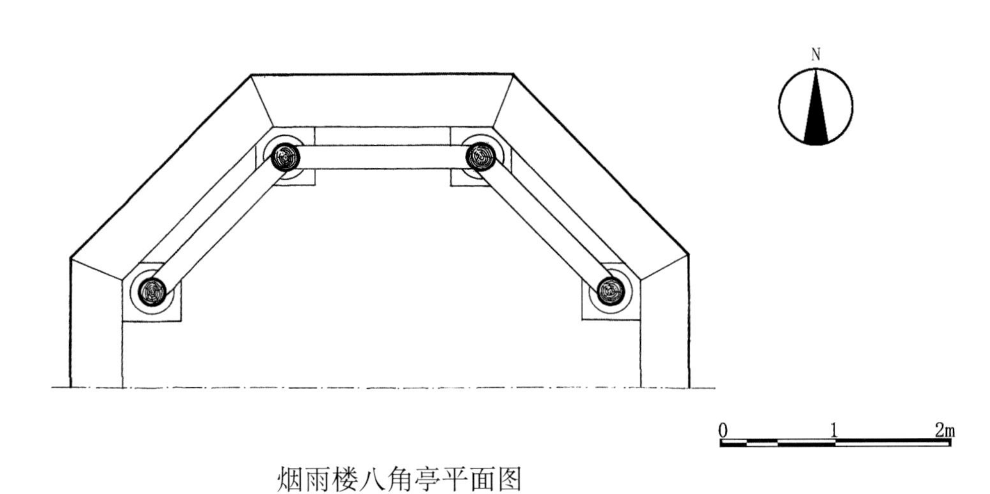

烟雨楼八角亭平面图
Plan of bajiaoting of Yanyulou

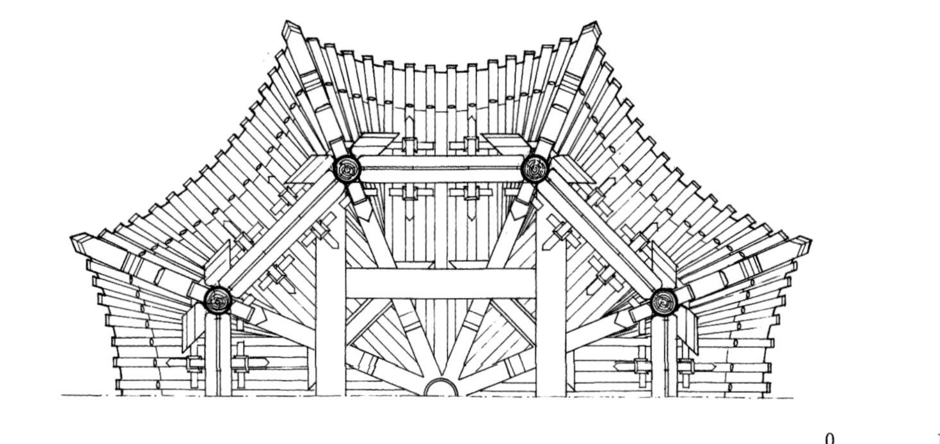

烟雨楼八角亭梁架仰视图
Plan of framework as seen from below of bajiaoting of Yanyulou

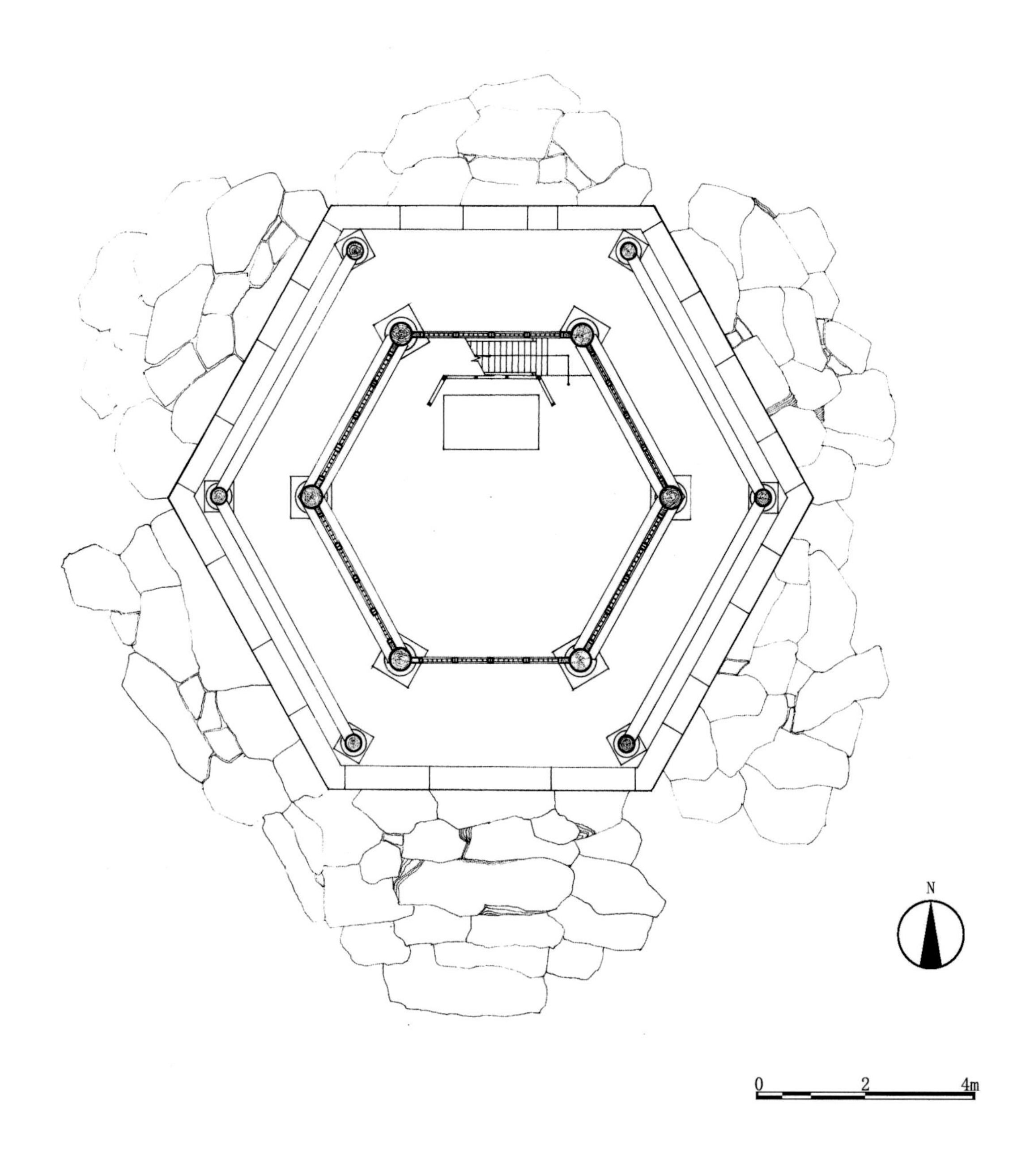

金山上帝阁一层平面图
Plan of first floor of Jinshan shangdige

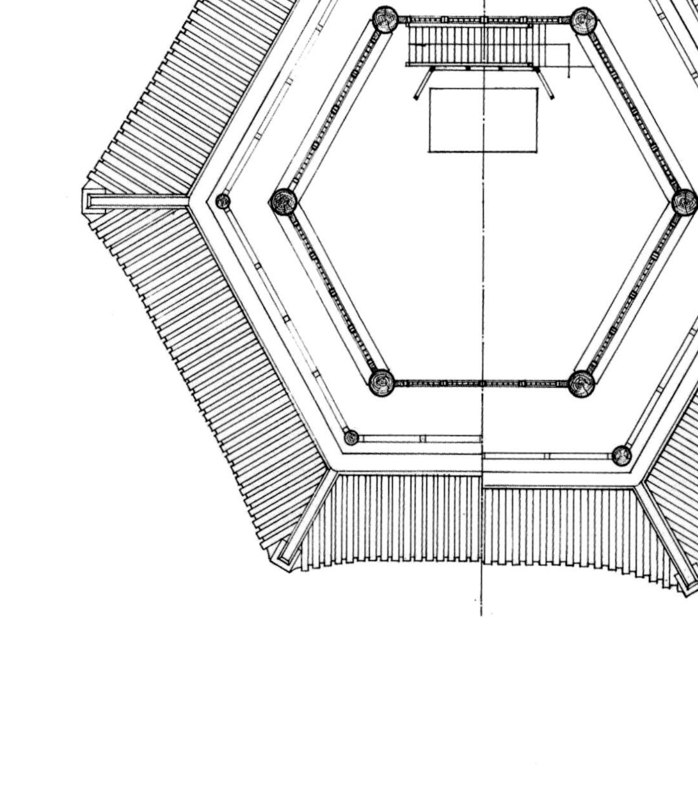

金山上帝阁二、三层平面图
Plan of second and third floor of Jinshan shangdige

金山上帝阁横剖面图
Cross-section of Jinshan shangdige

金山上帝阁正立面图
Front elevation of Jinshan shangdige

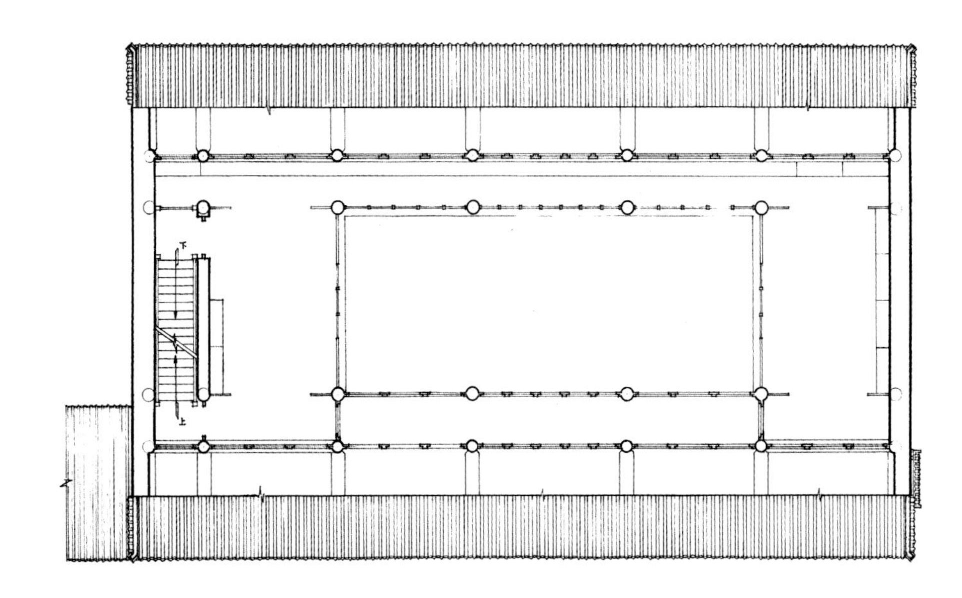

文津阁夹层平面复原图
Restored plan of mezzanine of Wenjinge

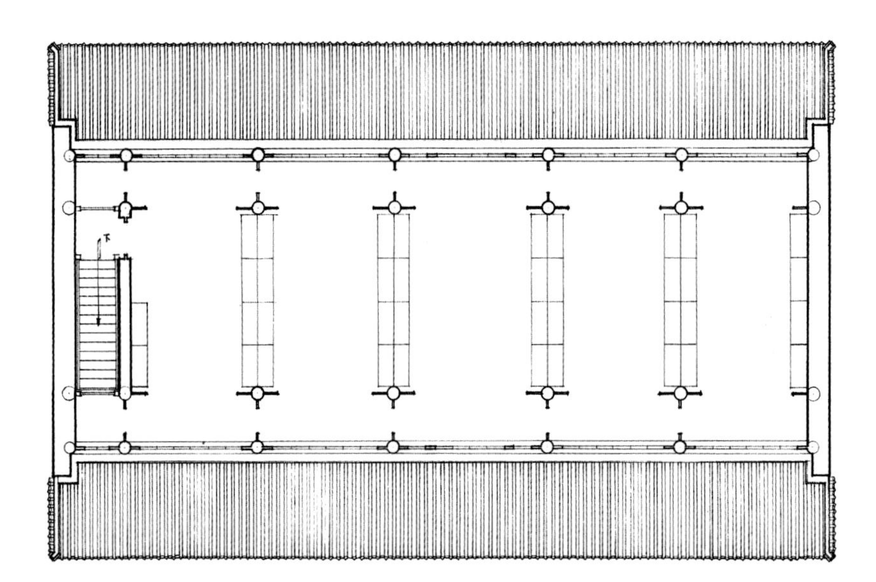

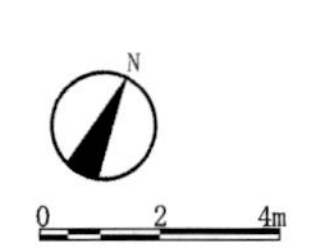

文津阁上层平面复原图
Restored plan of upper floor of Wenjinge

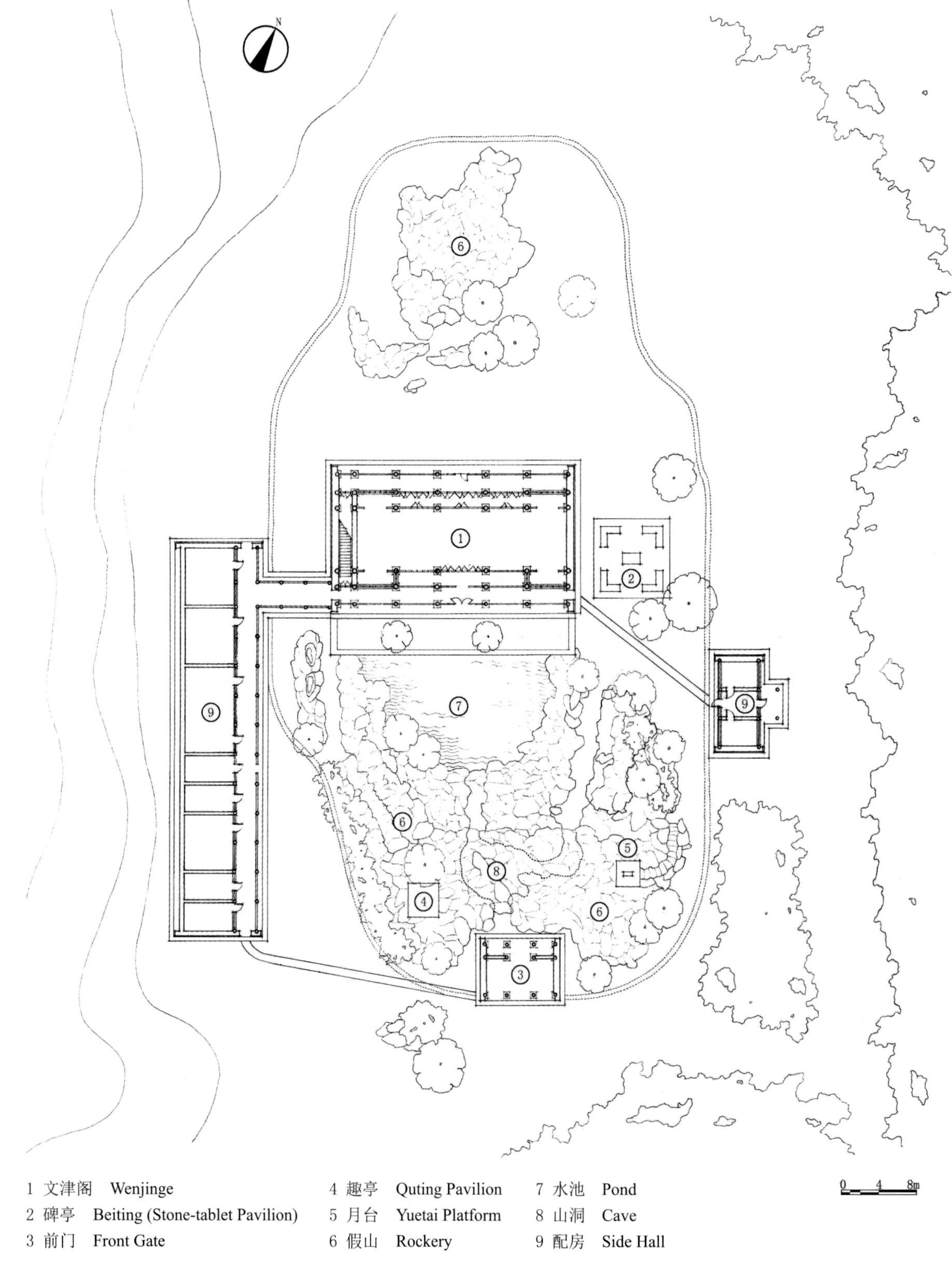

文津阁总平面图
Site plan of Wenjinge

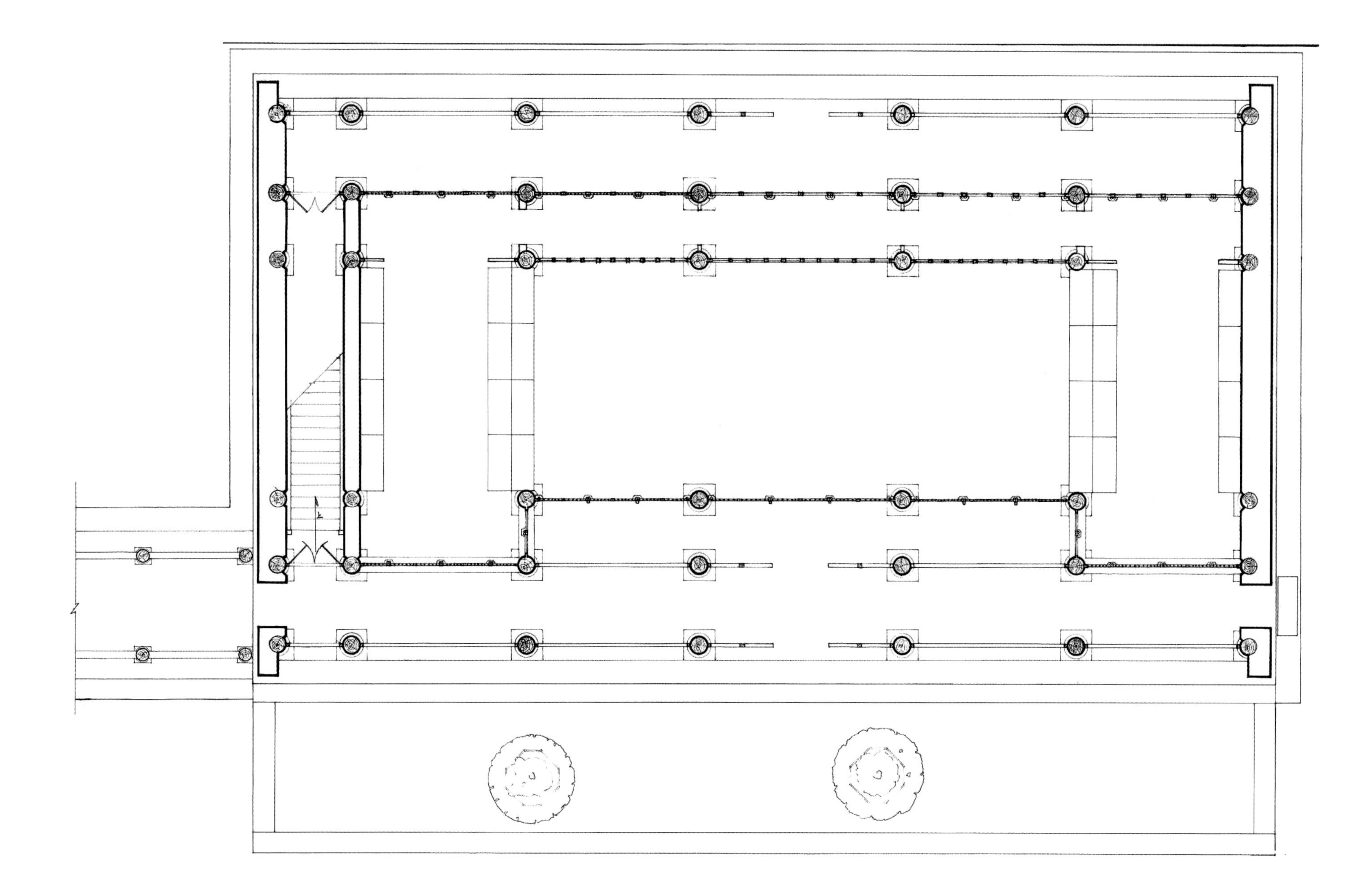

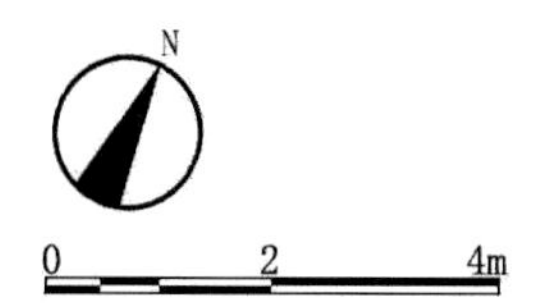

文津阁一层平面复原图
Restored plan of first floor of Wenjinge

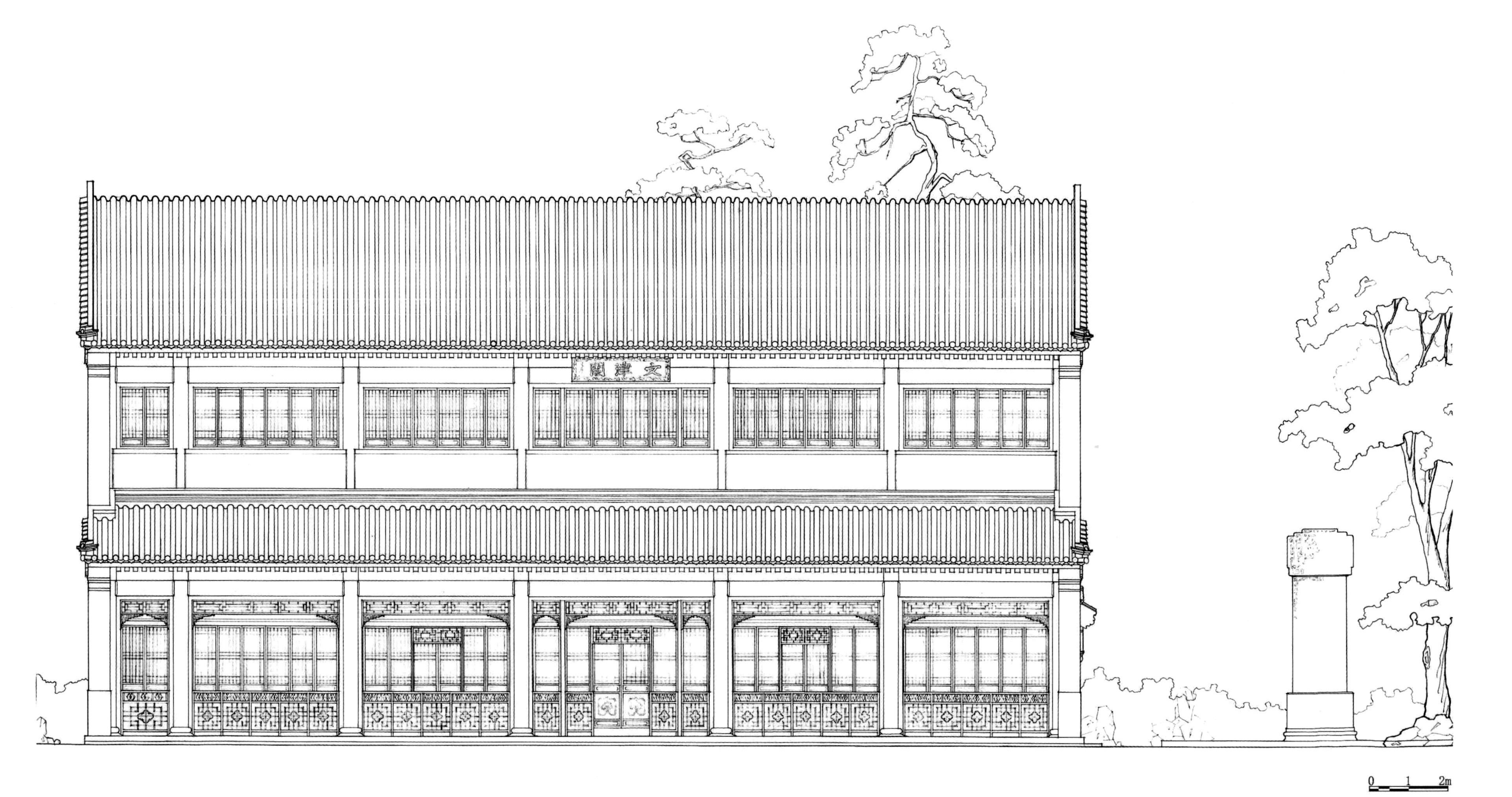

文津阁南立面图
South elevation of Wenjinge

文津阁东立面图
East elevation of Wenjinge

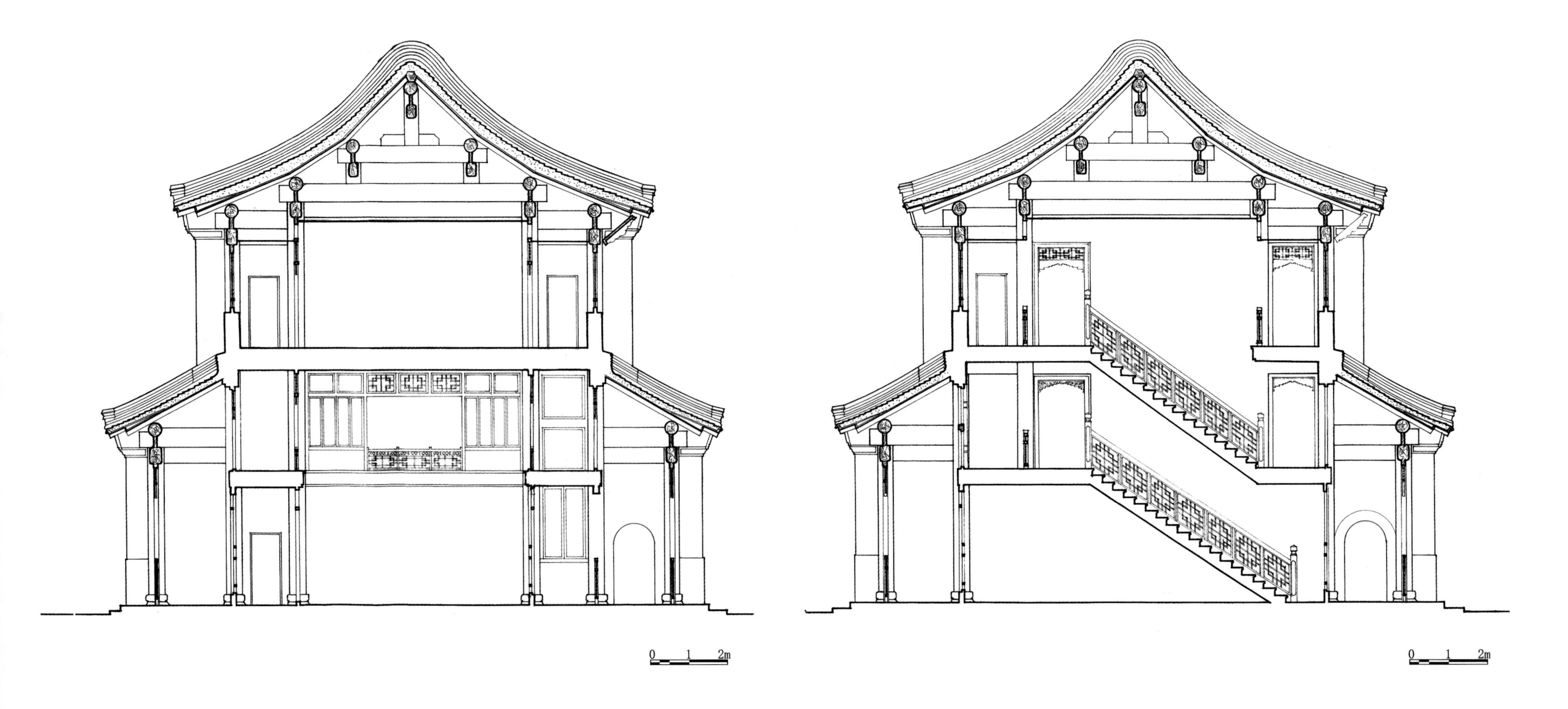

文津阁明间横剖面复原图
Restored cross-section of central-bay of Wenjinge

文津阁梢间横剖面复原图
Restored corss-section of second-to-last-bay of Wenjinge

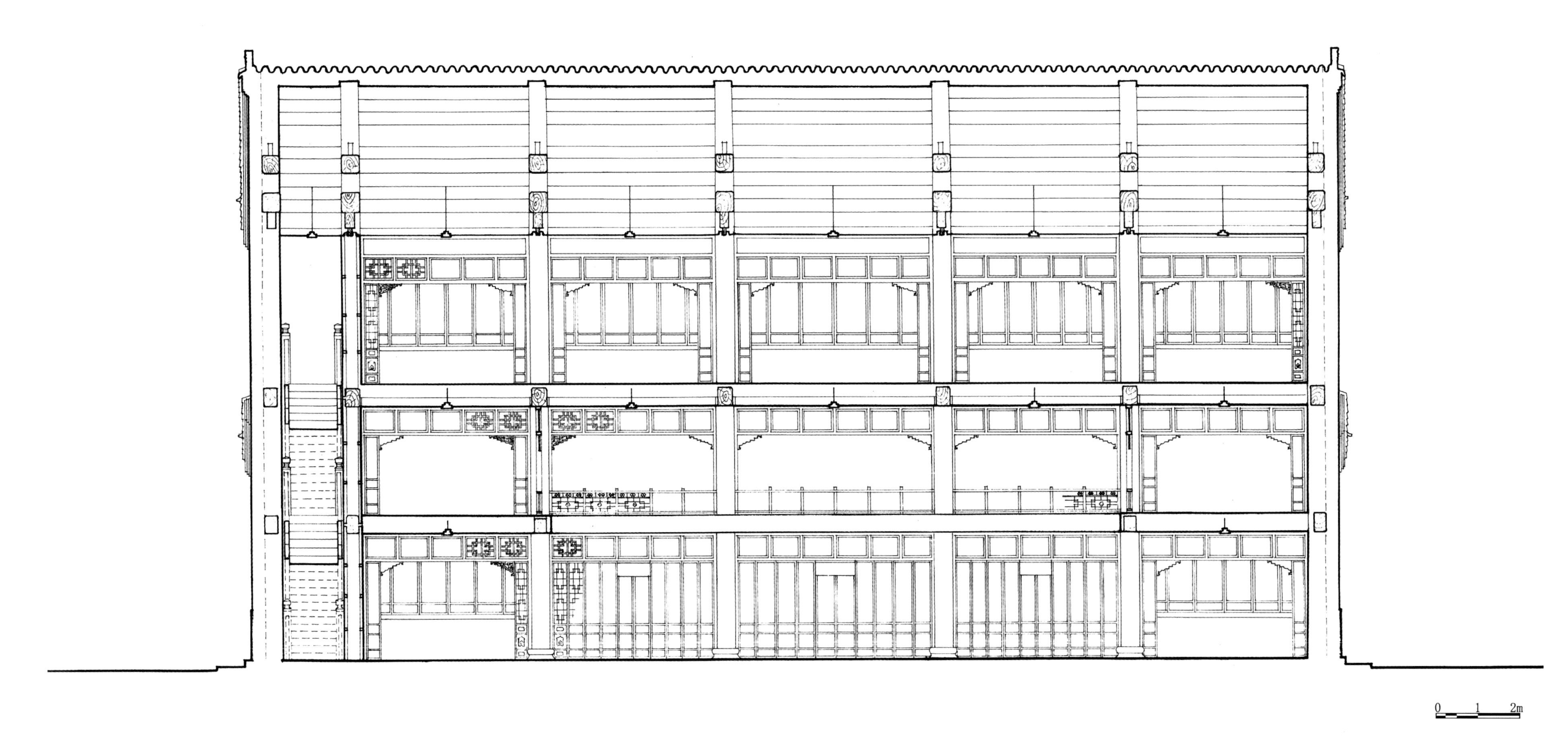

文津阁纵剖面复原图
Restored longitudinal section of Wenjinge

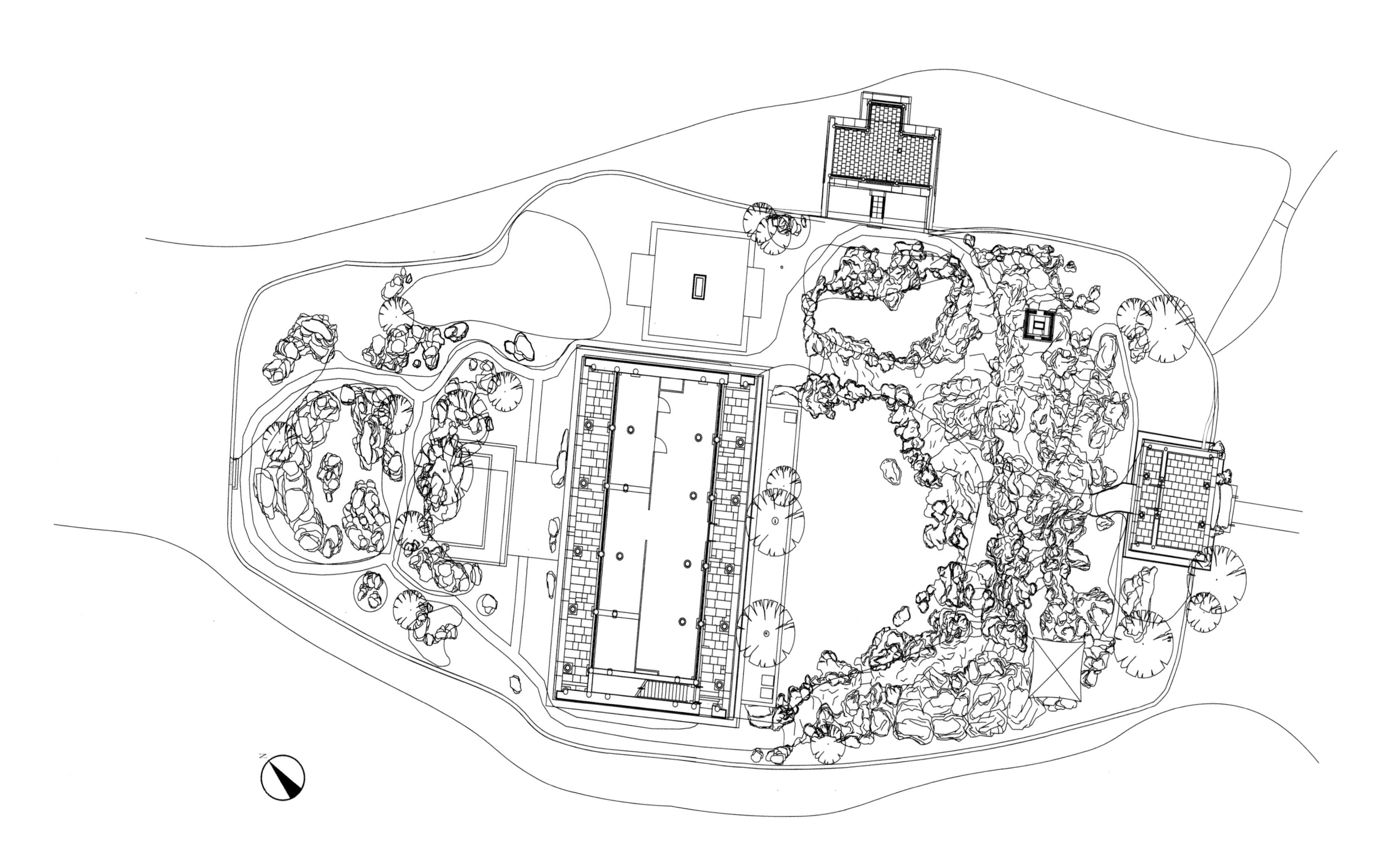

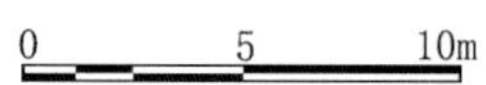

文津阁总平面图（2012 年测绘）
Site plan of Wenjinge (2012 survey)

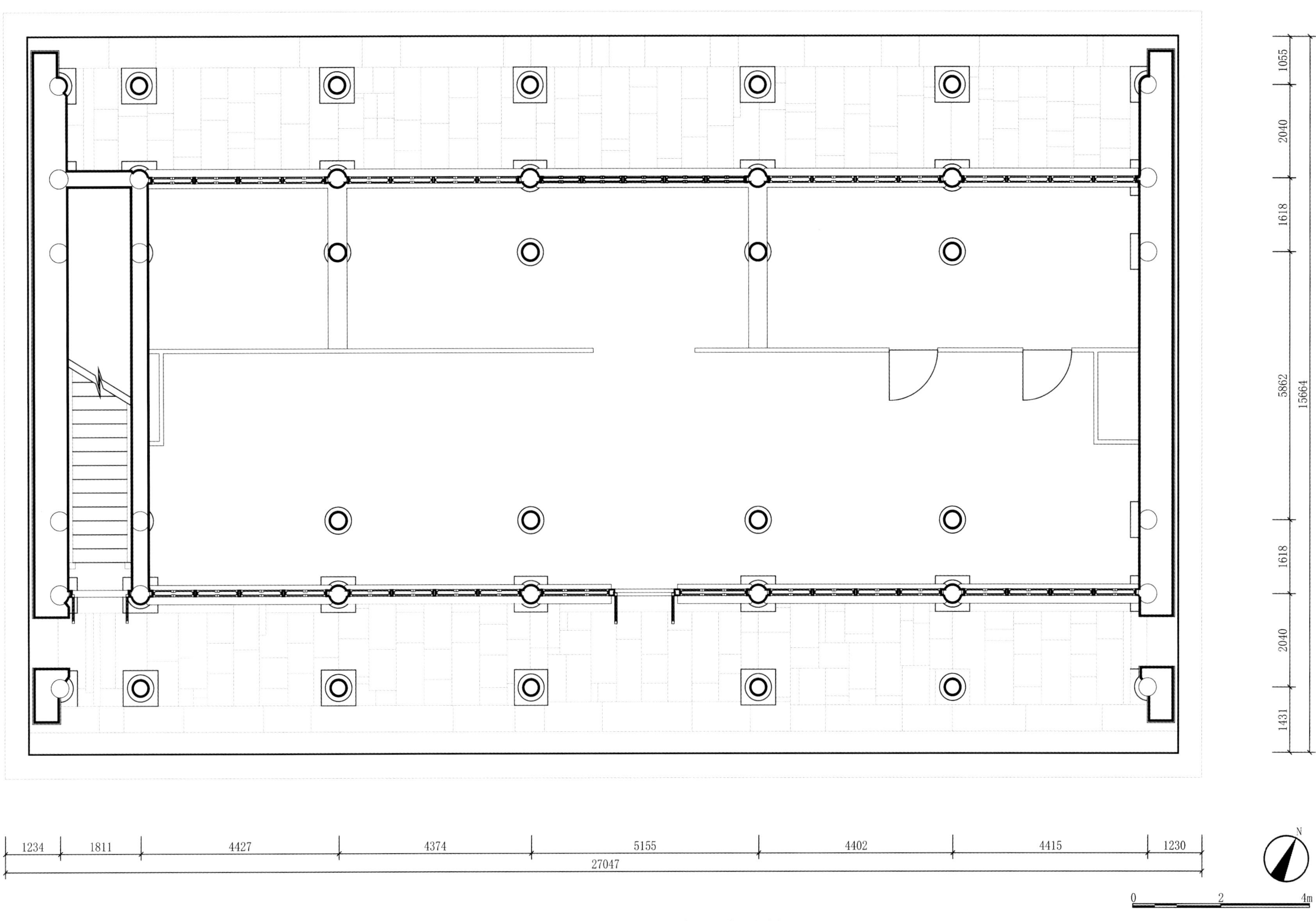

文津阁一层平面图（2012 年测绘）
Plan of first floor of Wenjinge (2012 survey)

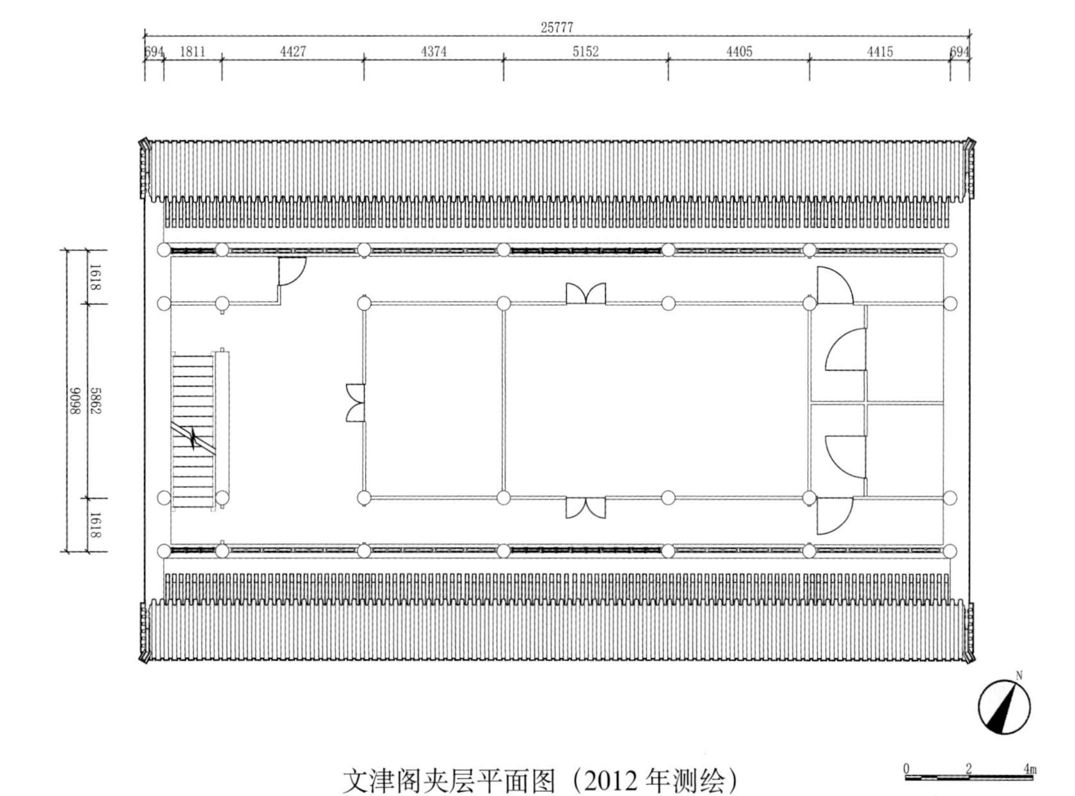

文津阁夹层平面图（2012 年测绘）
Plan of mezzanine of Wenjinge (2012 survey)

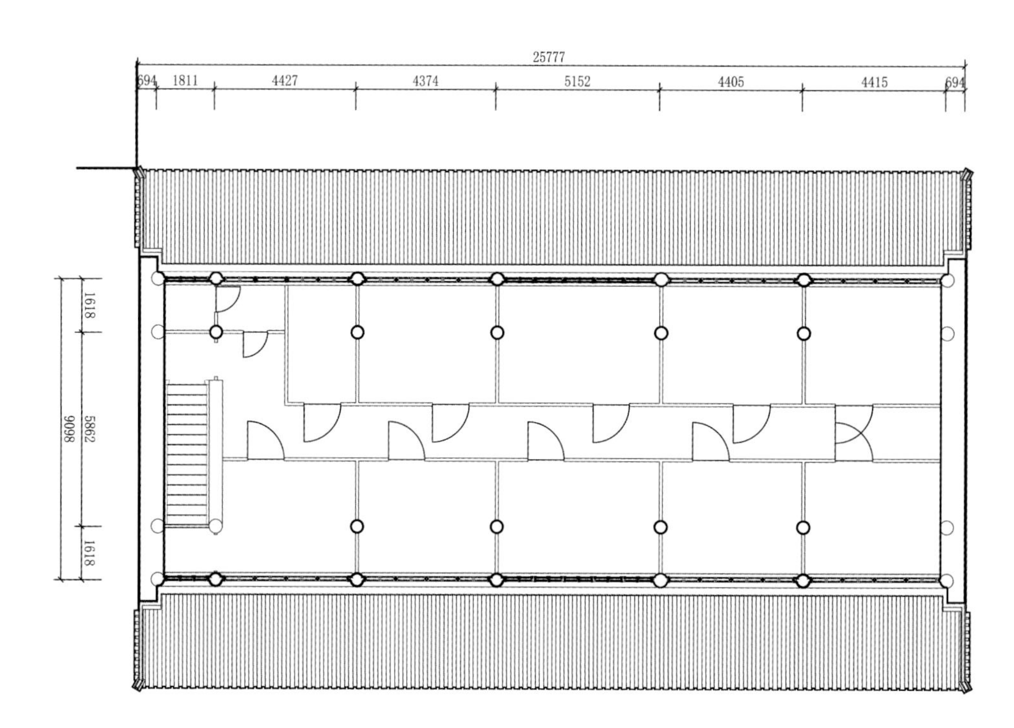

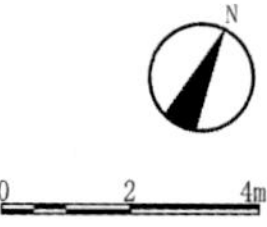

文津阁上层面图（2012 年测绘）
Plan of upper floor of Wenjinge (2012 survey)

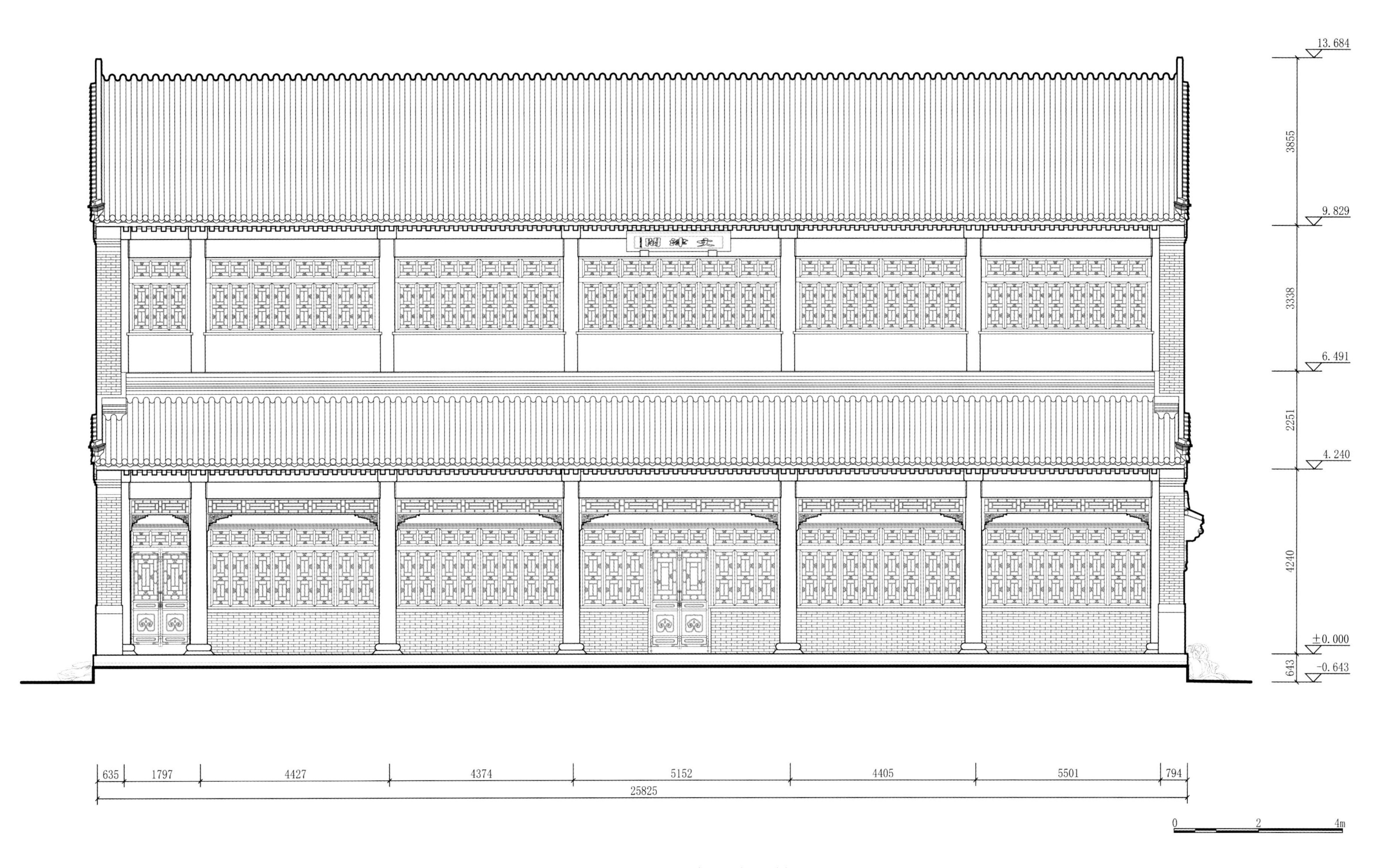

文津阁正立面图（2012 年测绘）
Front elevation of Wenjinge (2012 survey)

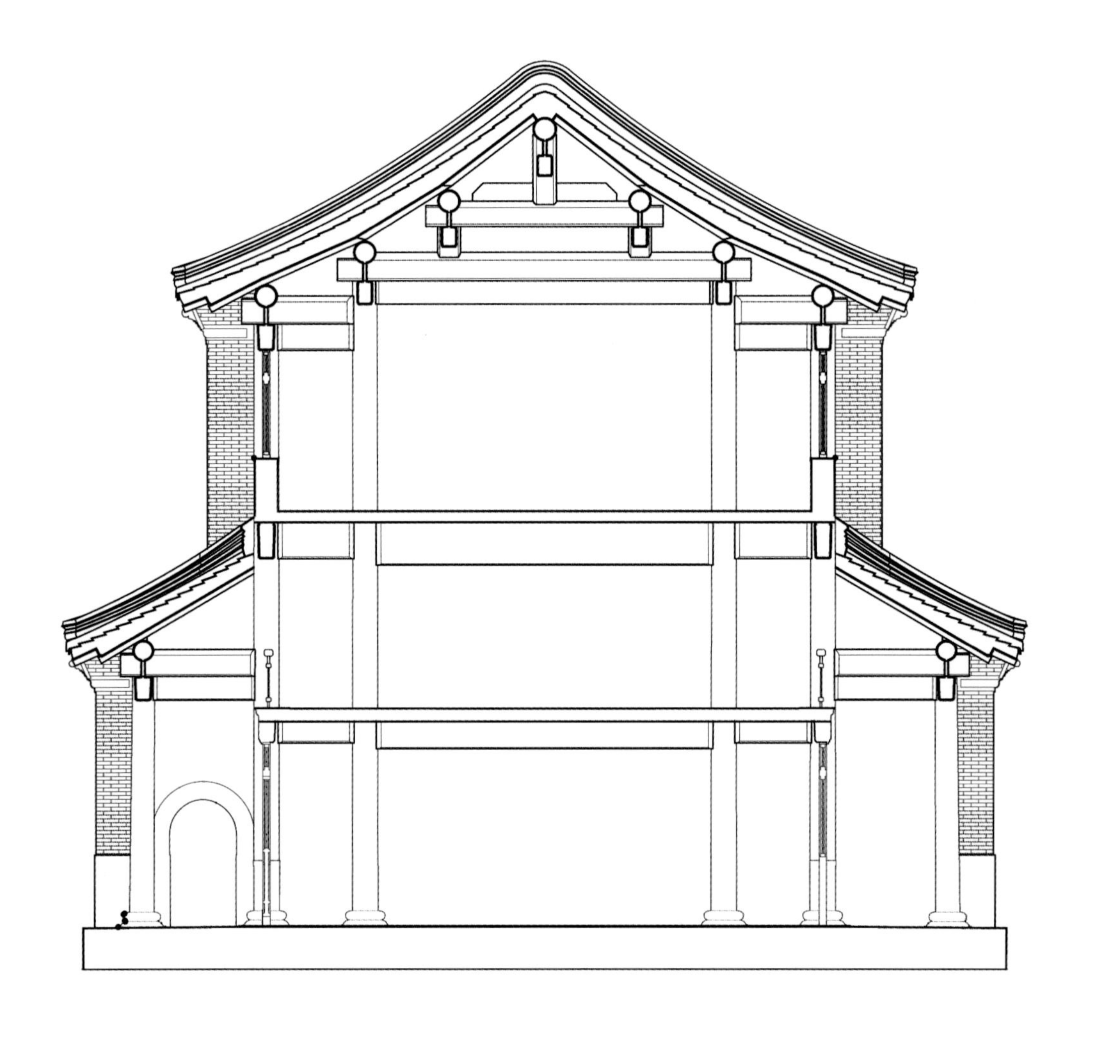

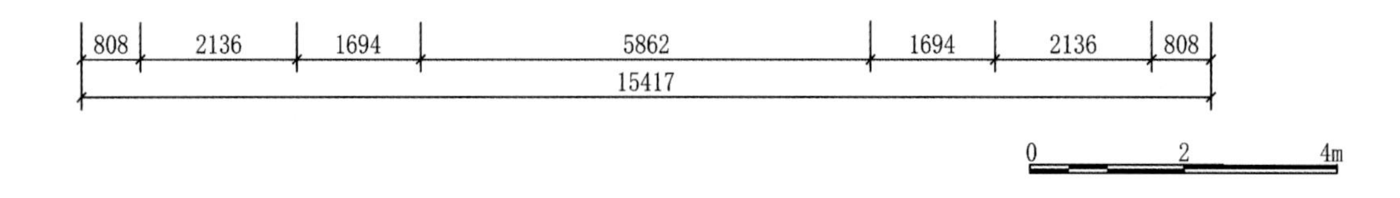

文津阁明间横剖面图（2012 年测绘）
Cross-section of central-bay of Wenjinge (2012 survey)

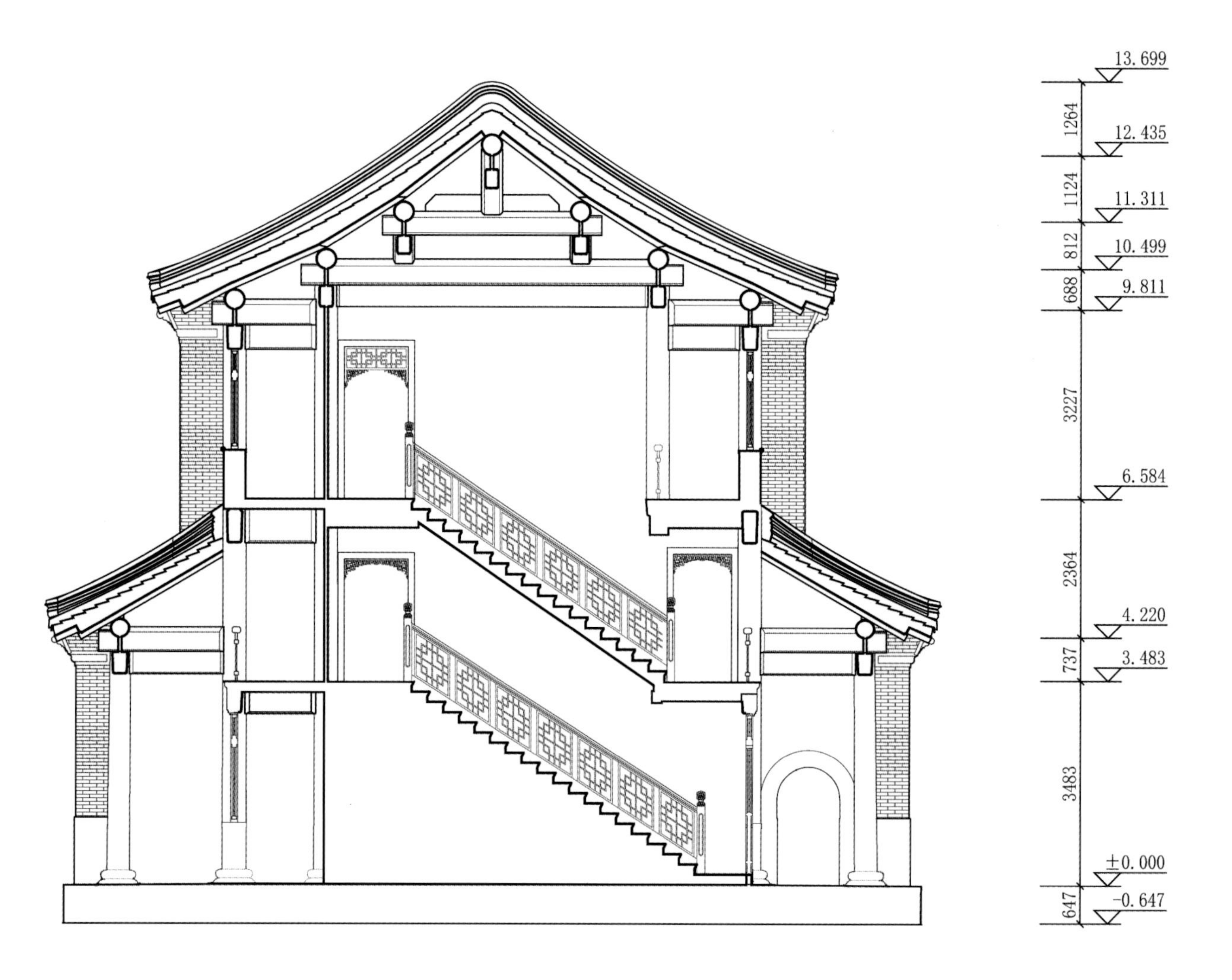

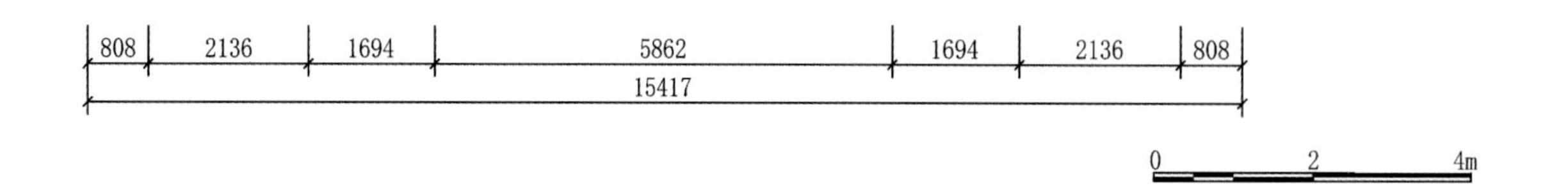

文津阁楼梯间横剖面图（2012 年测绘）
Cross-section of staircase-bay of Wenjinge (2012 survey)

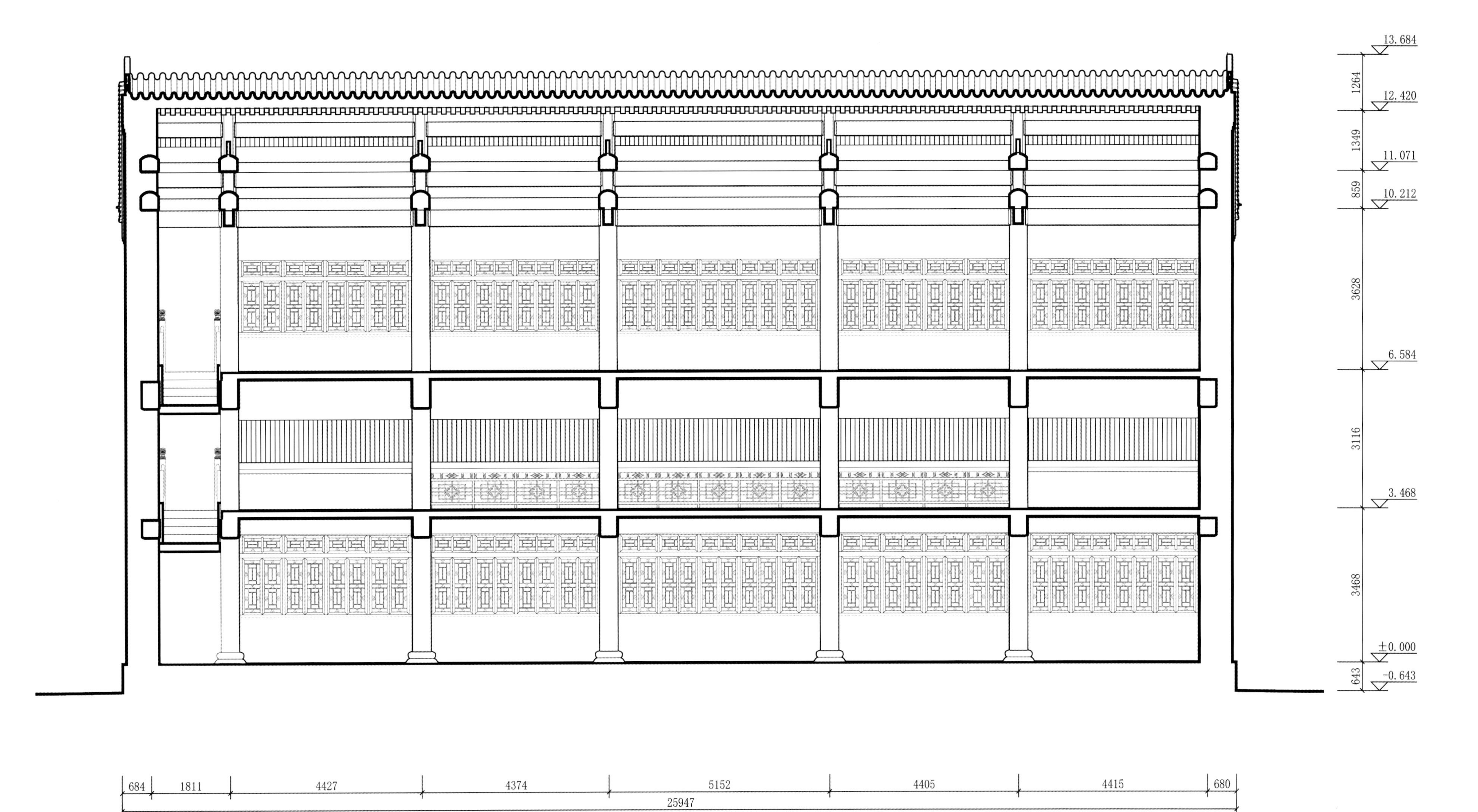

文津阁纵剖面图（2012年测绘）
Longitudinal section of Wenjinge (2012 survey)

芳渚临流立面图
Elevation of Fangzhu linliu

0 1 2m

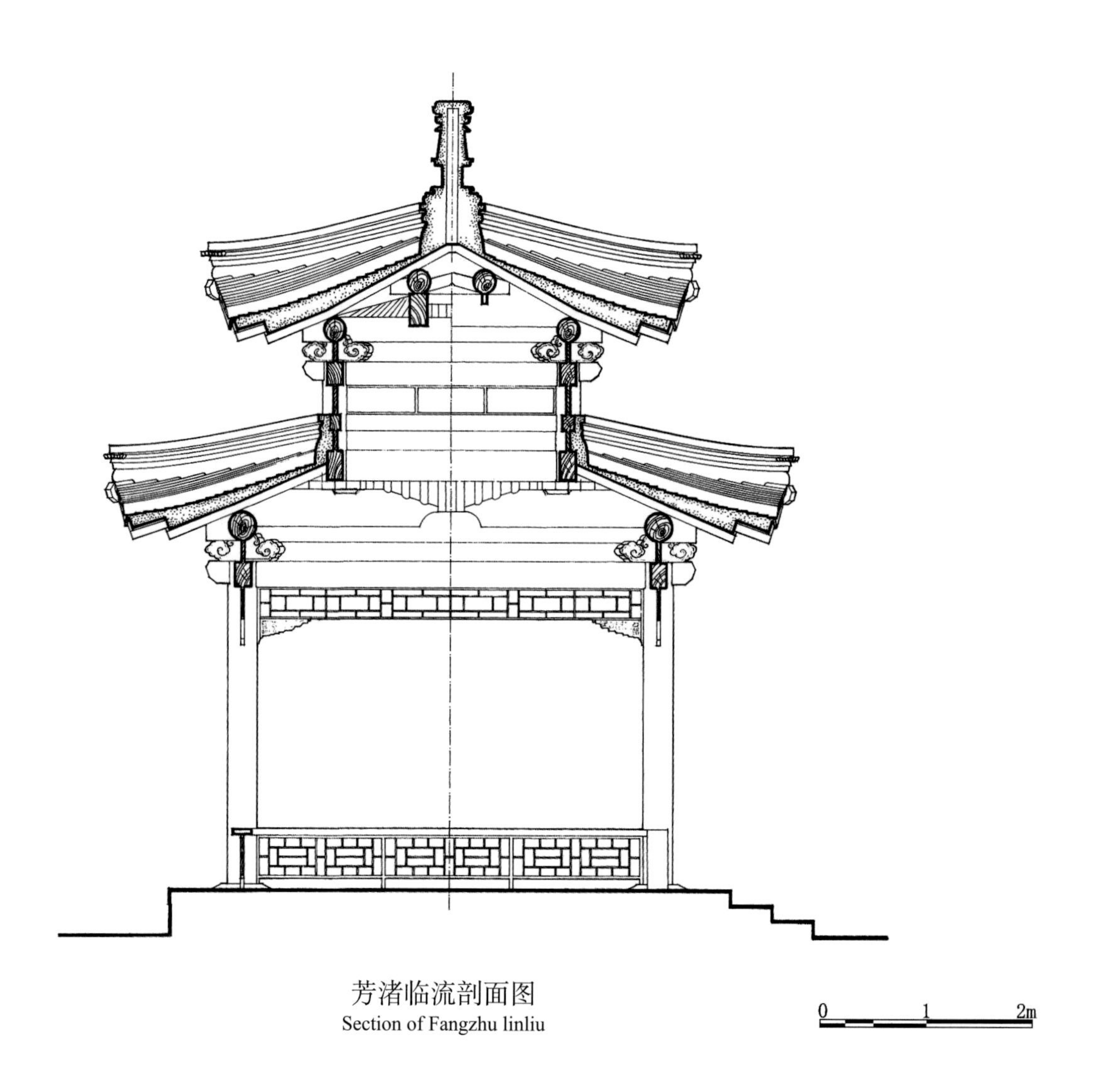

芳渚临流剖面图
Section of Fangzhu linliu

0 1 2m

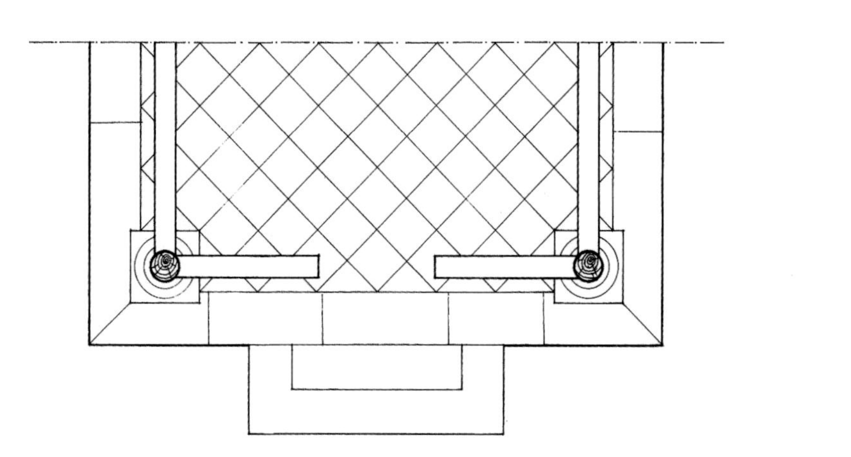

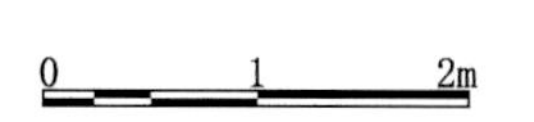

芳渚临流平面图
Plan of Fangzhu linliu

0 1 2m

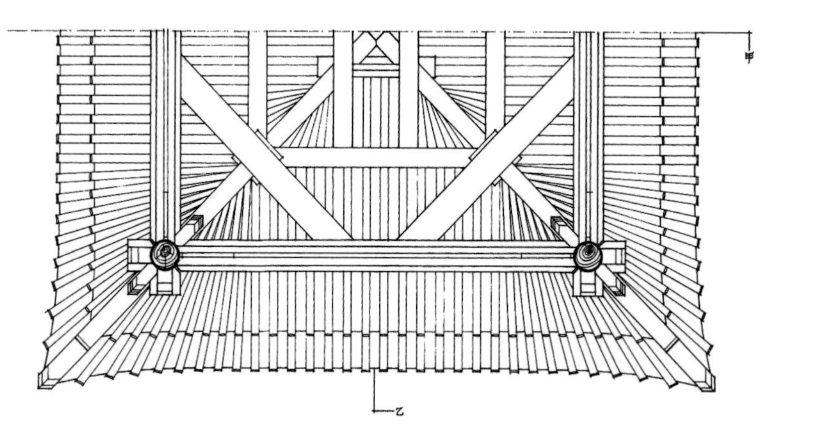

芳渚临流梁架仰视图
Plan of framework as seen from below of Fangzhu linliu

0 1 2m

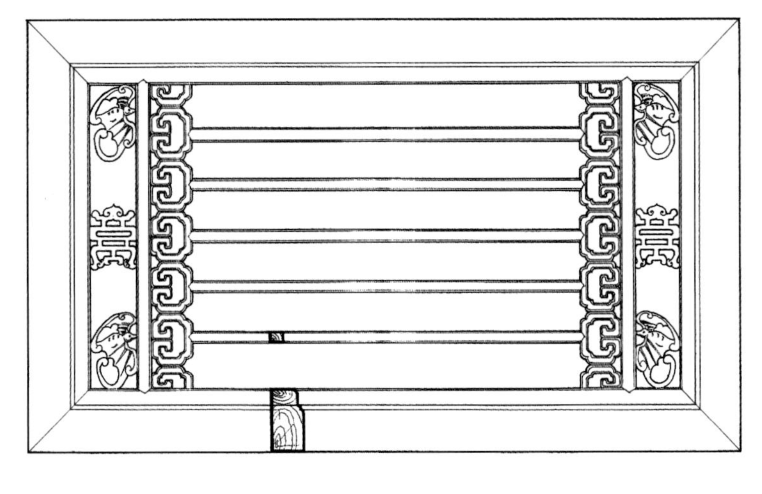

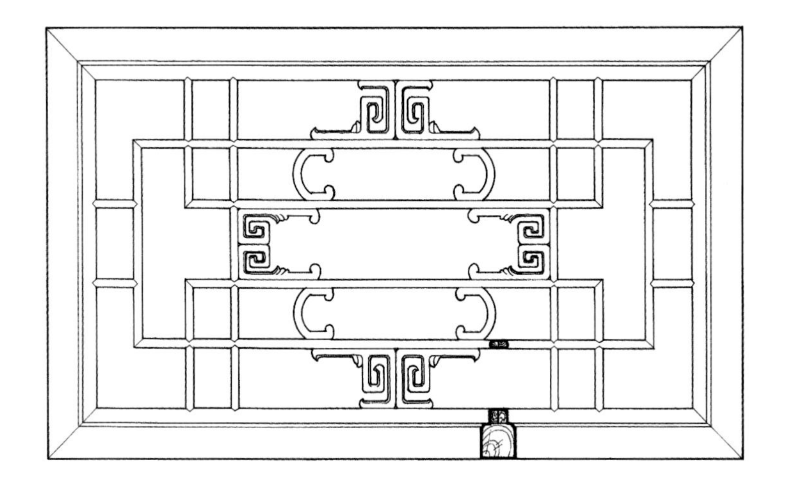

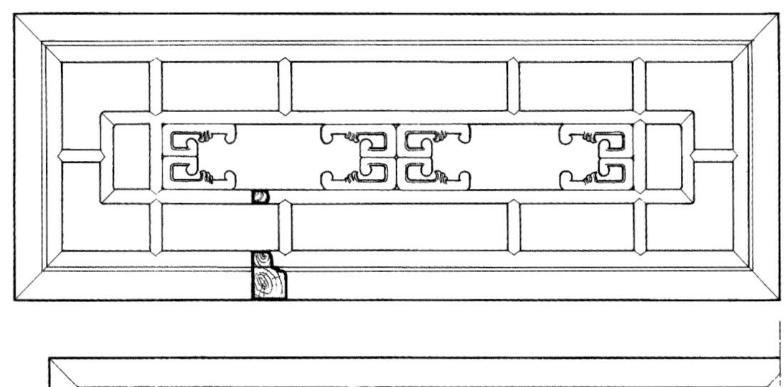

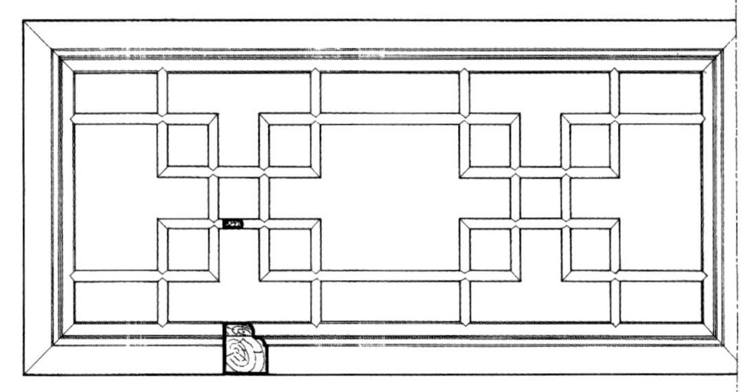

建筑物横披大样图
Hengpi of buildings

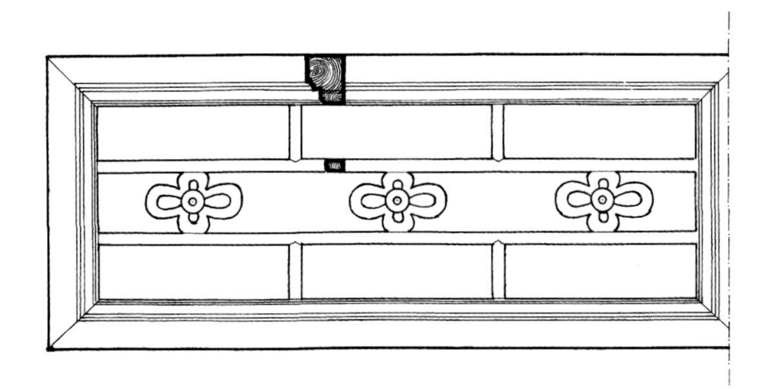

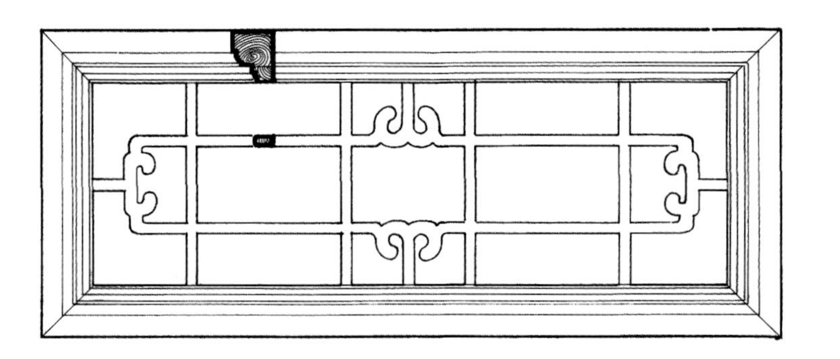

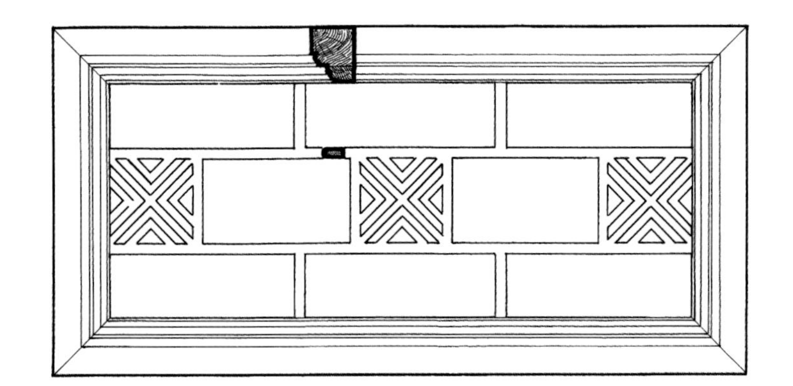

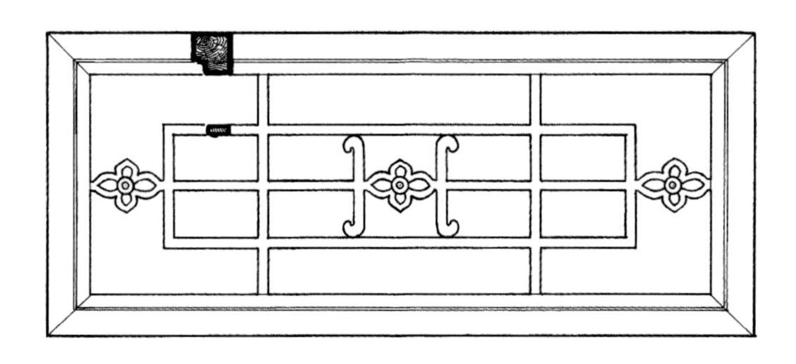

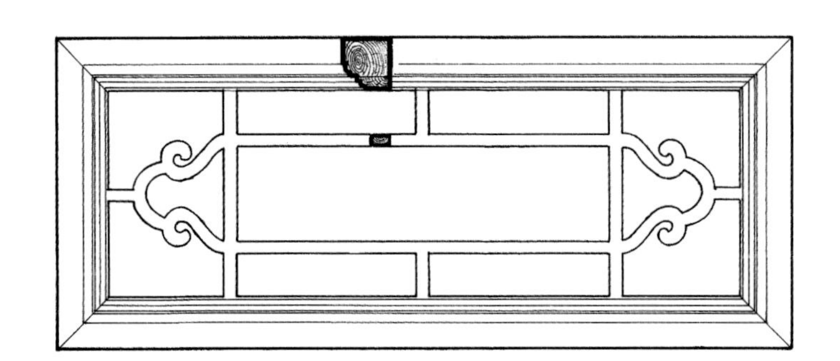

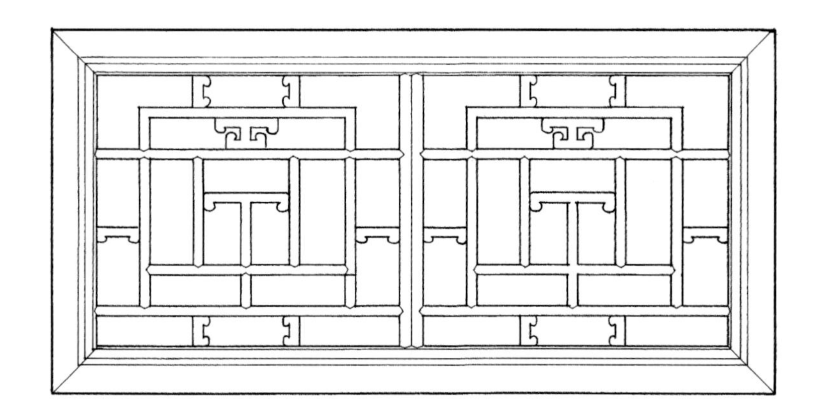

0 0.2 0.4m

建筑物横披大样图

Hengpi of buildings

普宁寺
Puningsi

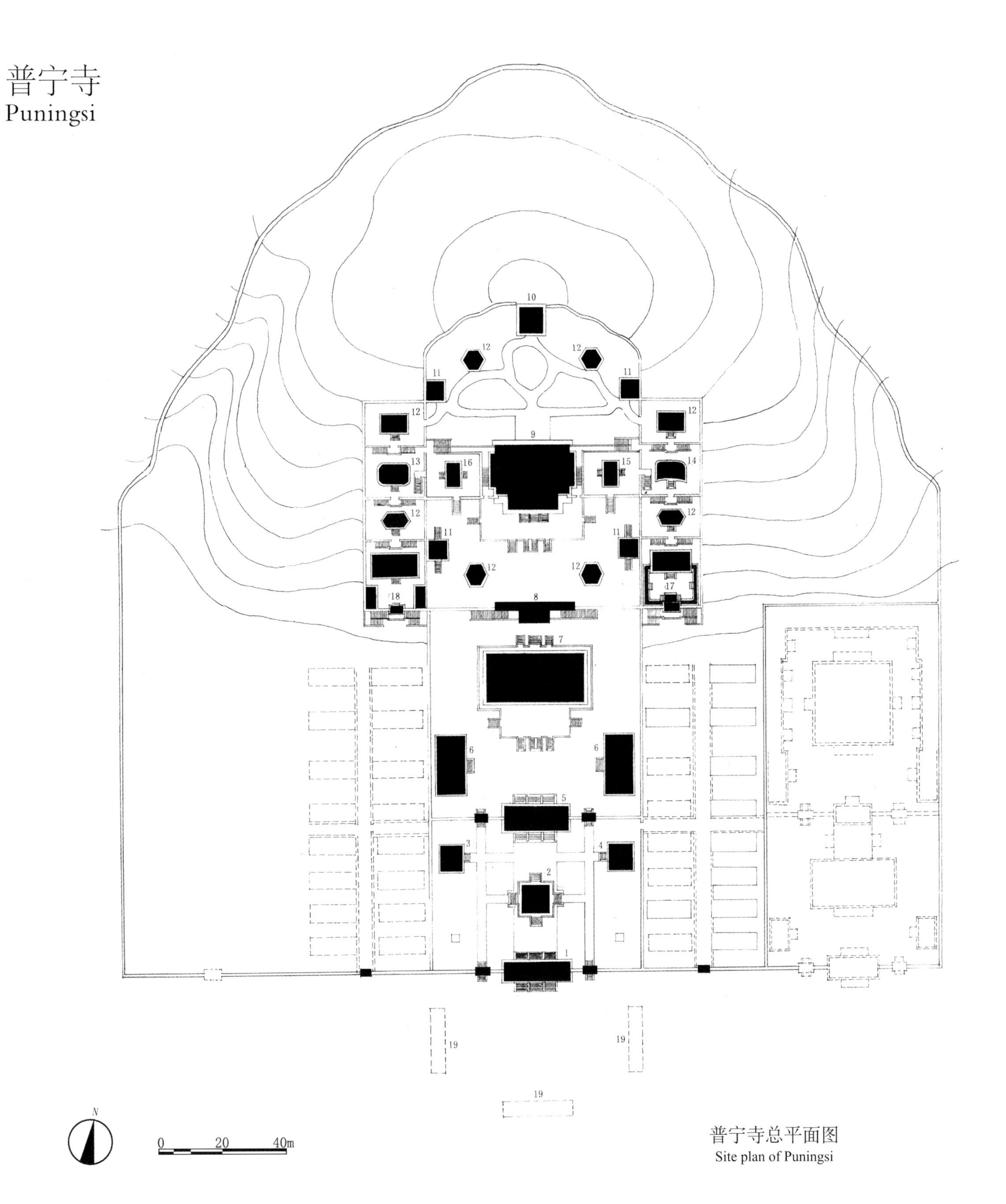

普宁寺总平面图
Site plan of Puningsi

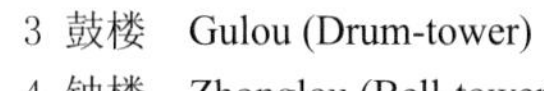

1 山门 Shanmen (Gate Hall)
2 碑亭 Beiting (Stone-tablet Pavilion)
3 鼓楼 Gulou (Drum-tower)
4 钟楼 Zhonglou (Bell-tower)
5 天王殿 Tian wangdian (Hall of Heavenly King)
6 配殿 Peidian (Side Halls)
7 大雄宝殿 Daxiong baodian (Mahavira Hall)
8 南瞻部洲 Nanzhan buzhou (Jambudvīpa Pavilion)
9 大乘之阁 Dacheng zhige
10 北俱卢洲 Beiju luzhou (Uttarakuru Pavilion)
13 西牛贺洲 Xiniu hezhou (Aparagodānīya Pavilion)
14 东胜神州 Dongsheng shenzhou (Pūrvavideha Pavilion)
15 日殿 Sun-shape Pavilion
16 月殿 Moon-shape Pavilion
17 妙严室 Miaoyanshi
18 讲经堂 Jiangjingtang
19 牌楼遗址 Site of pailou (Decorated Archways)

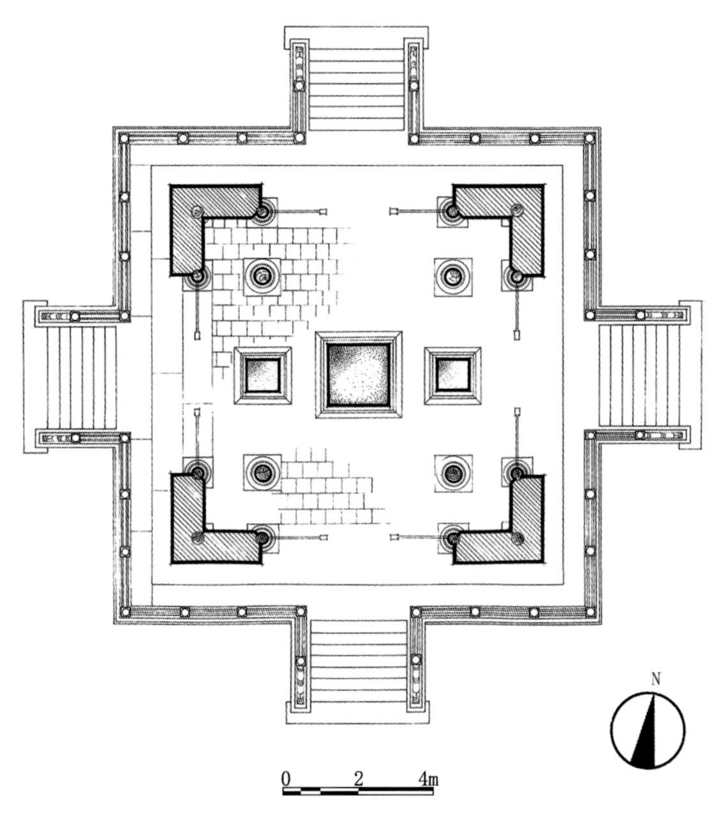

普宁寺碑亭平面图
Plan of beiting of Puningsi

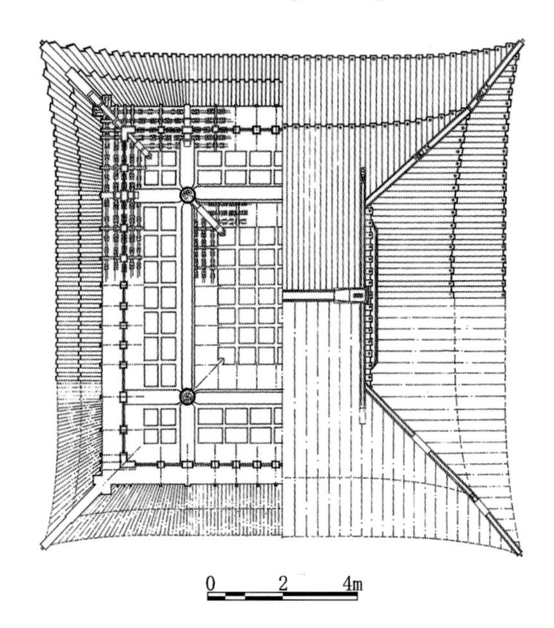

普宁寺碑亭梁架仰视图、屋顶平面图
Plan of framework as seen from below and roof plan of beiting of Puningsi

普宁寺碑亭正立面图
Front elevation of beiting of Puningsi

*原图已失，仅存照片
The original picture has been lost,
only photos are left.

普宁寺碑亭侧立面图*
Side elevation of beiting of Puningsi*

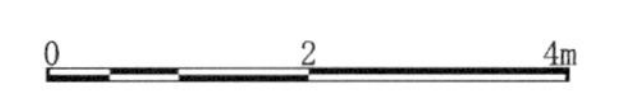

普宁寺碑亭石碑大样图
Shibei of beiting of Puningsi

0 0.2 0.4m

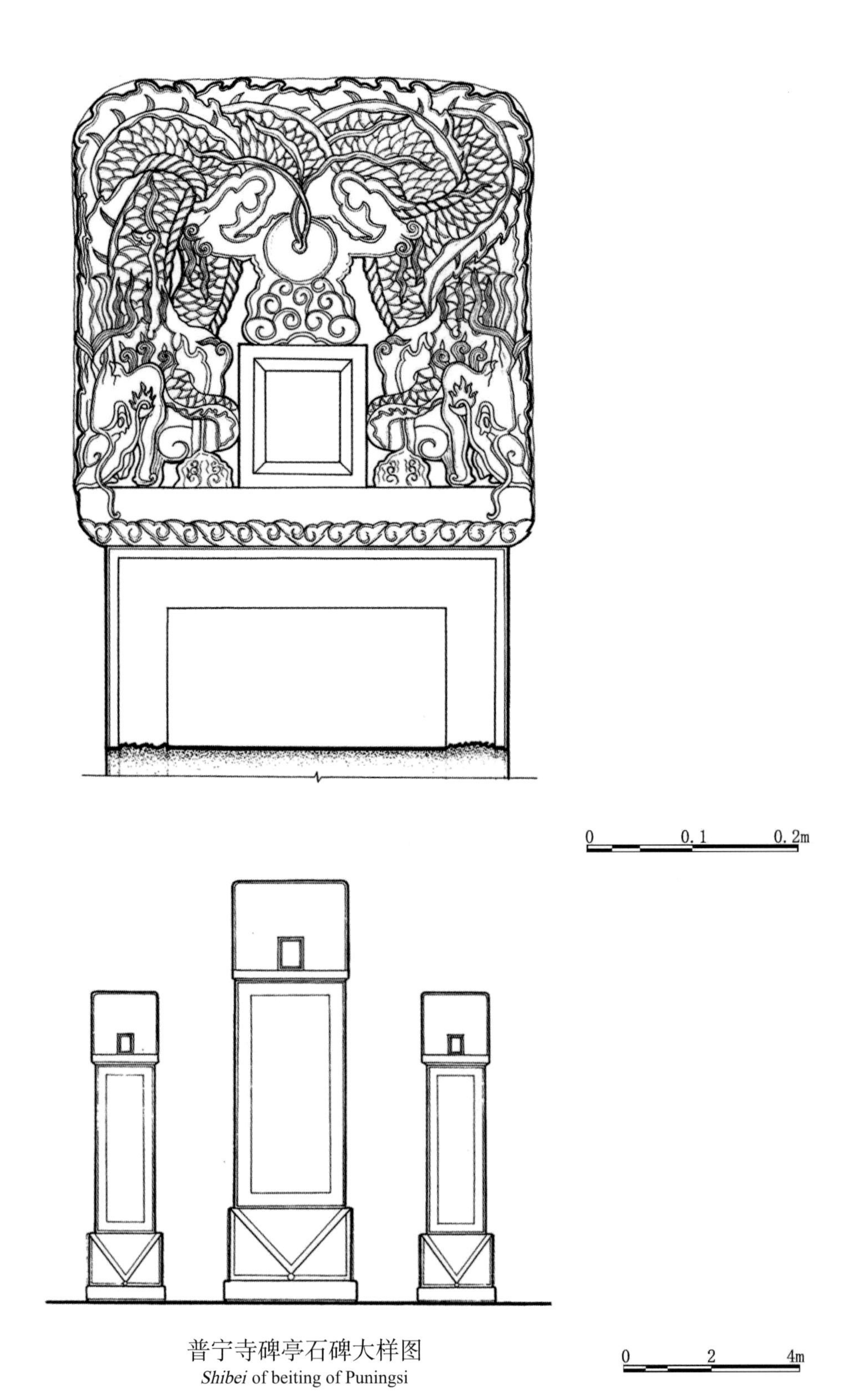

普宁寺碑亭石碑大样图

Shibei of beiting of Puningsi

普宁寺碑亭碑座大样图

Beizuo of beiting of Puningsi

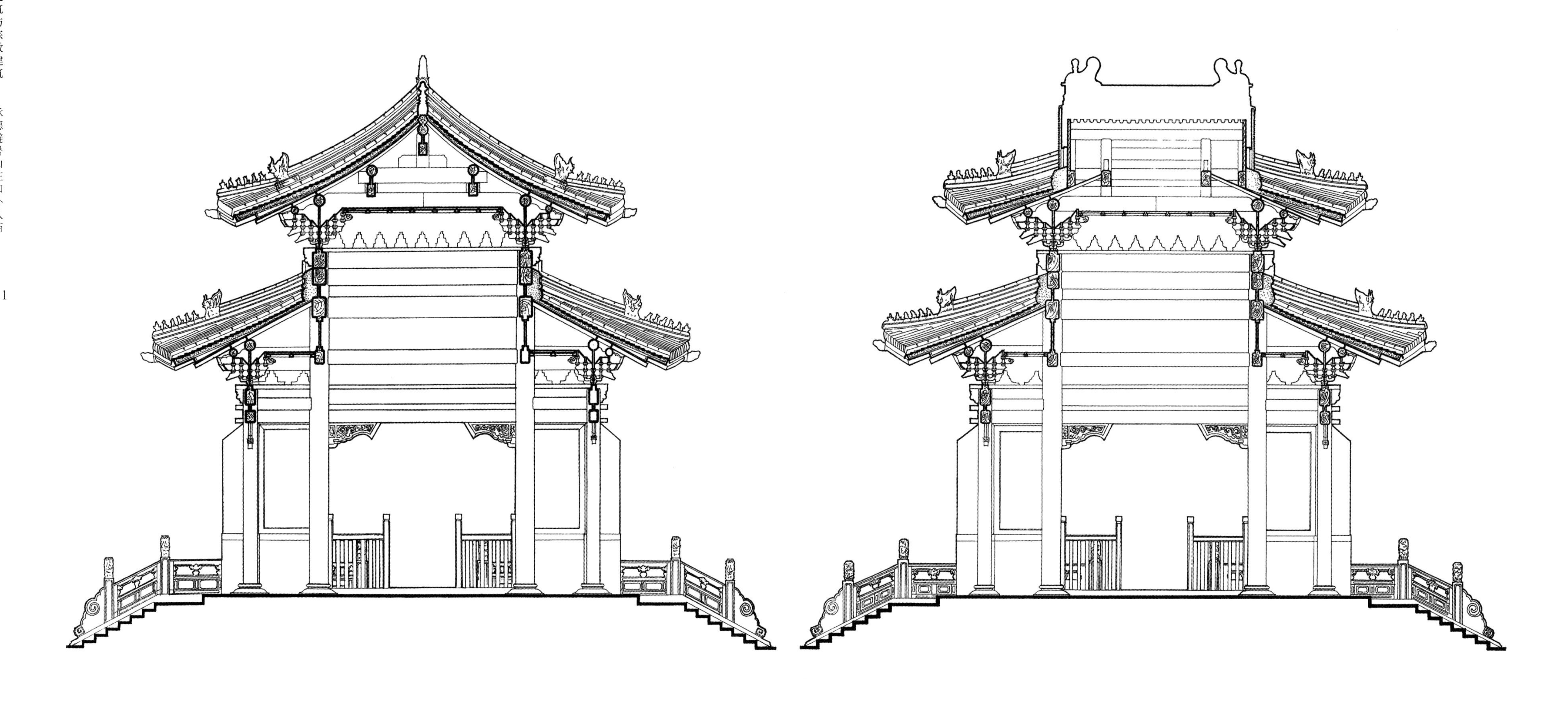

普宁寺碑亭横剖面图
Cross-section of beiting of Puningsi

普宁寺碑亭纵剖面图
Longitudinal section of beiting of Puningsi

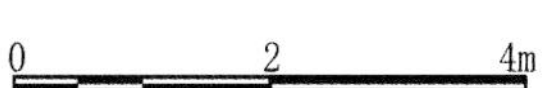

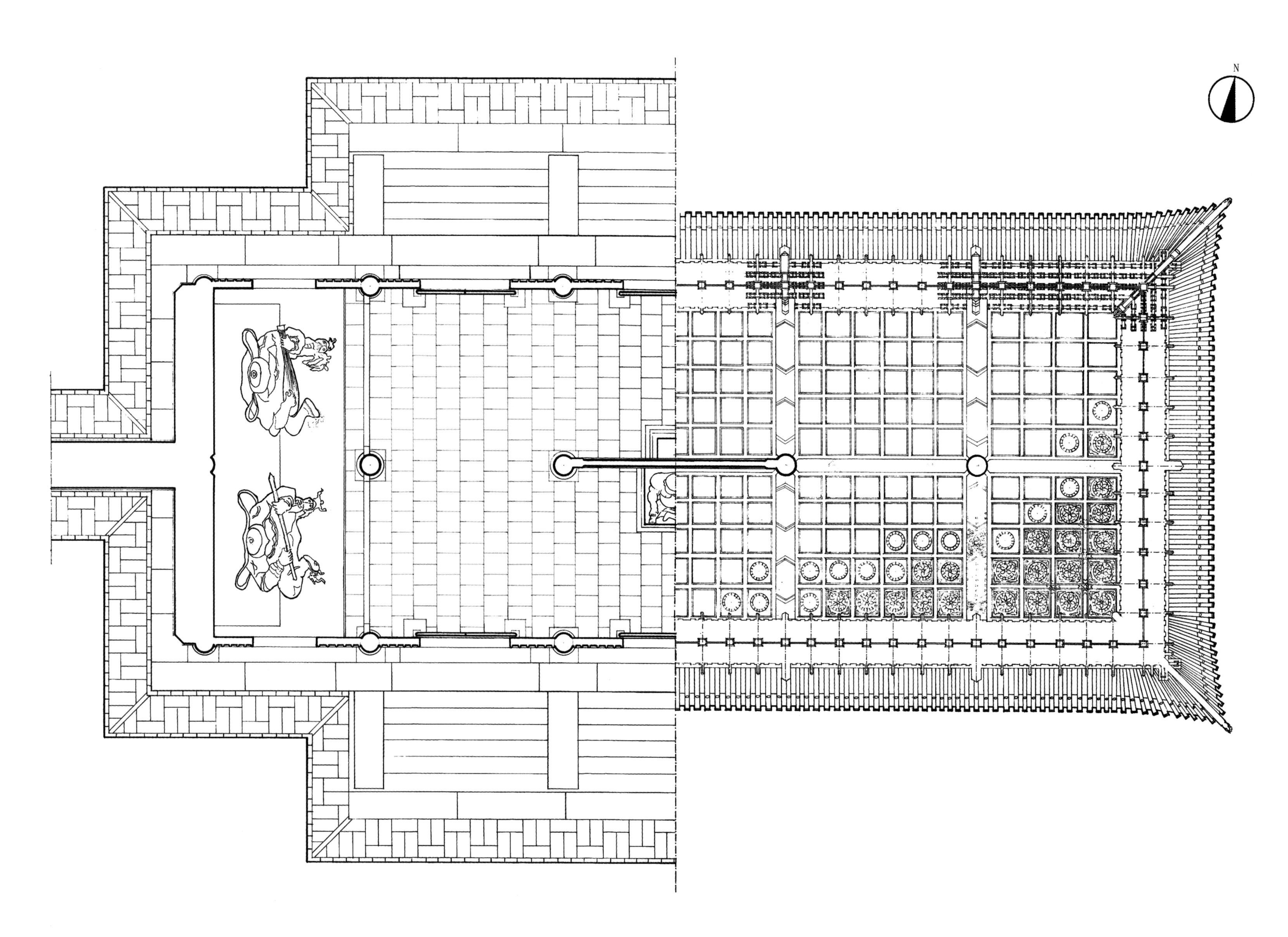

普宁寺天王殿平面图、天花仰视图
Plan and reflected ceiling plan of Tianwangdian of Puningsi

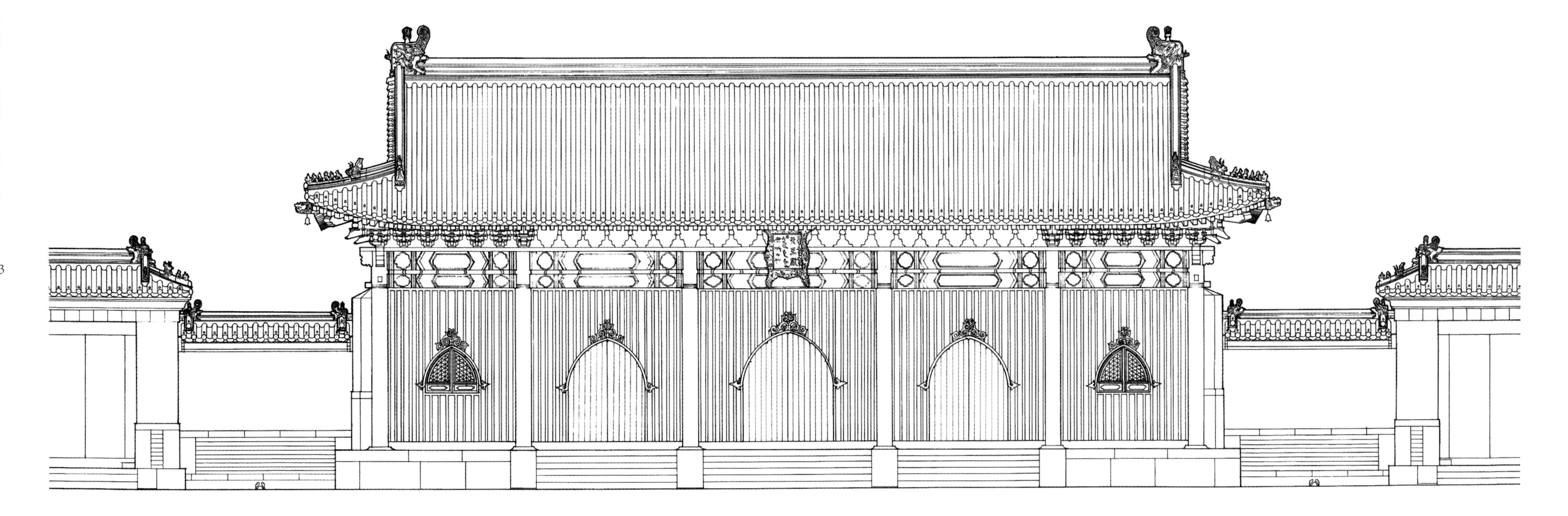

普宁寺天王殿正立面图
Front elevation of Tianwangdian of Puningsi

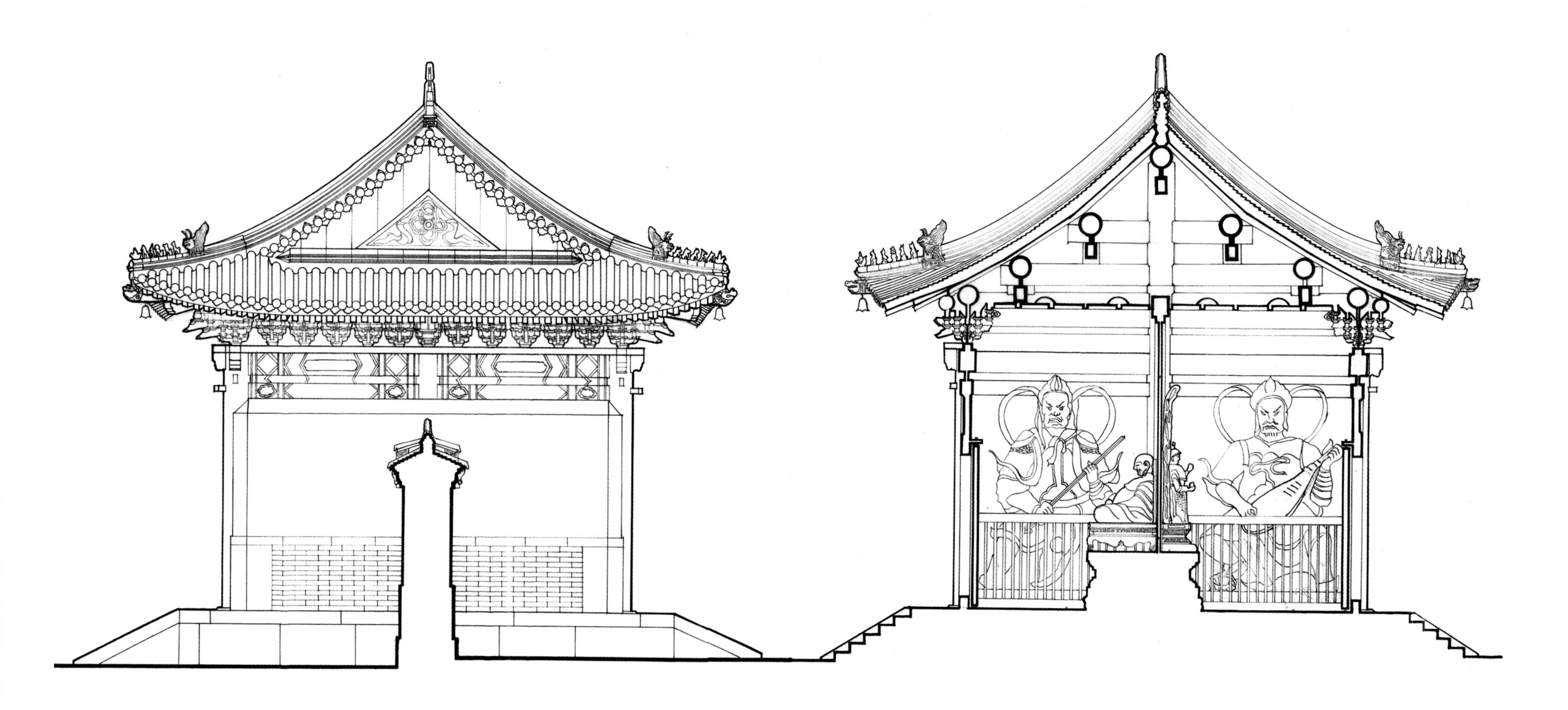

普宁寺天王殿侧立面图
Side elevation of Tianwangdian of Puningsi

0 2 4m

普宁寺天王殿横剖面图
Cross-section of Tianwangdian of Puningsi

0 2 4m

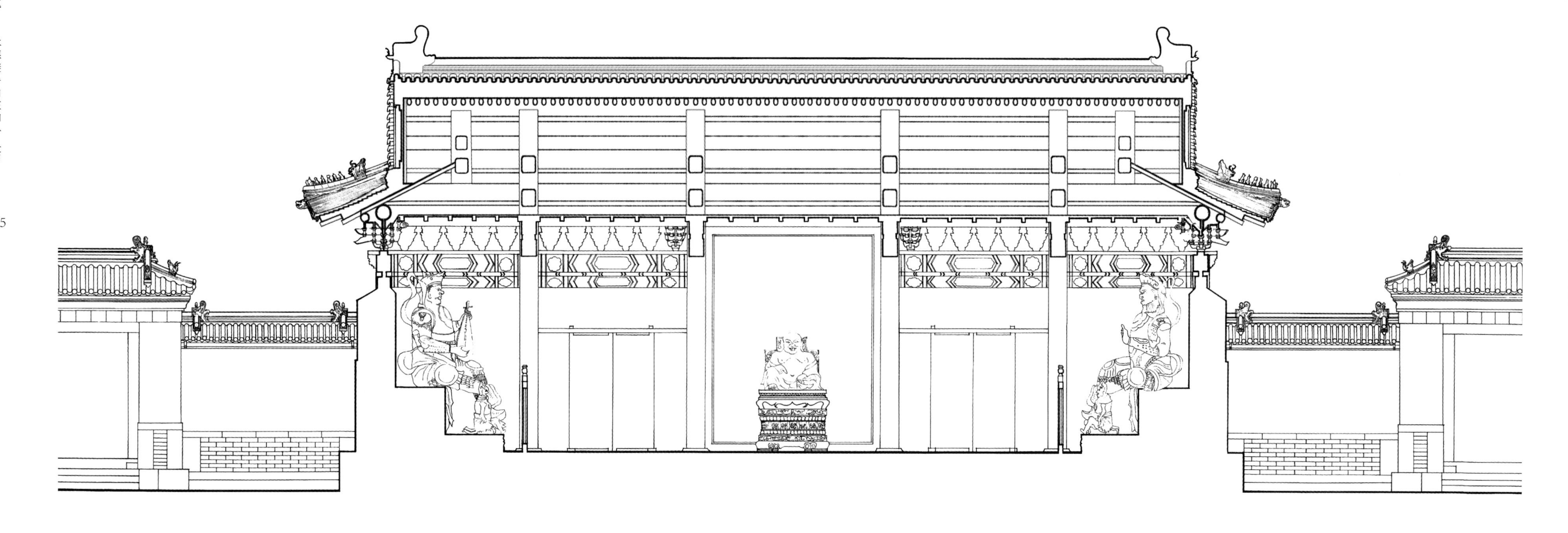

普宁寺天王殿纵剖面图
Longitudinal section of Tianwangdian of Puningsi

0 2 4m

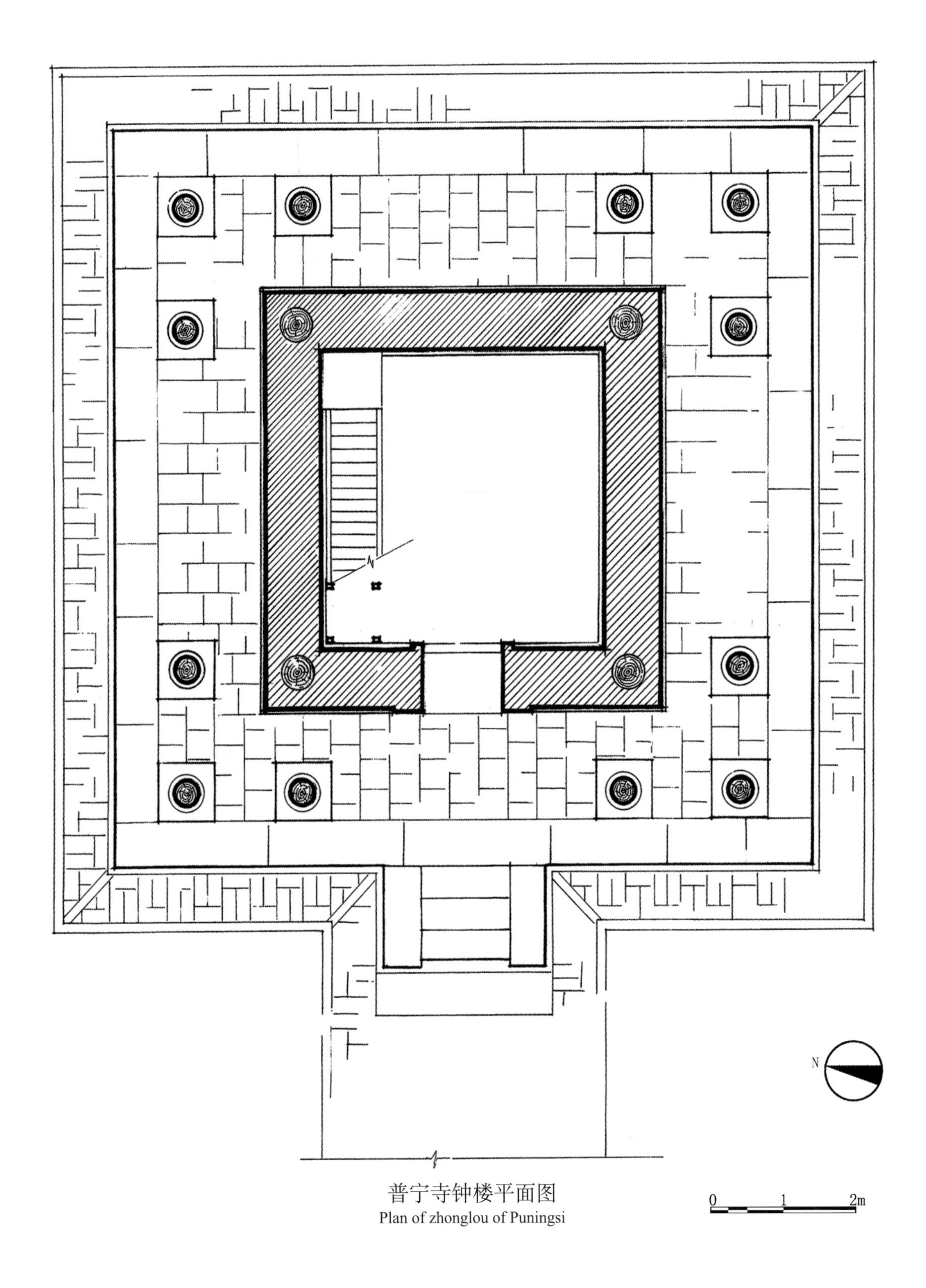

普宁寺钟楼平面图
Plan of zhonglou of Puningsi

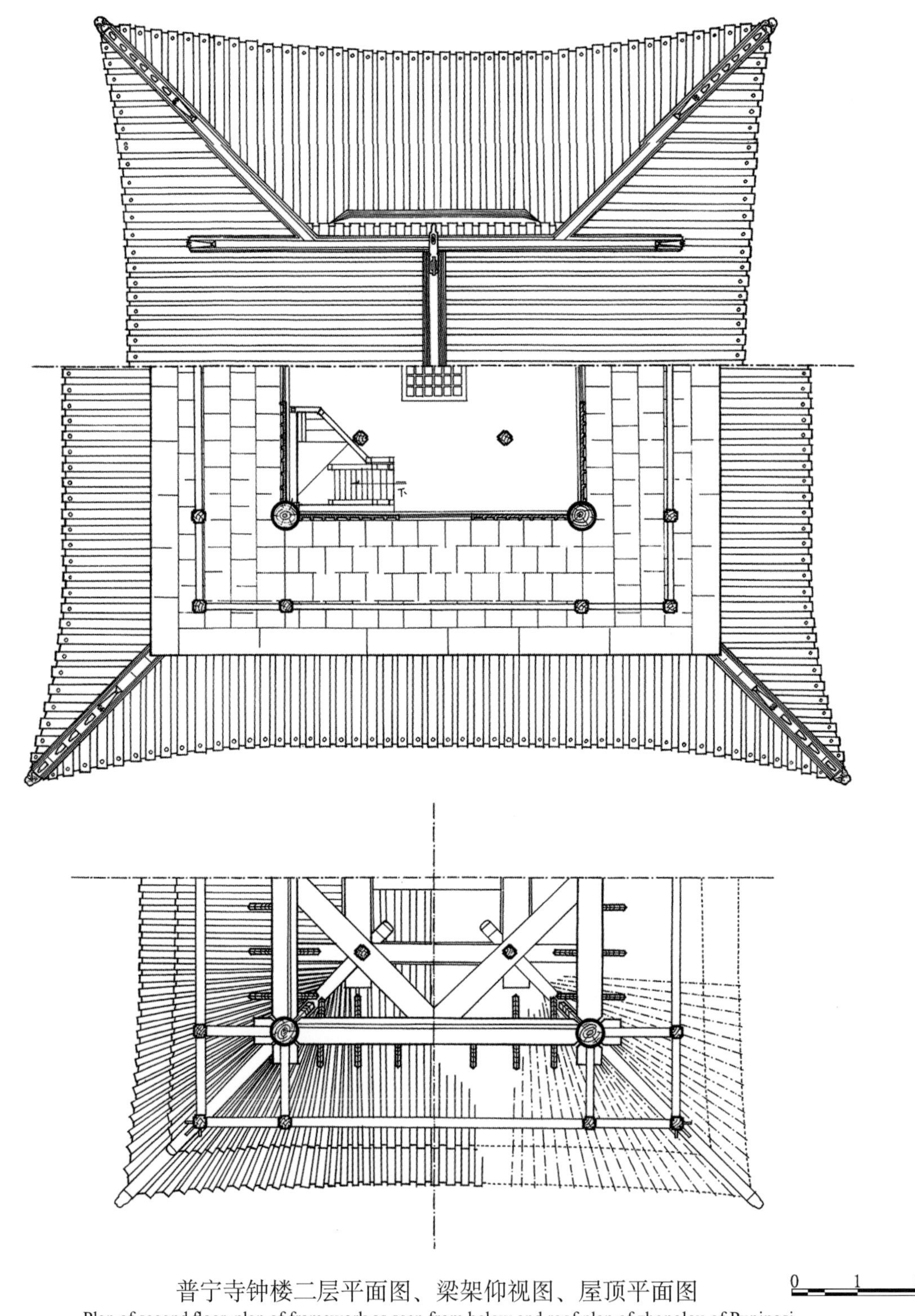

普宁寺钟楼二层平面图、梁架仰视图、屋顶平面图
Plan of second floor, plan of framework as seen from below and roof plan of zhonglou of Puningsi

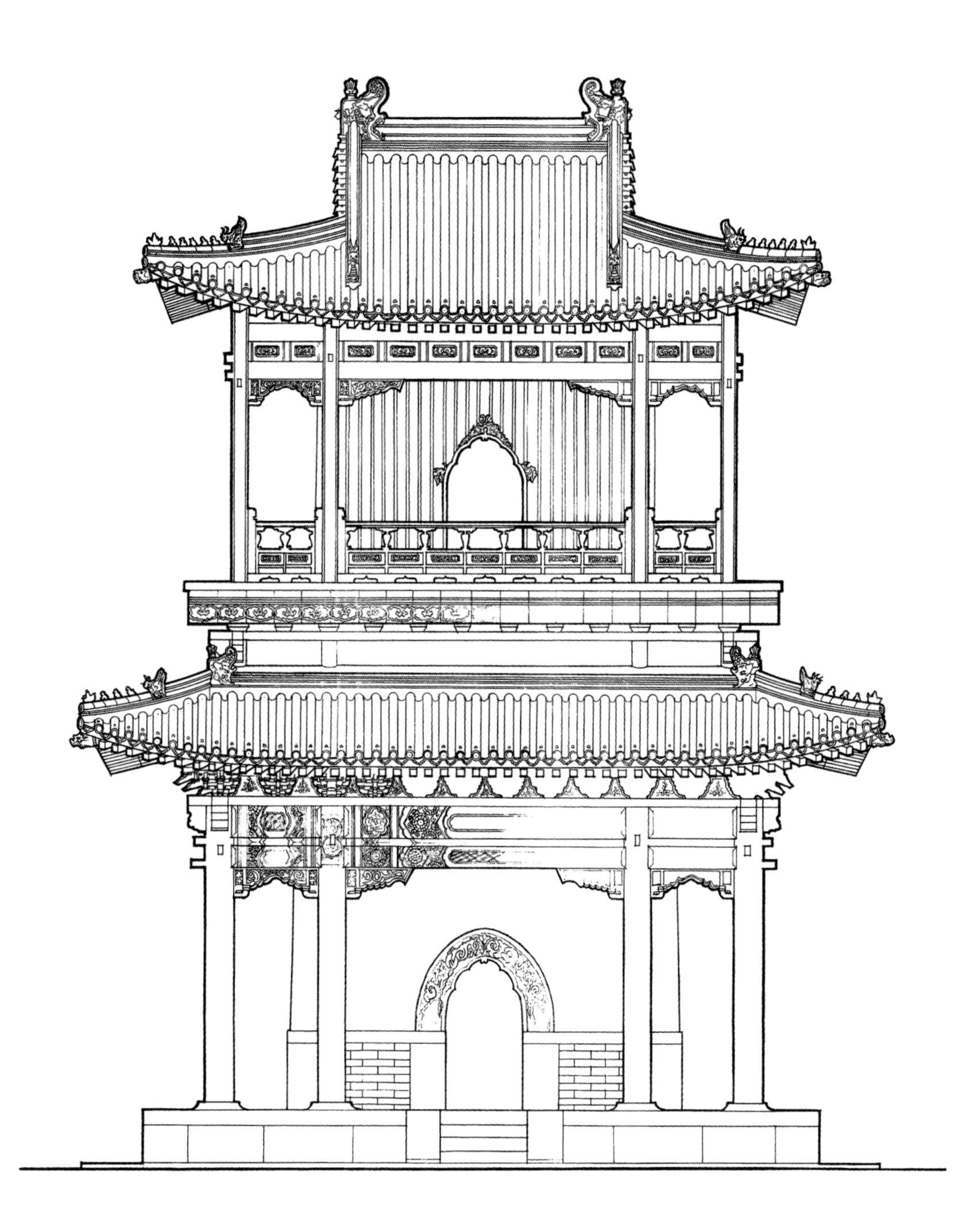

普宁寺钟楼正立面图
Front elevation of zhonglou of Puningsi

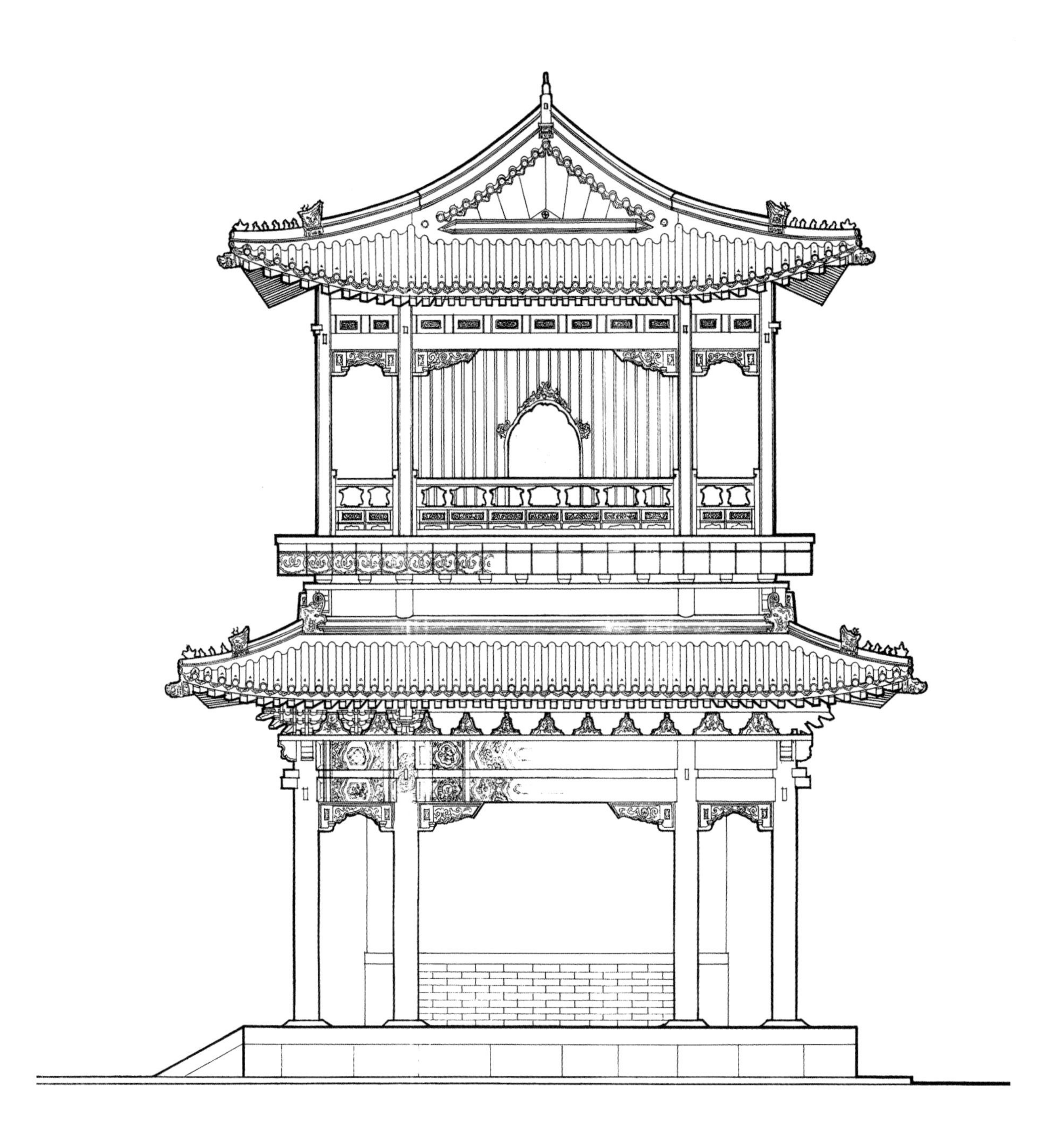

普宁寺钟楼侧立面图
Side elevation of zhonglou of Puningsi

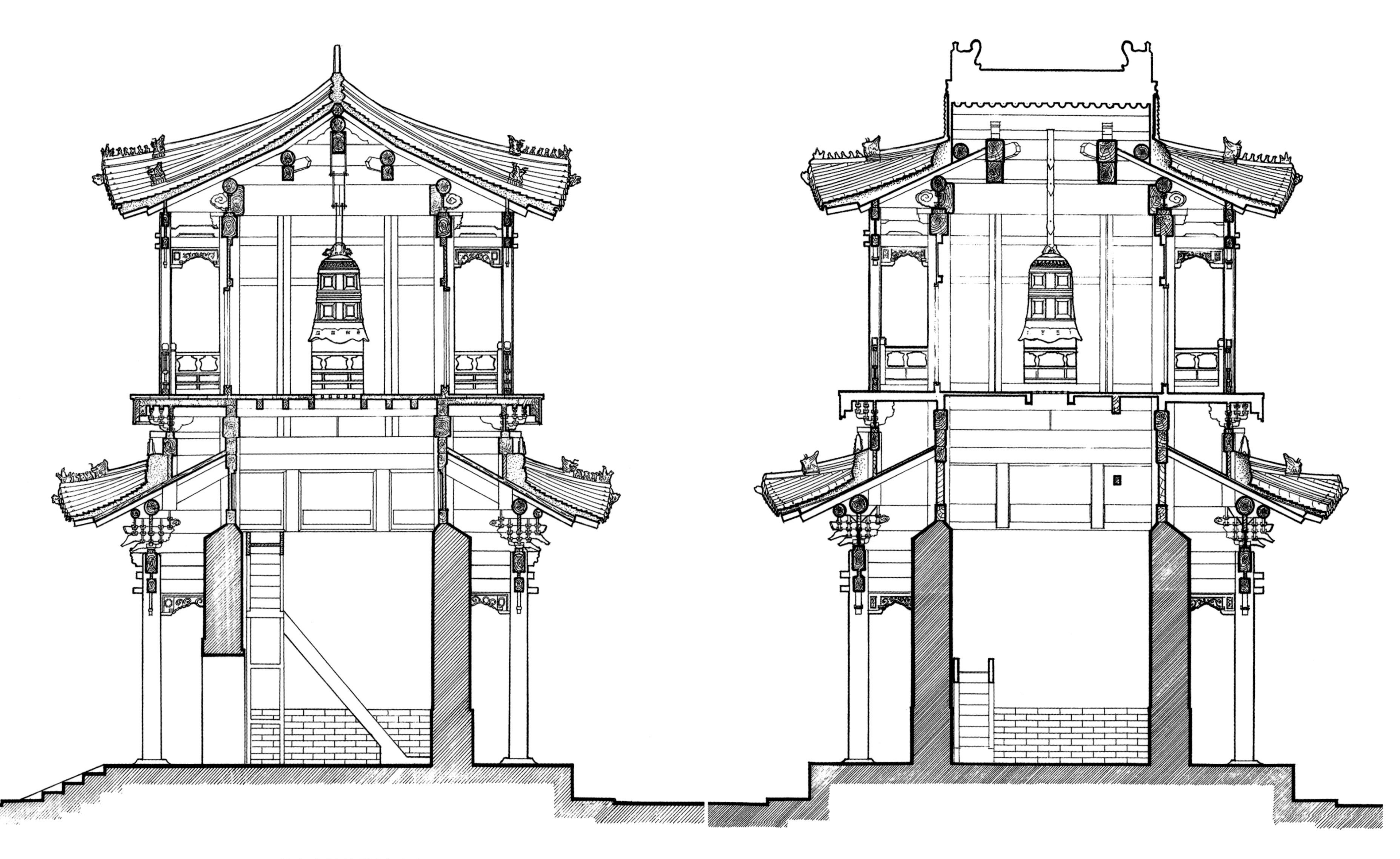

普宁寺钟楼横剖面图

Cross-section of zhonglou of Puningsi

普宁寺钟楼纵剖面图

Longitudinal section of zhonglou of Puningsi

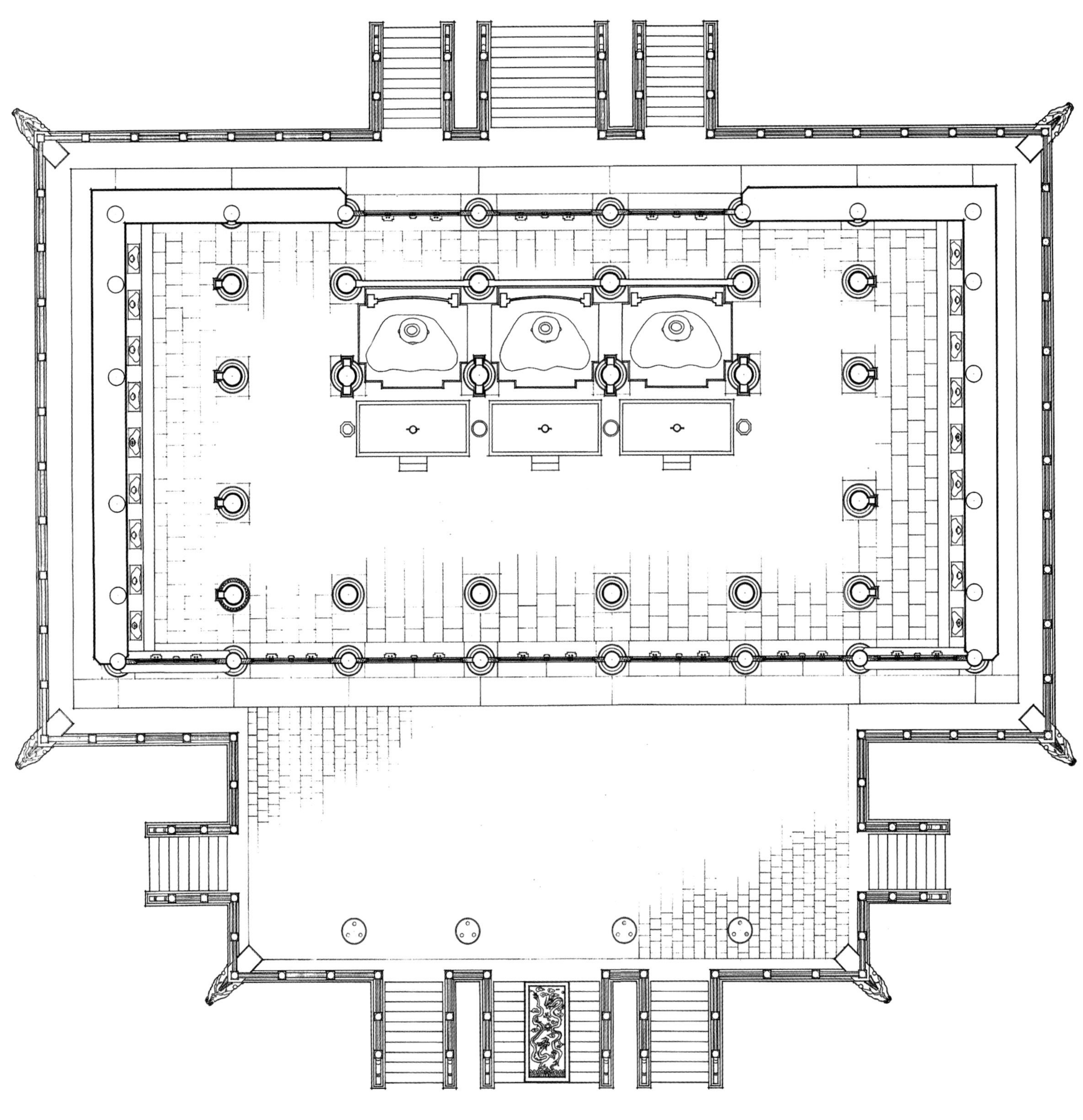

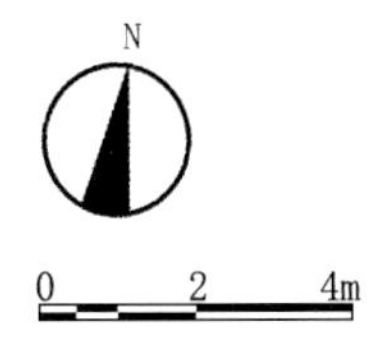

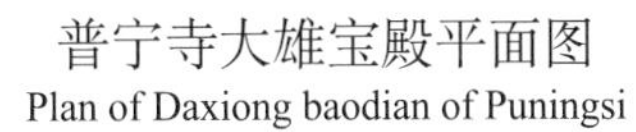

普宁寺大雄宝殿平面图

Plan of Daxiong baodian of Puningsi

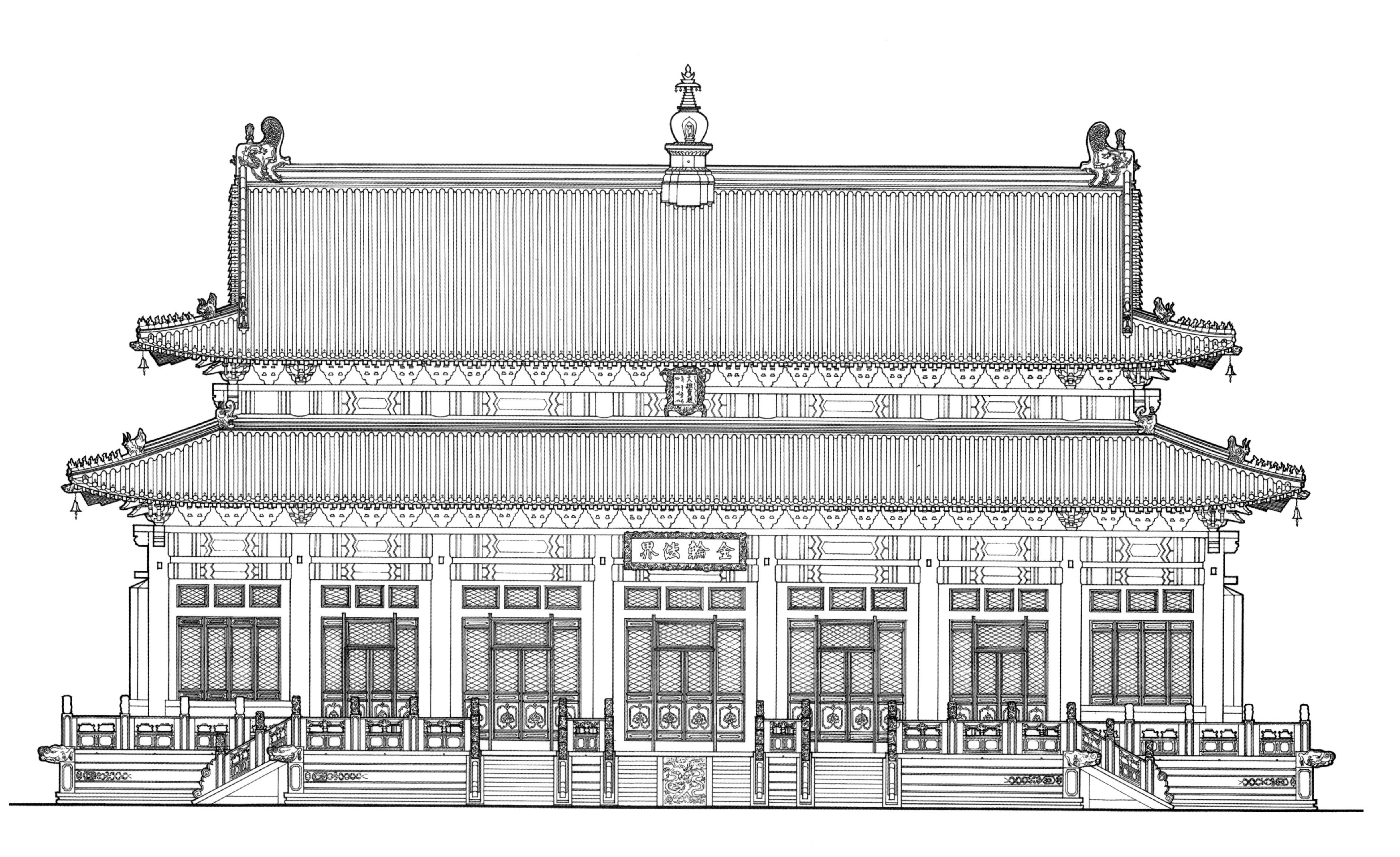

普宁寺大雄宝殿正立面图
Front elevation of Daxiong baodian of Puningsi

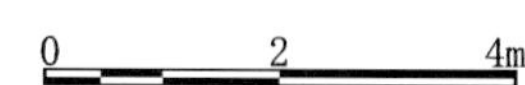

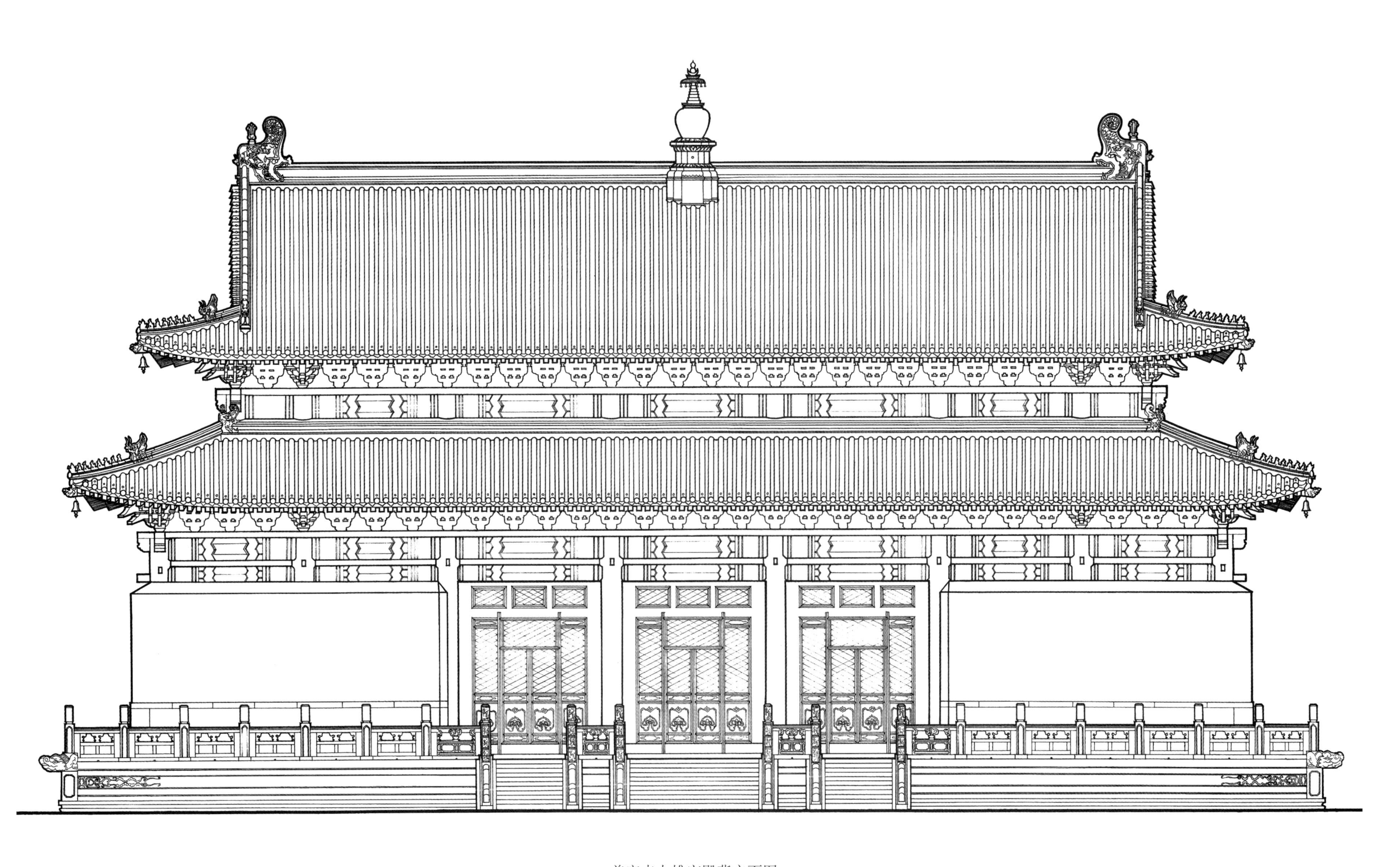

普宁寺大雄宝殿背立面图
Rear elevation of Daxiong baodian of Puningsi

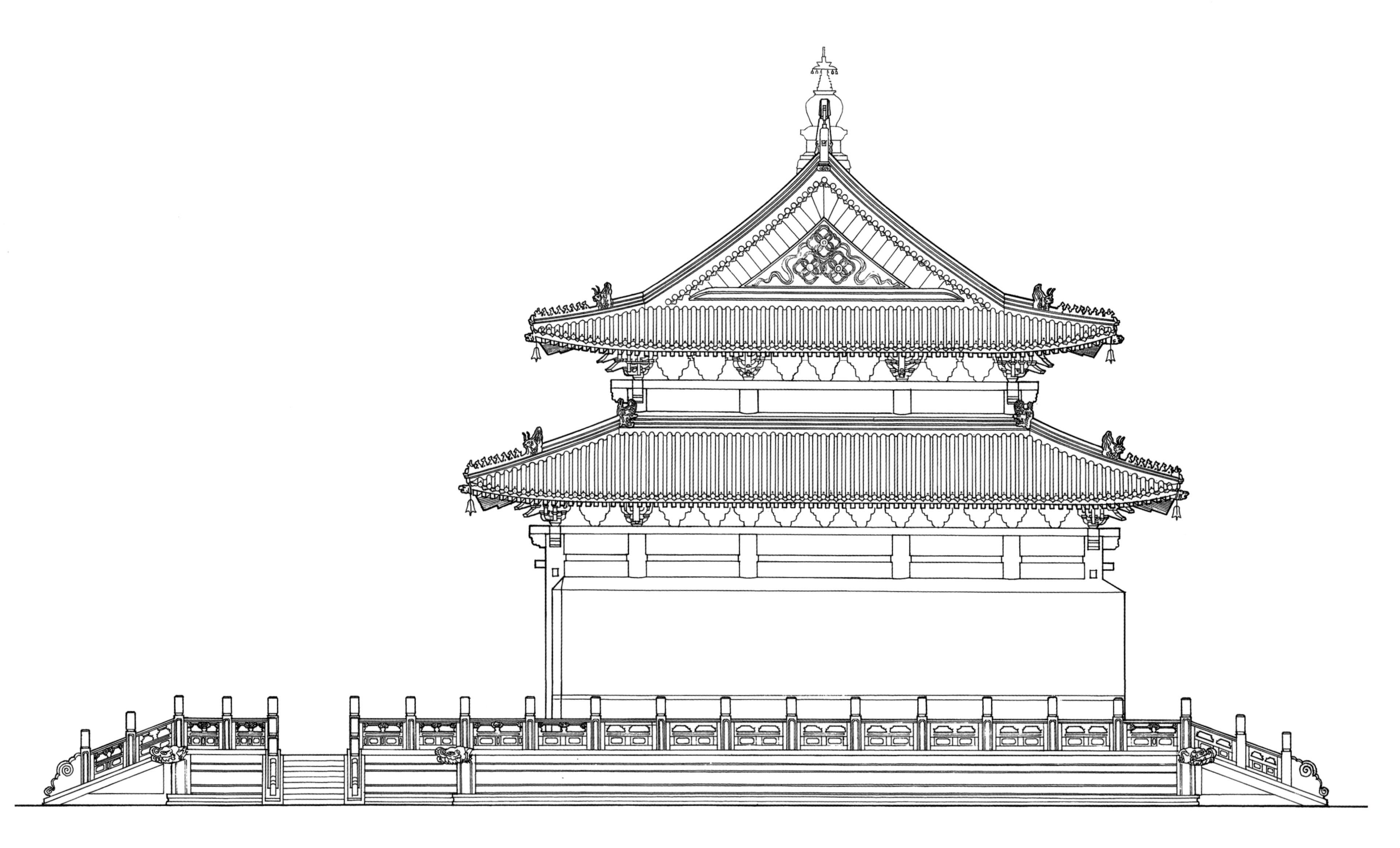

普宁寺大雄宝殿侧立面图
Side elevation of Daxiong baodian of Puningsi

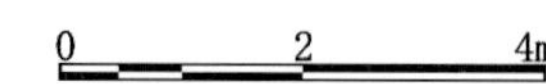

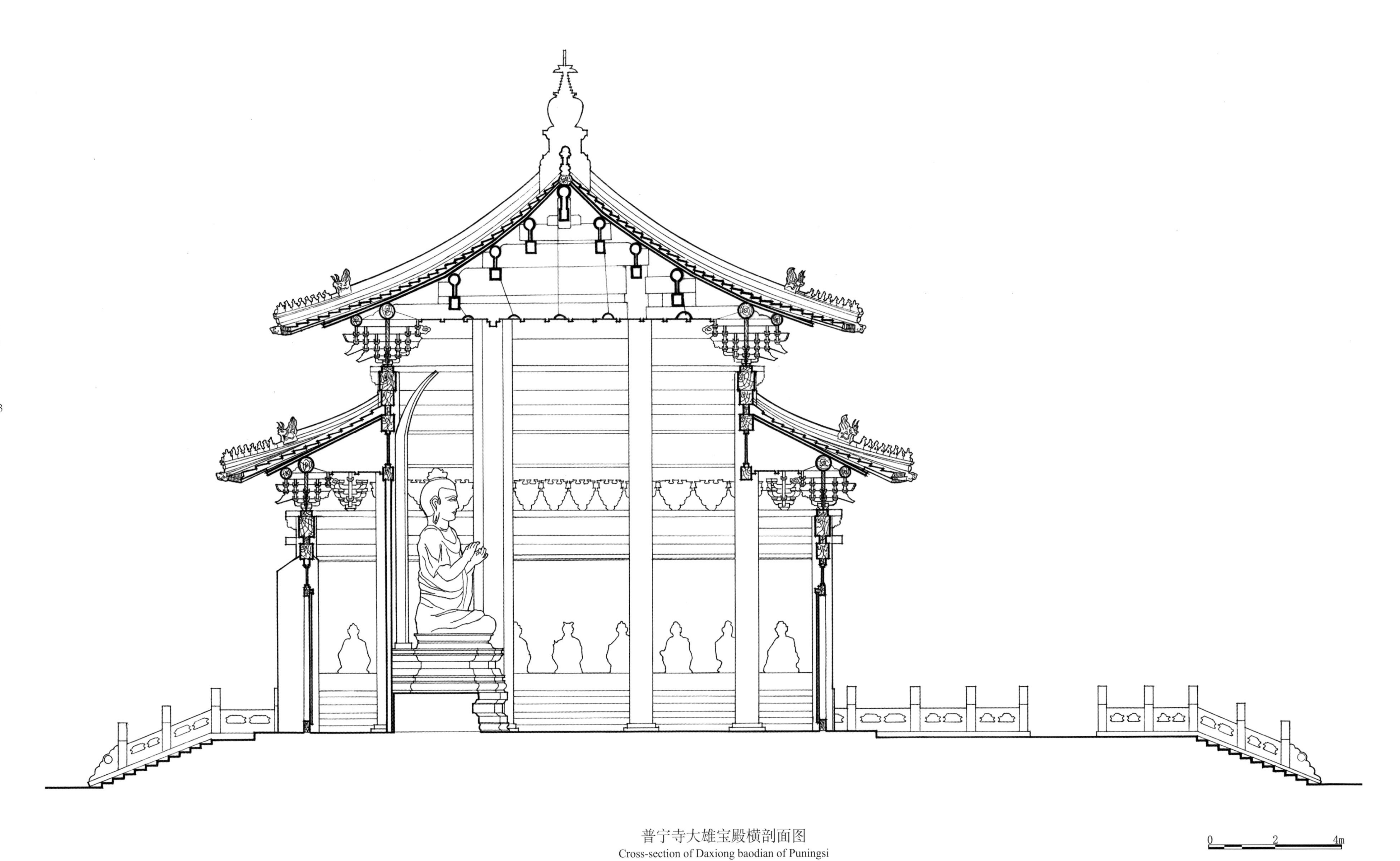

普宁寺大雄宝殿横剖面图
Cross-section of Daxiong baodian of Puningsi

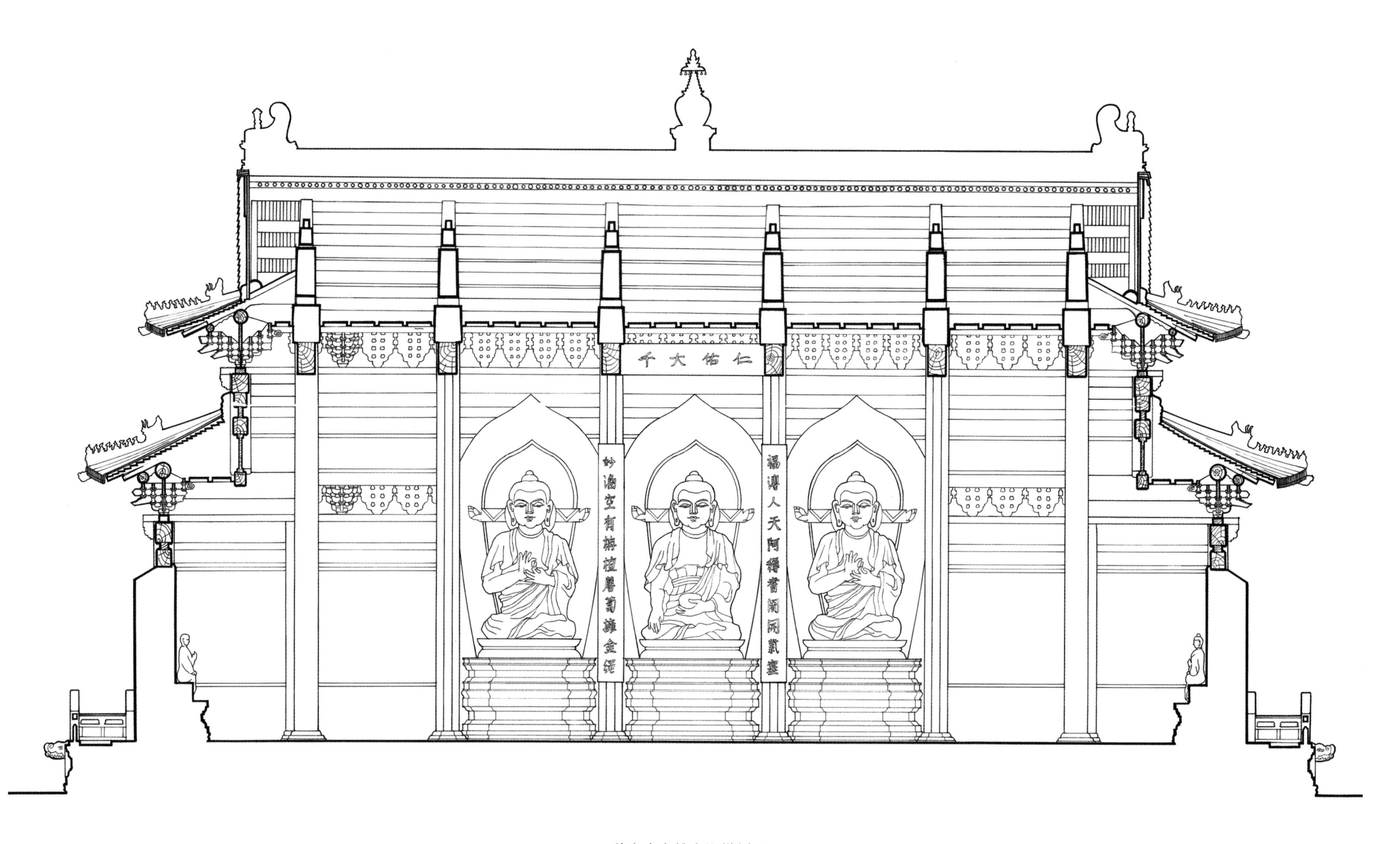

普宁寺大雄宝殿纵剖面图
Longitudinal section of Daxiong baodian of Puningsi

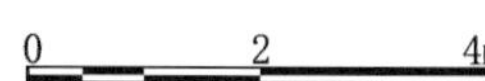

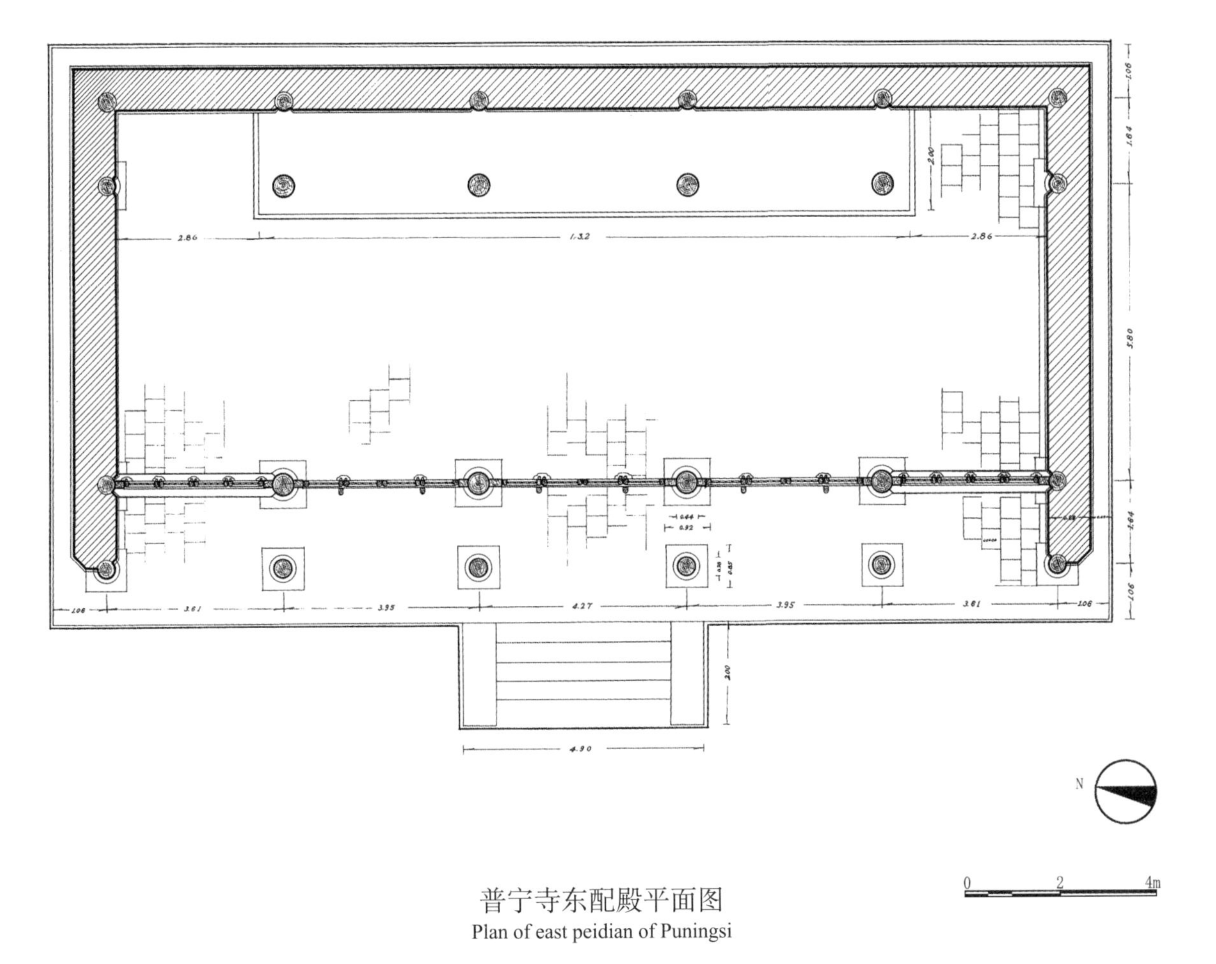

普宁寺东配殿平面图

Plan of east peidian of Puningsi

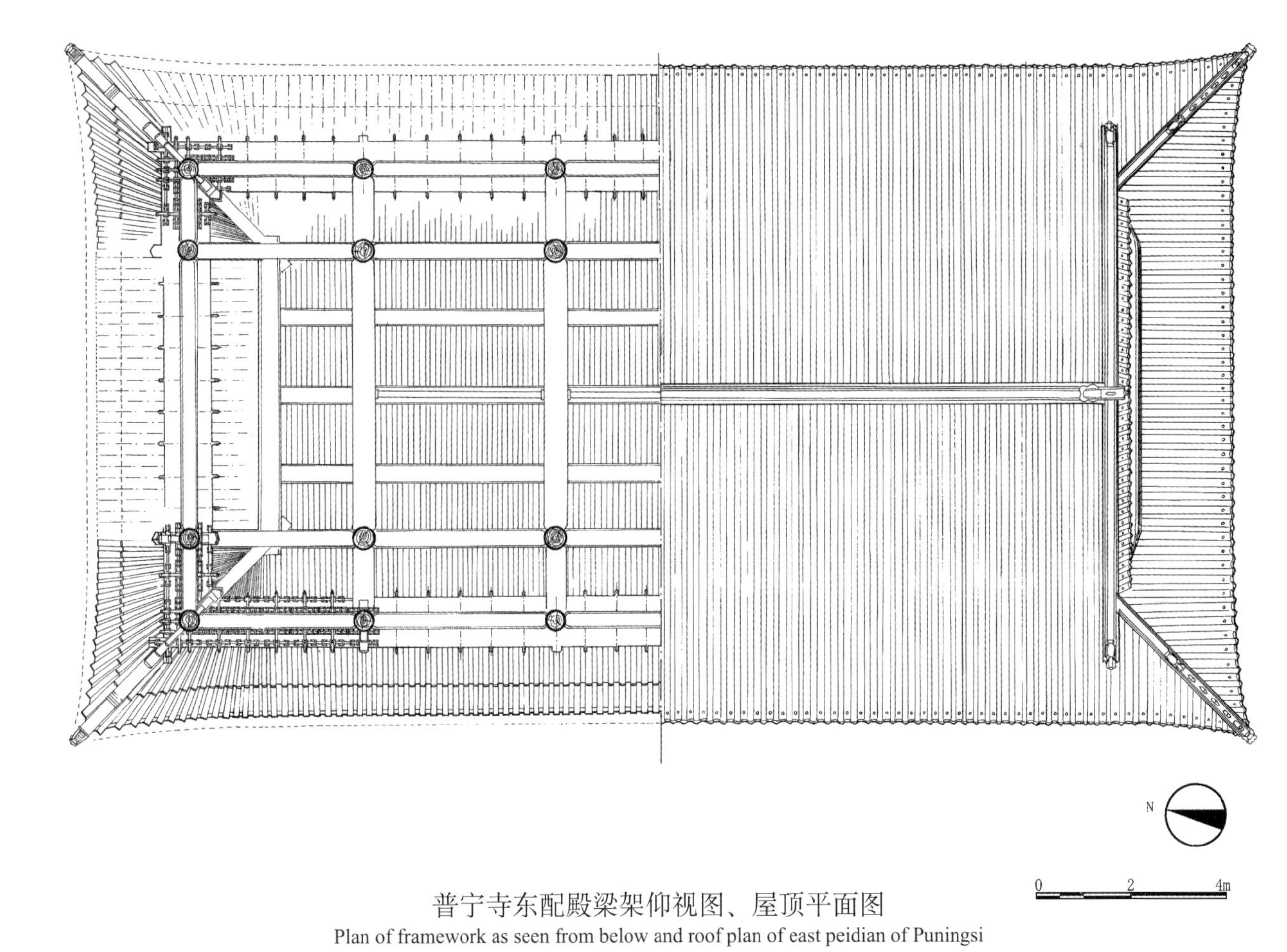

普宁寺东配殿梁架仰视图、屋顶平面图

Plan of framework as seen from below and roof plan of east peidian of Puningsi

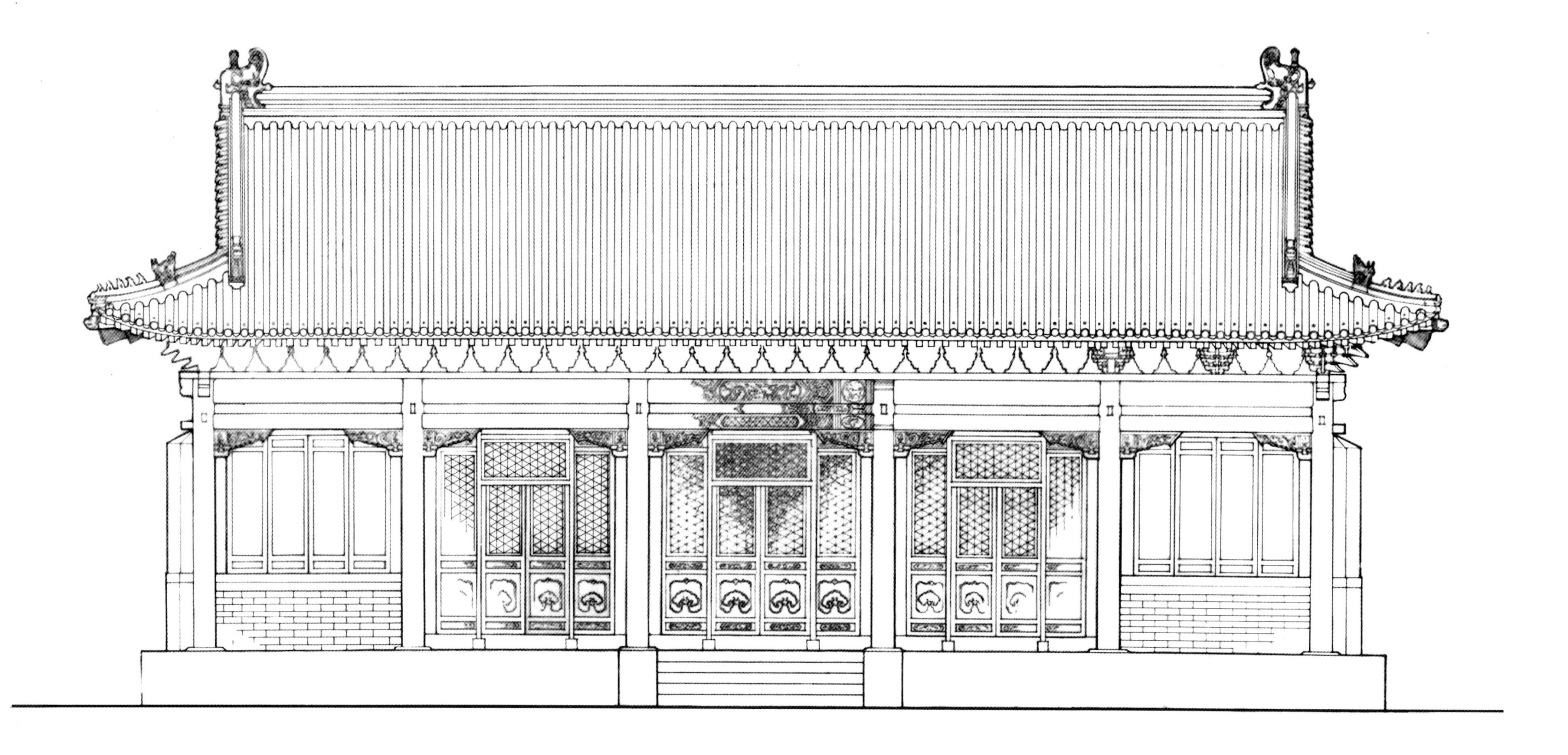

＊原图已失，仅存照片
The original picture has been lost,
only photos are left.

普宁寺东配殿正立面图＊
Front elevation of east peidian of Puningsi*

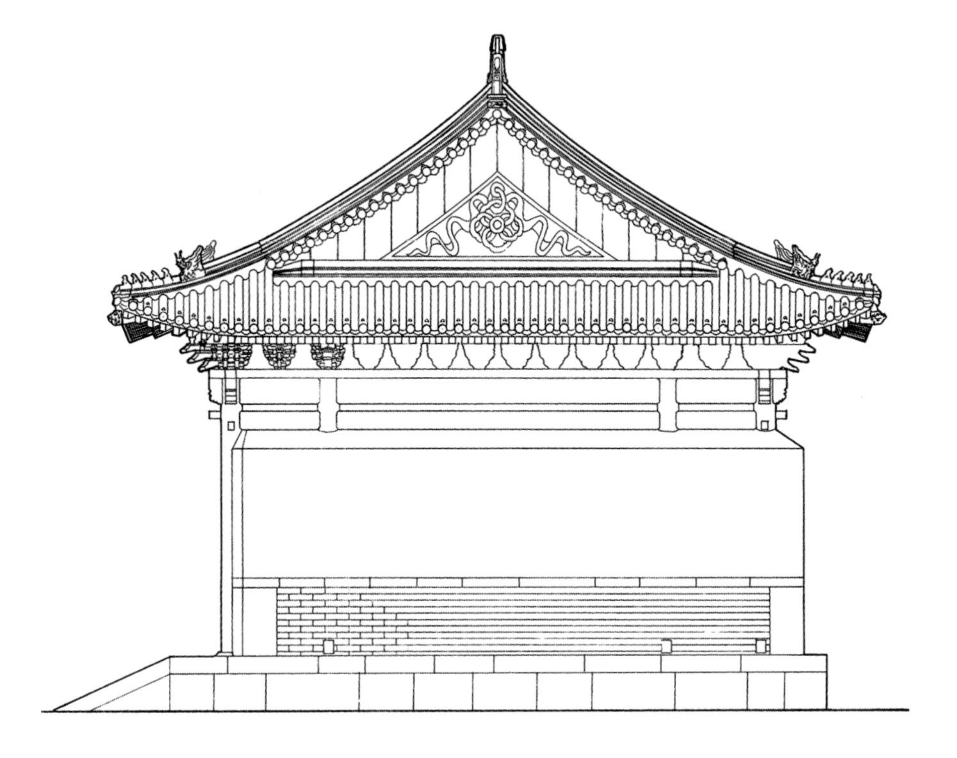

普宁寺东配殿侧立面图
Side elevation of east peidian of Puningsi

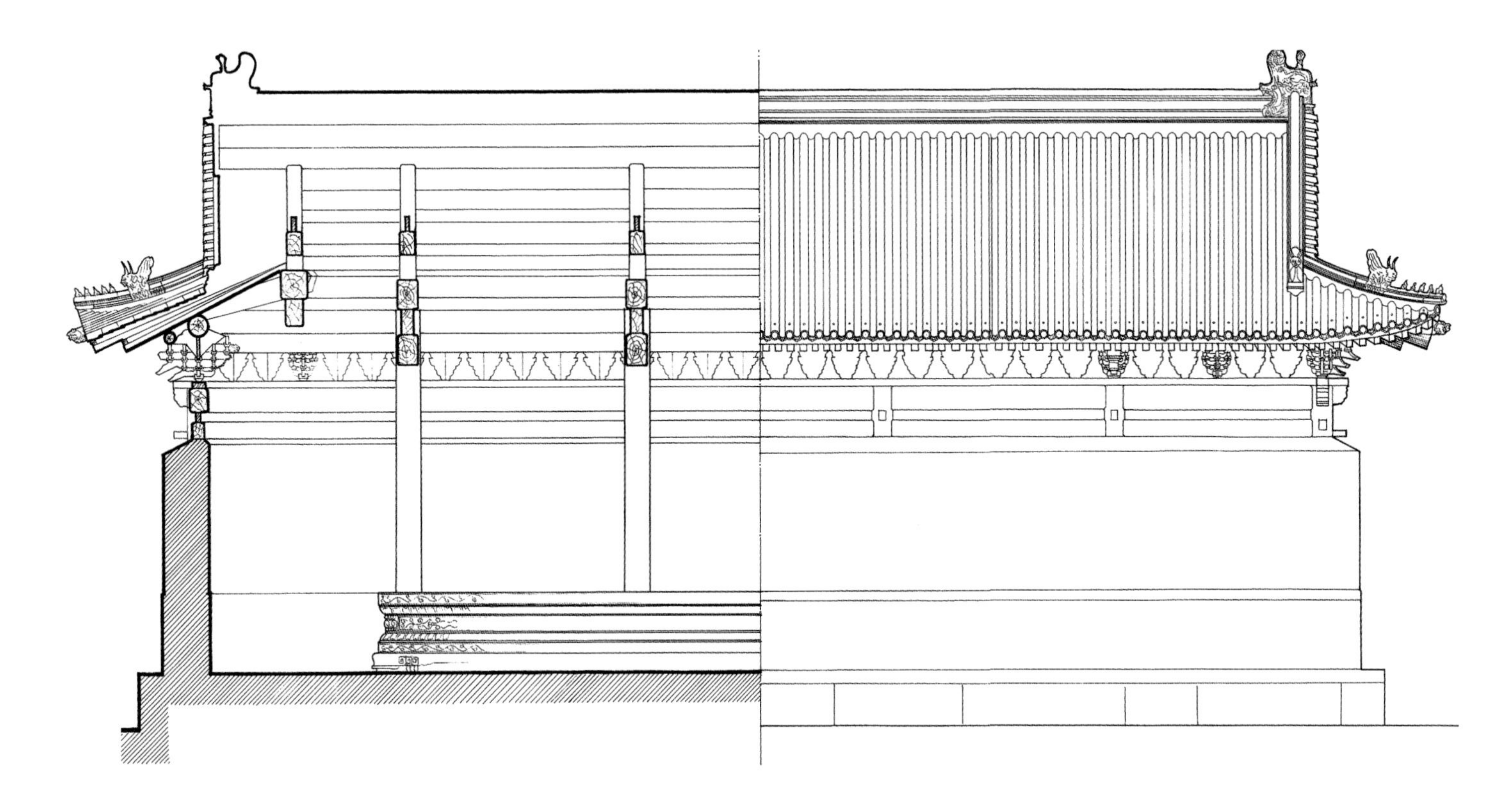

普宁寺东配殿纵剖面图、背立面图
Longitudinal section and rear elevation of east peidian of Puningsi

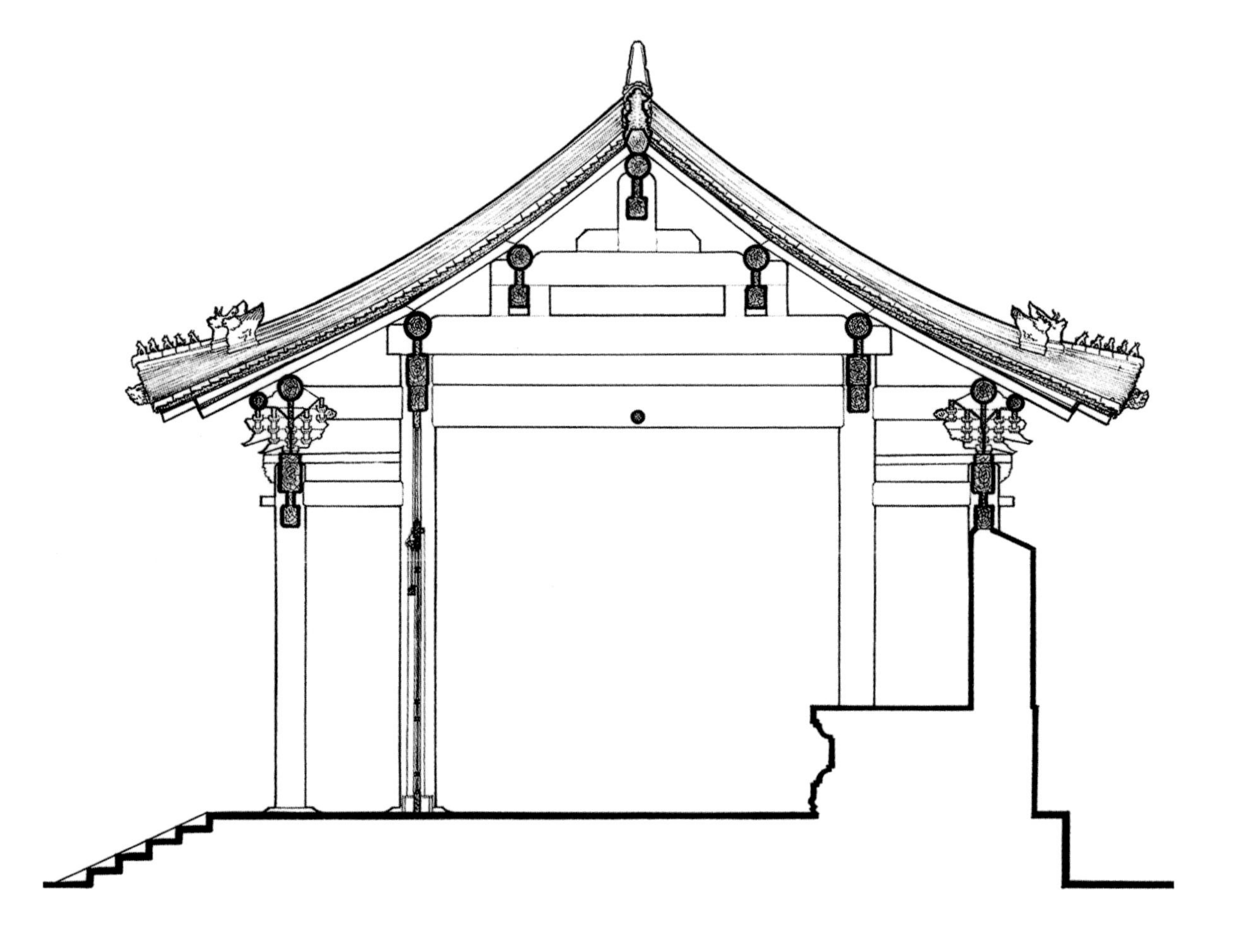

普宁寺东配殿明间横剖面图

Cross-section of central-bay of east peidian of Puningsi

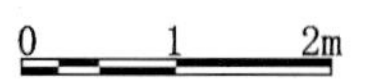

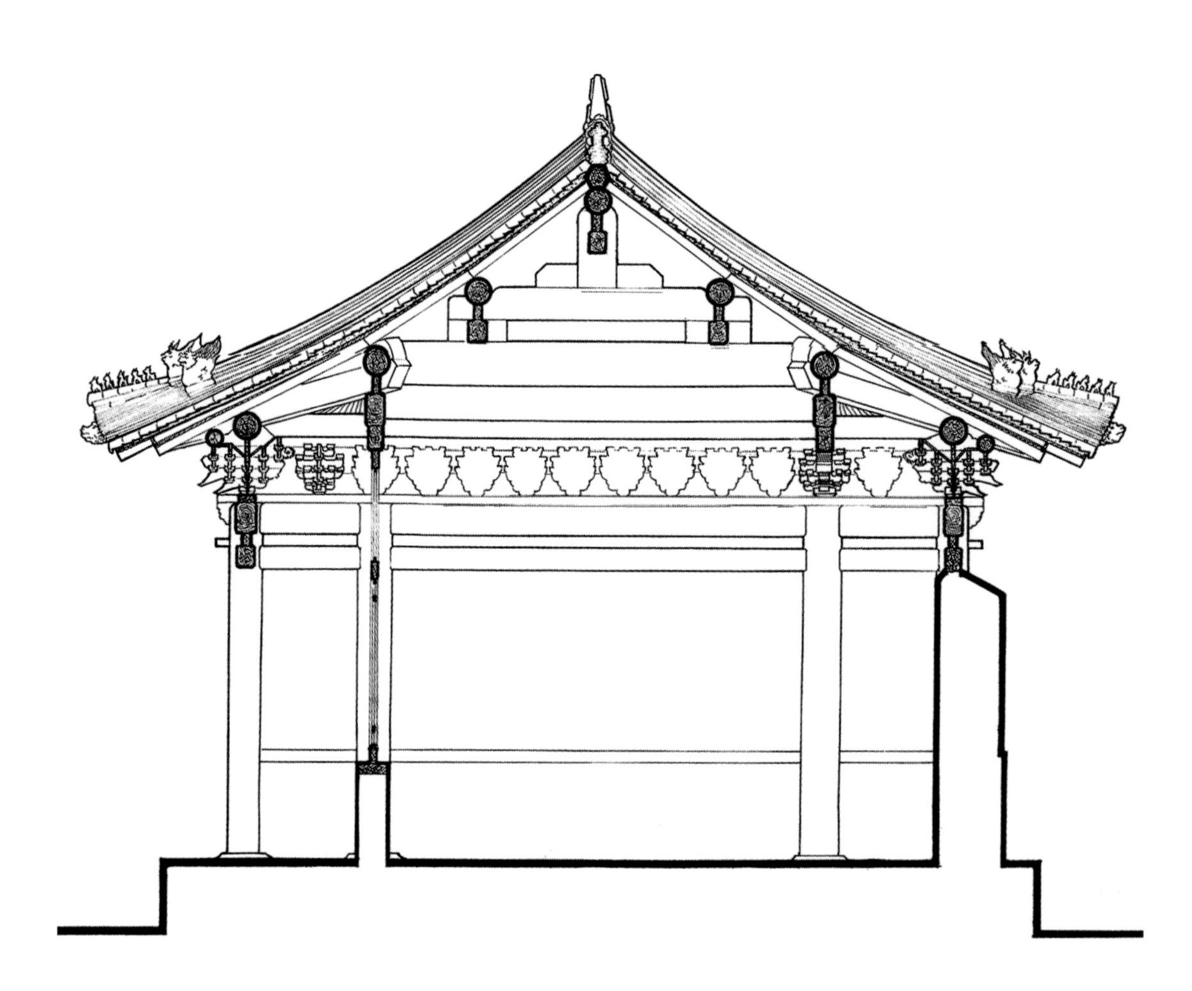

普宁寺东配殿梢间横剖面图

Cross-section of second-to-last-bay of east peidian of Puningsi

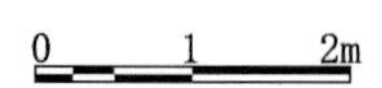

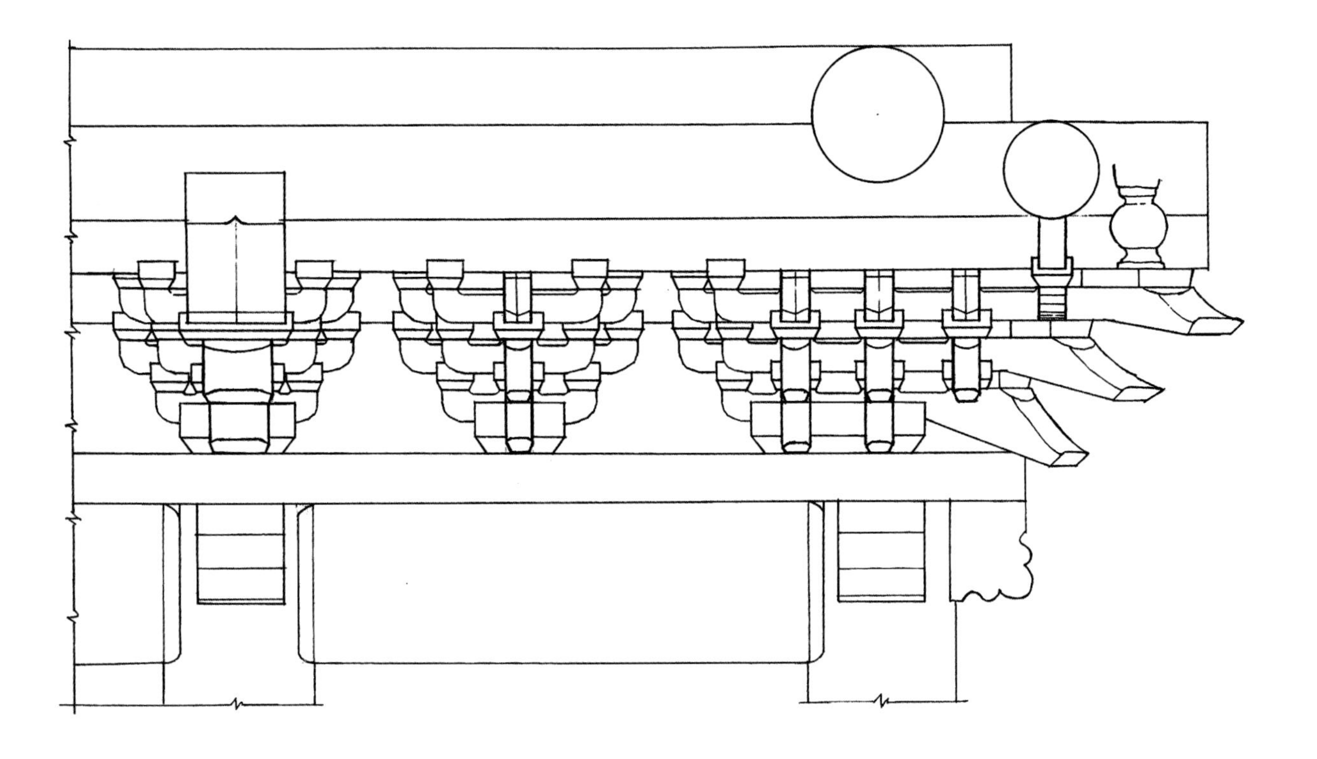

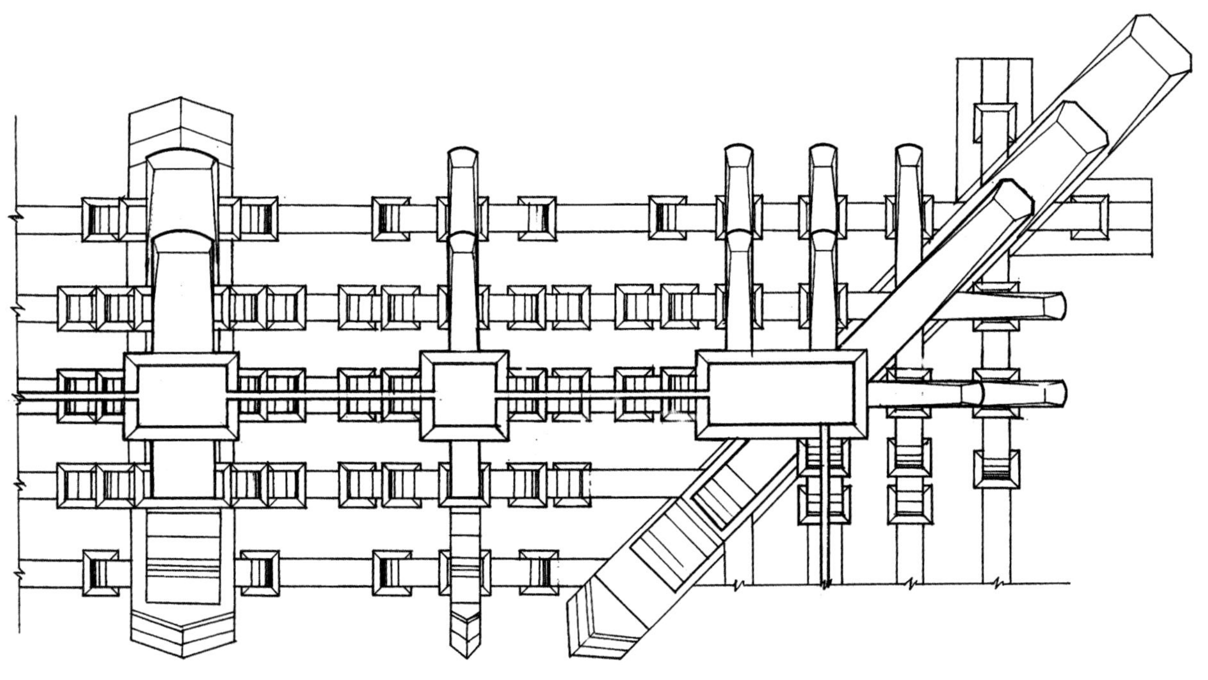

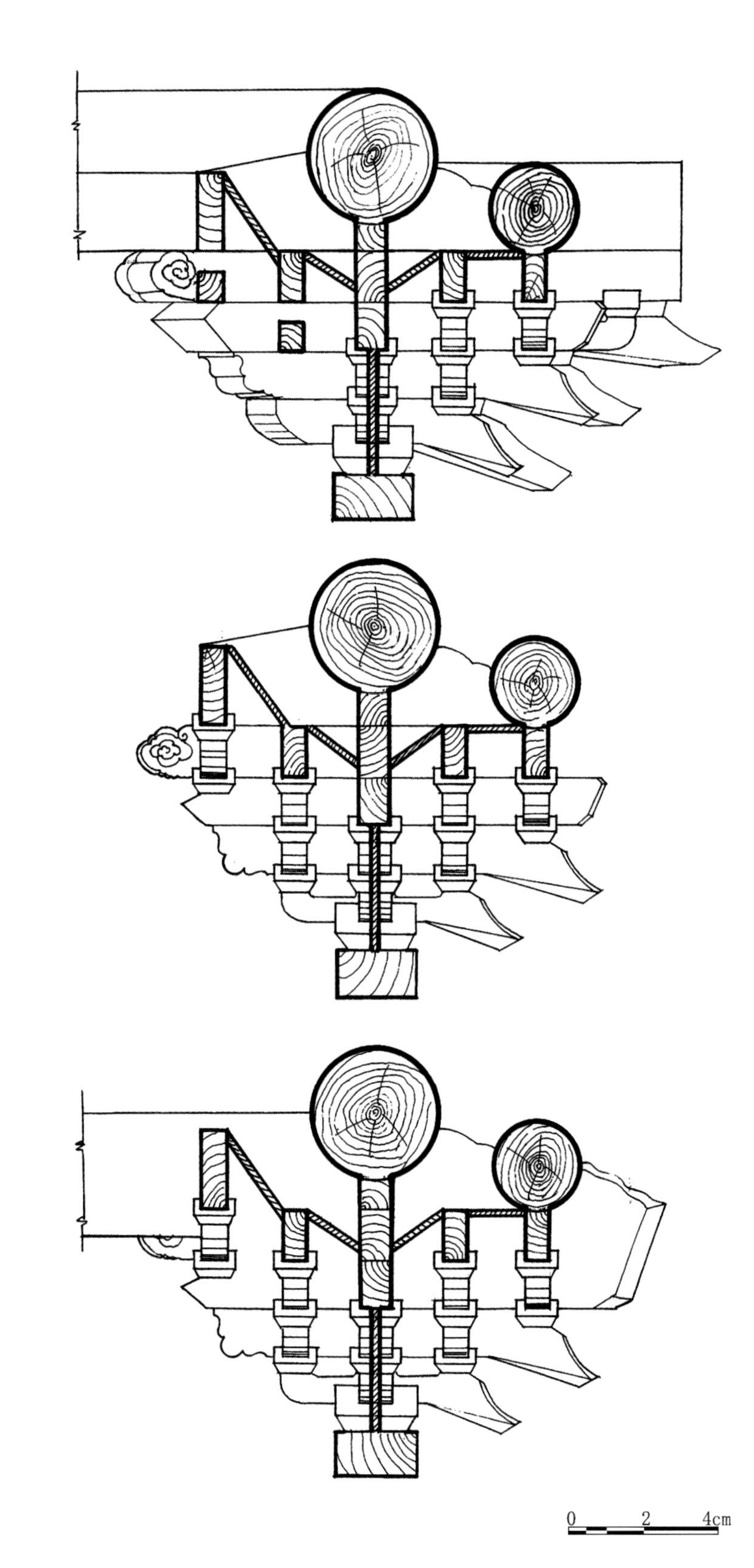

普宁寺东配殿斗栱大样图

Bracket set of east peidian of Puningsi

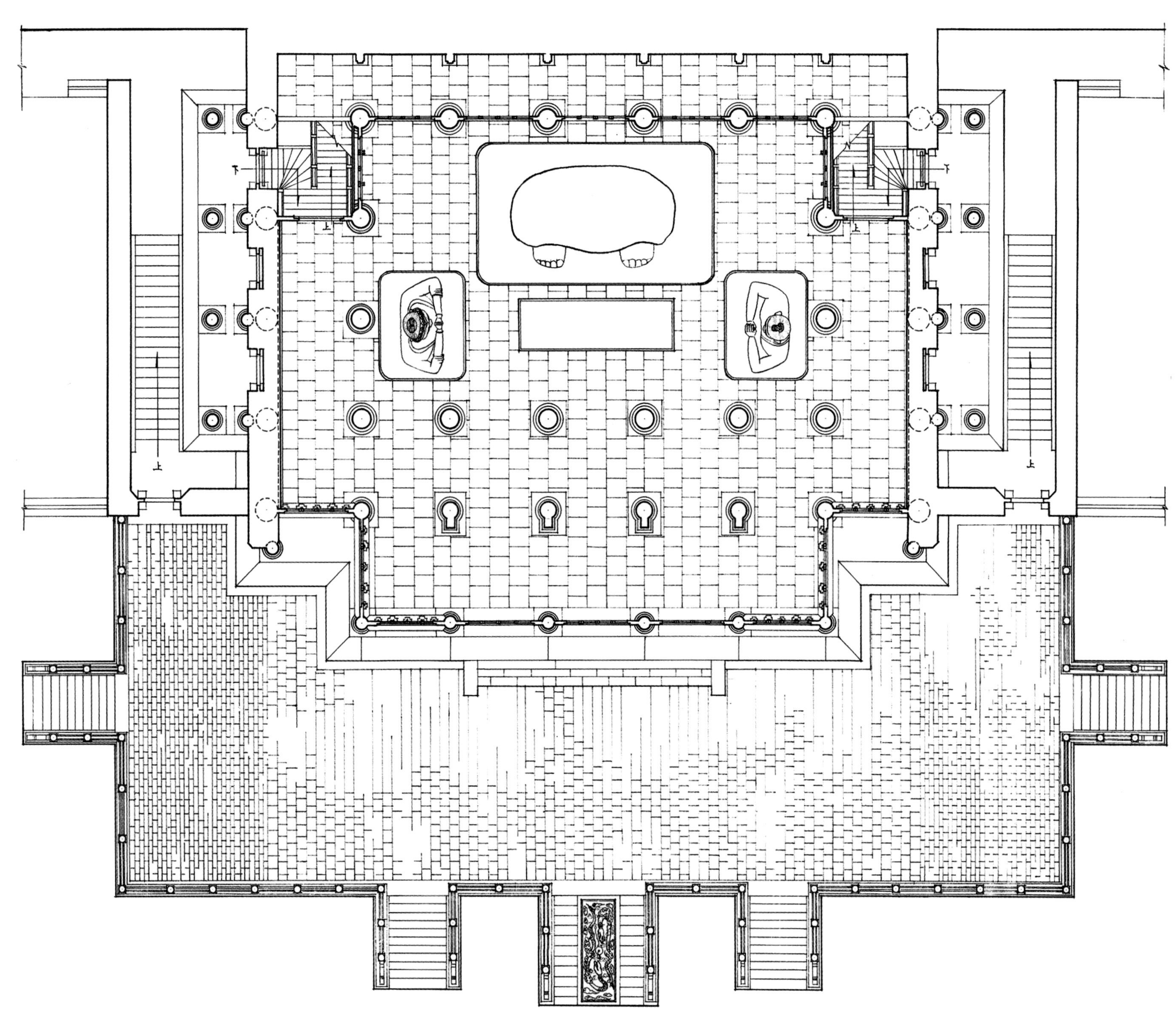

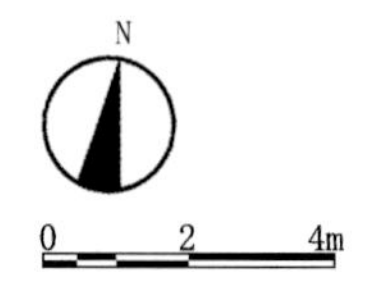

普宁寺大乘之阁一层平面图

Plan of first floor of Dacheng zhige of Puningsi

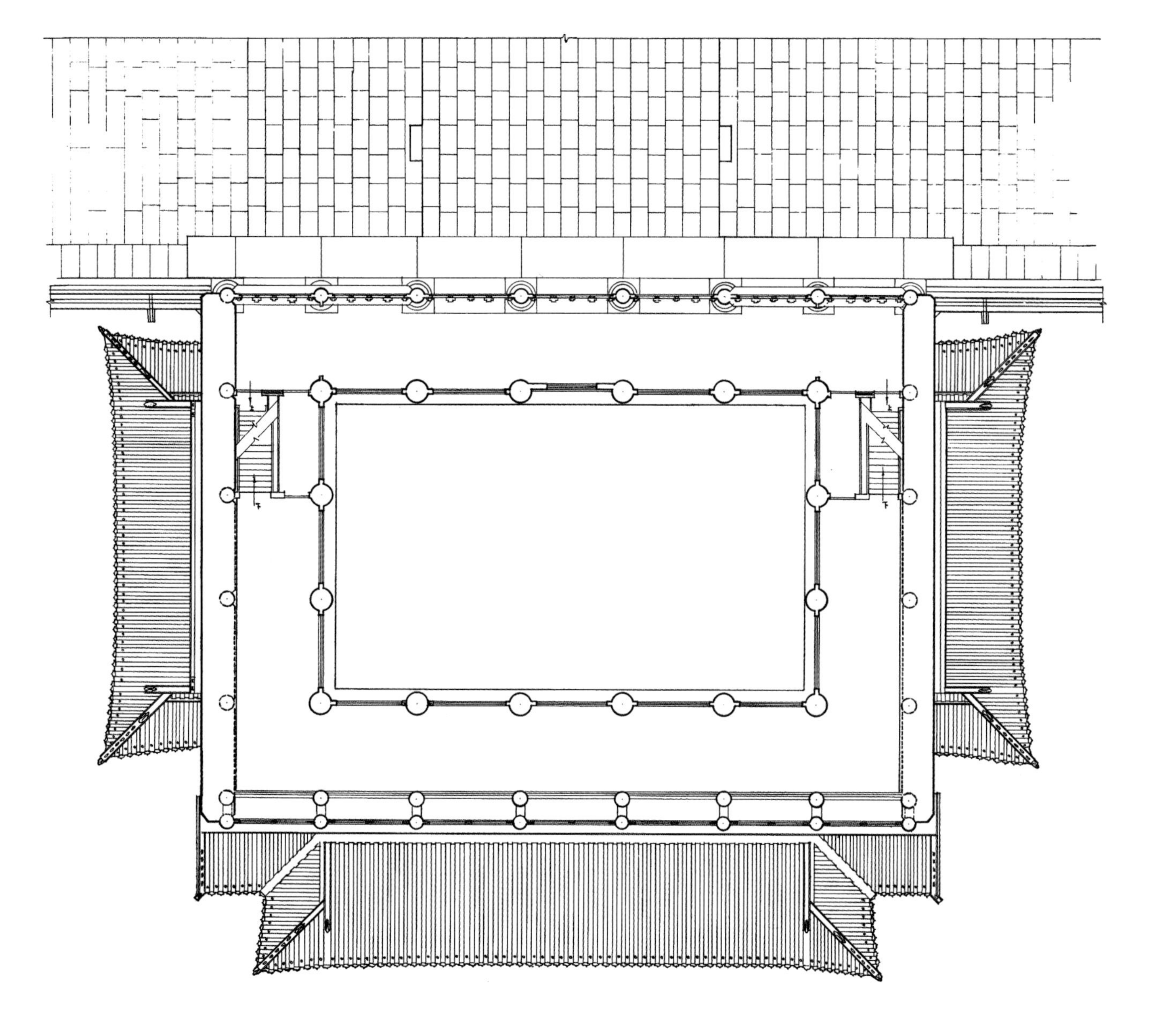

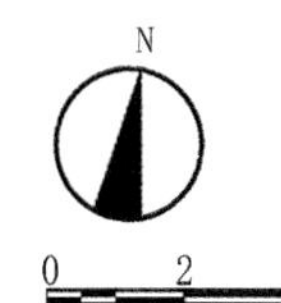

普宁寺大乘之阁二层平面图
Plan of second floor of Dacheng zhige of Puningsi

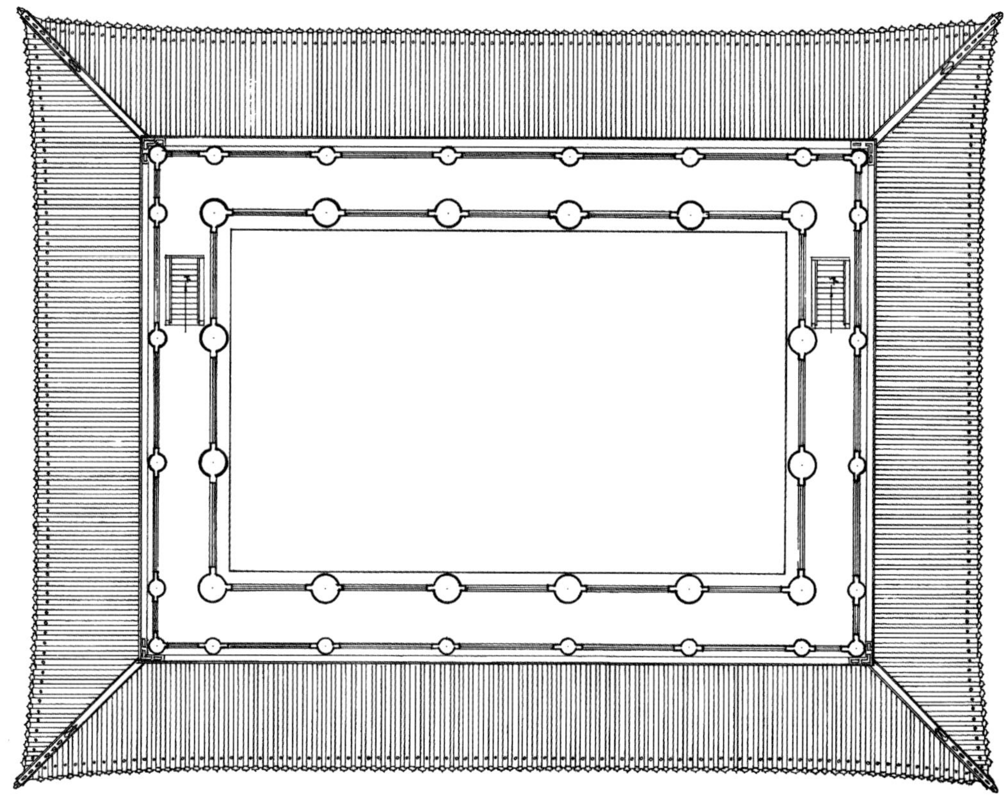

普宁寺大乘之阁三层平面图
Plan of third floor of Dacheng zhige of Puningsi

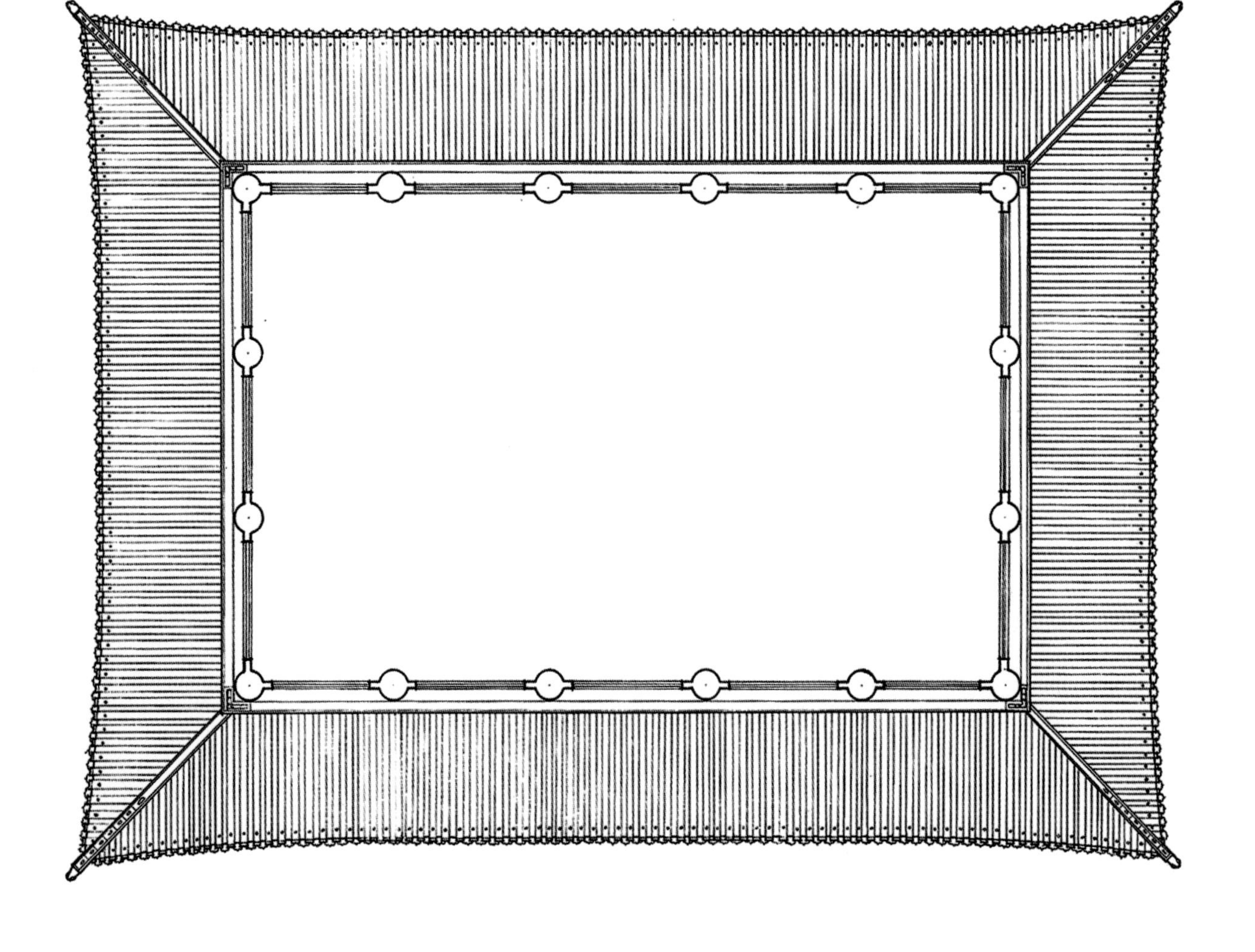

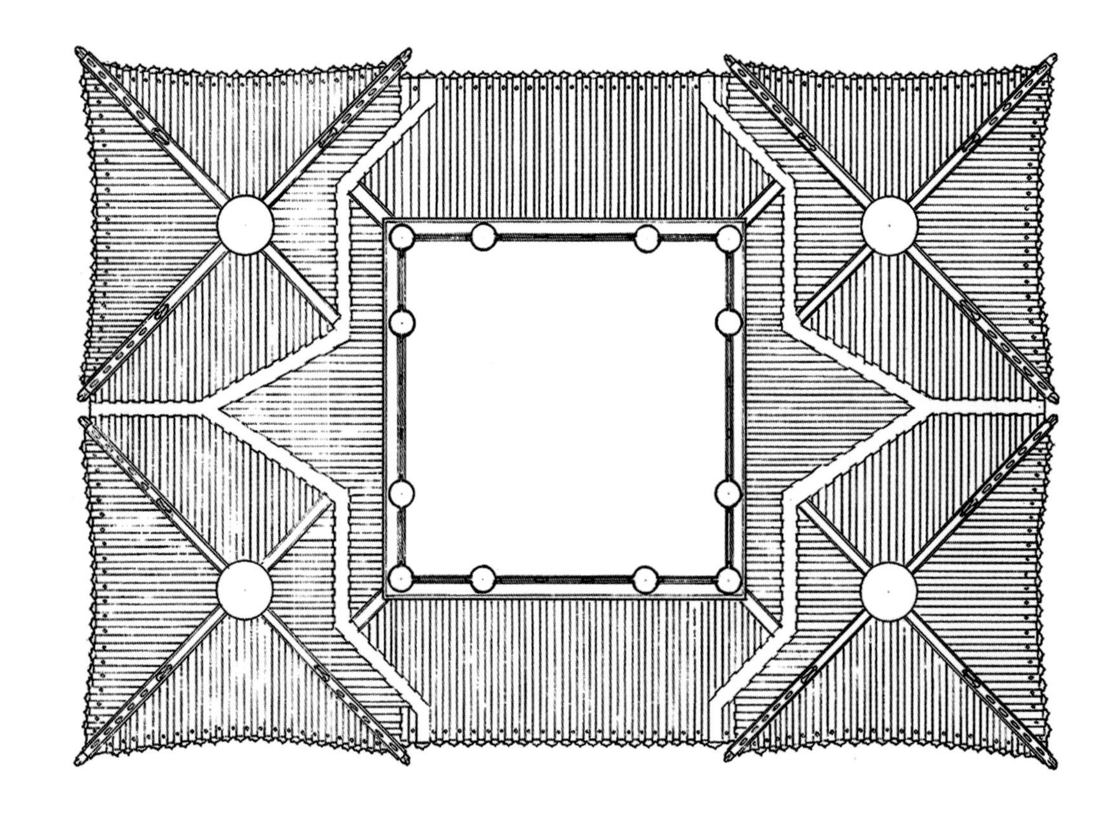

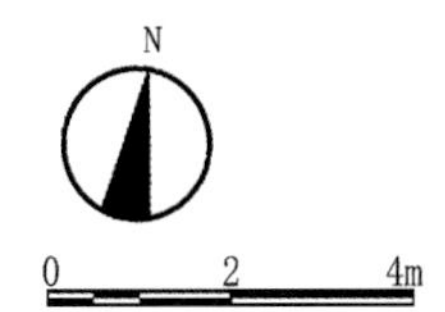

普宁寺大乘之阁四层平面图
Plan of fourth floor of Dacheng zhige of Puningsi

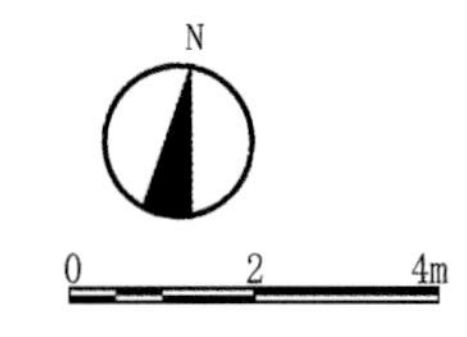

普宁寺大乘之阁五层平面图
Plan of fifth floor of Dacheng zhige of Puningsi

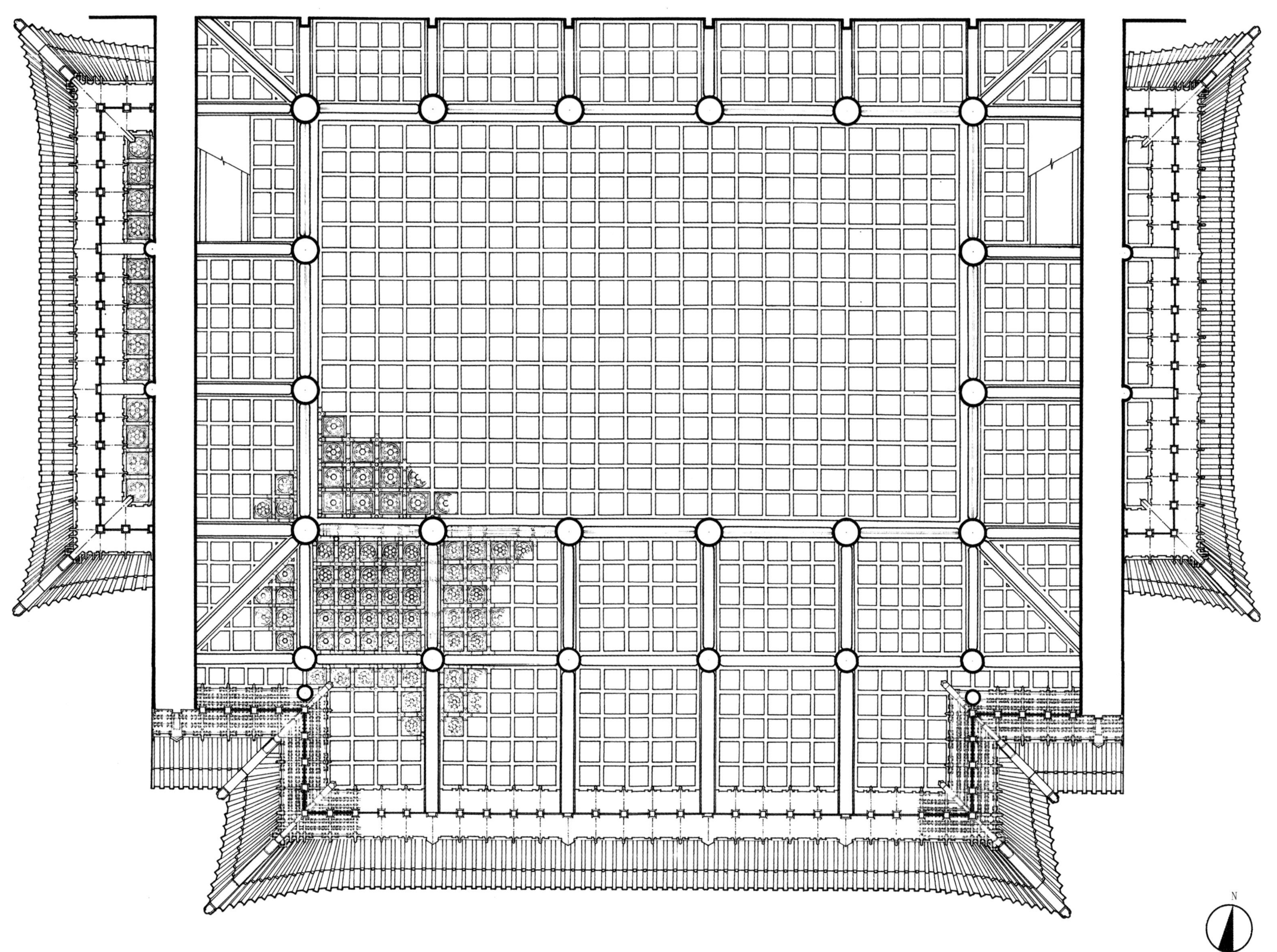

普宁寺大乘之阁天花仰视图
Reflected ceiling plan of Dacheng zhige of Puningsi

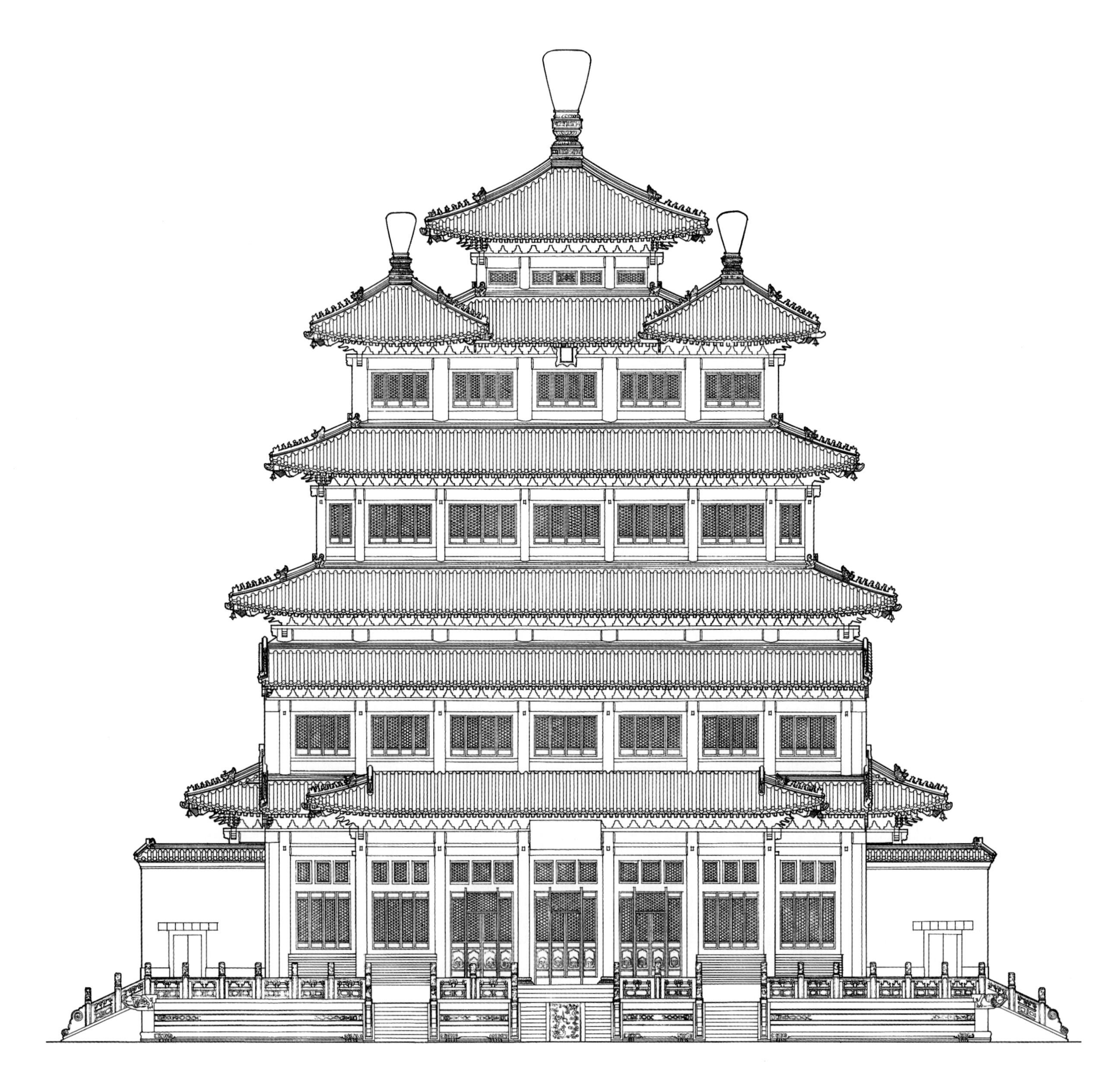

普宁寺大乘之阁正立面图

Front elevation of Dacheng zhige of Puningsi

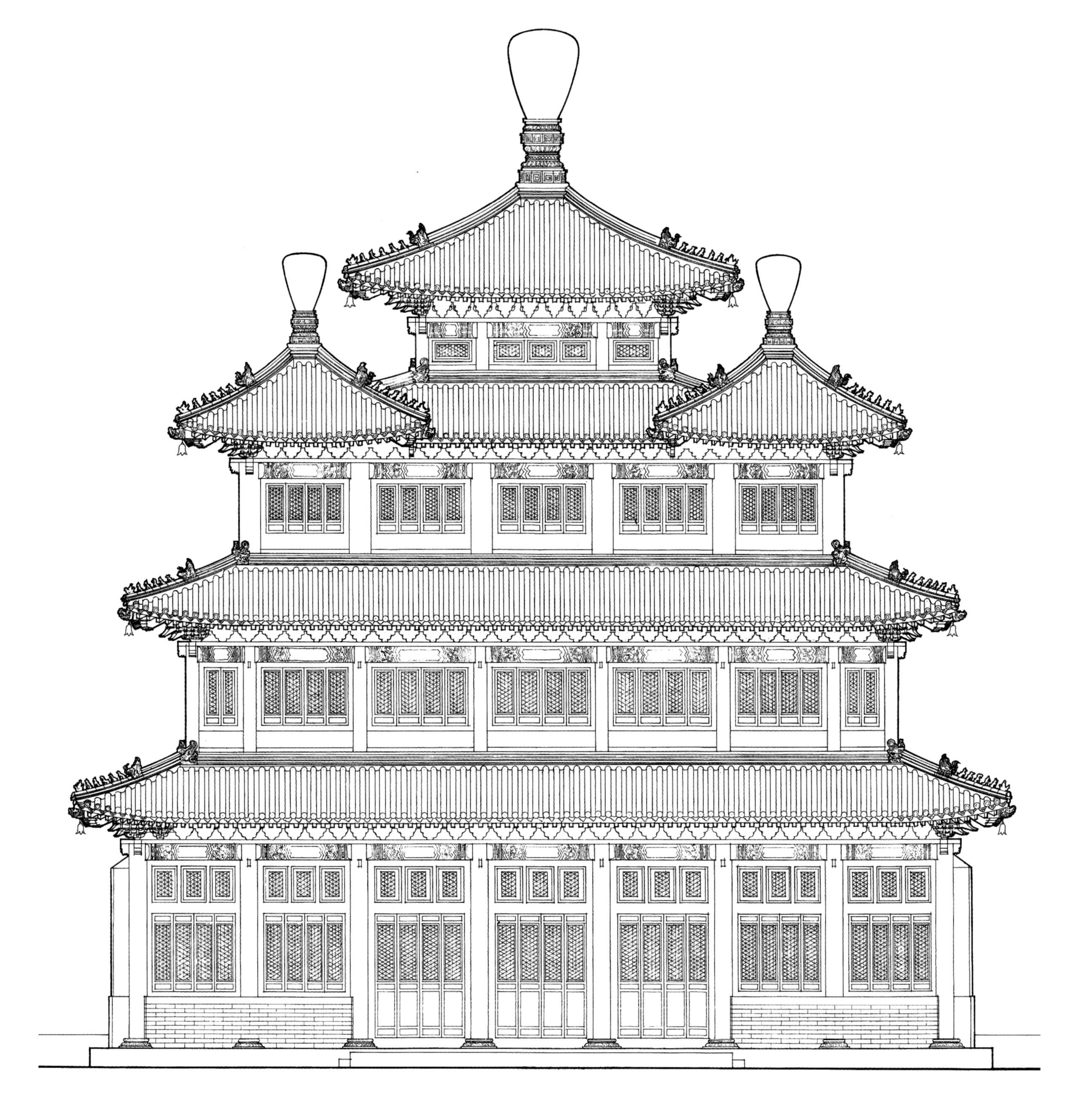

普宁寺大乘之阁背立面图

Rear elevation of Dacheng zhige of Puningsi

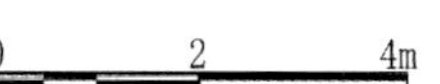

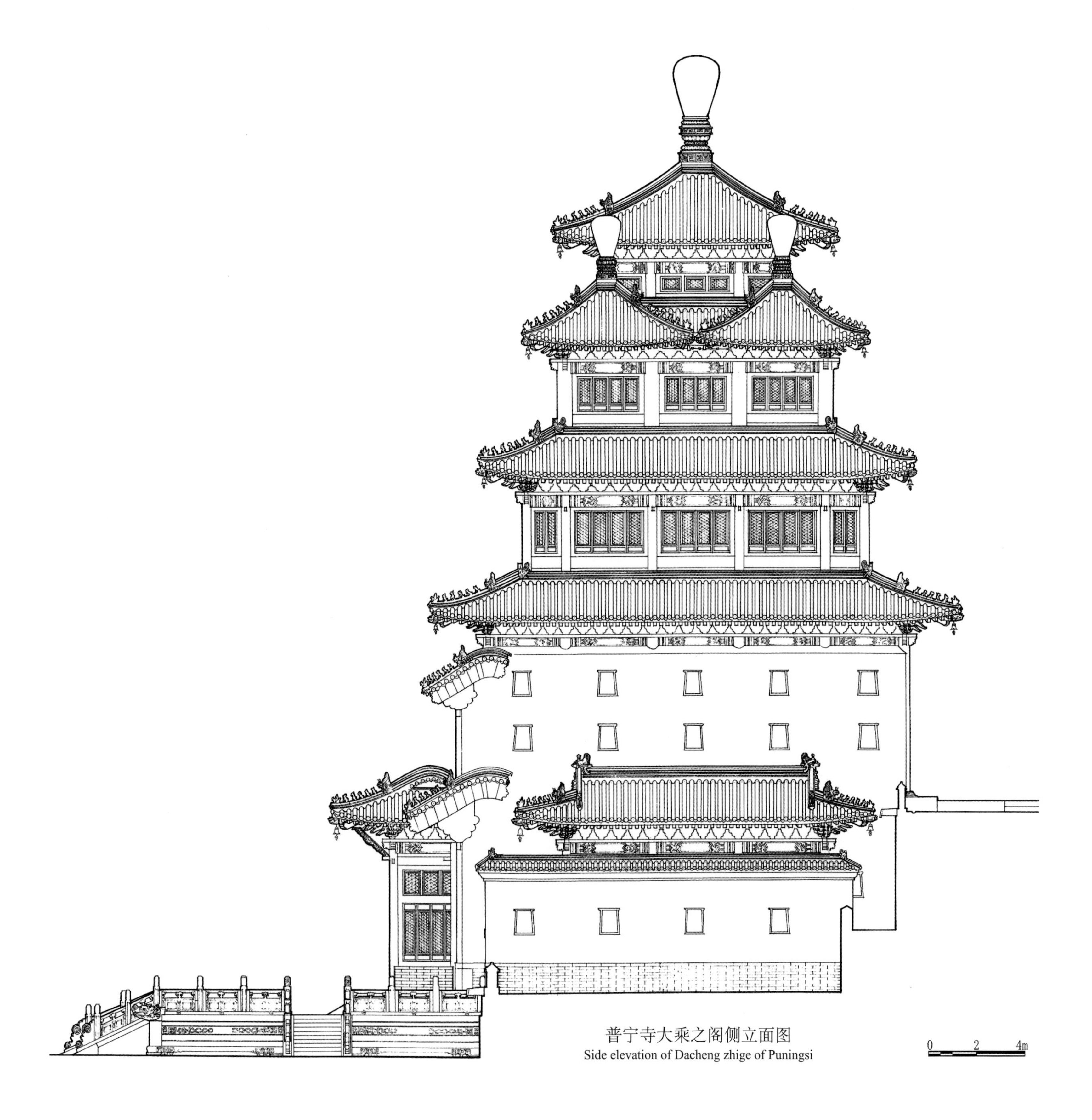

普宁寺大乘之阁侧立面图
Side elevation of Dacheng zhige of Puningsi

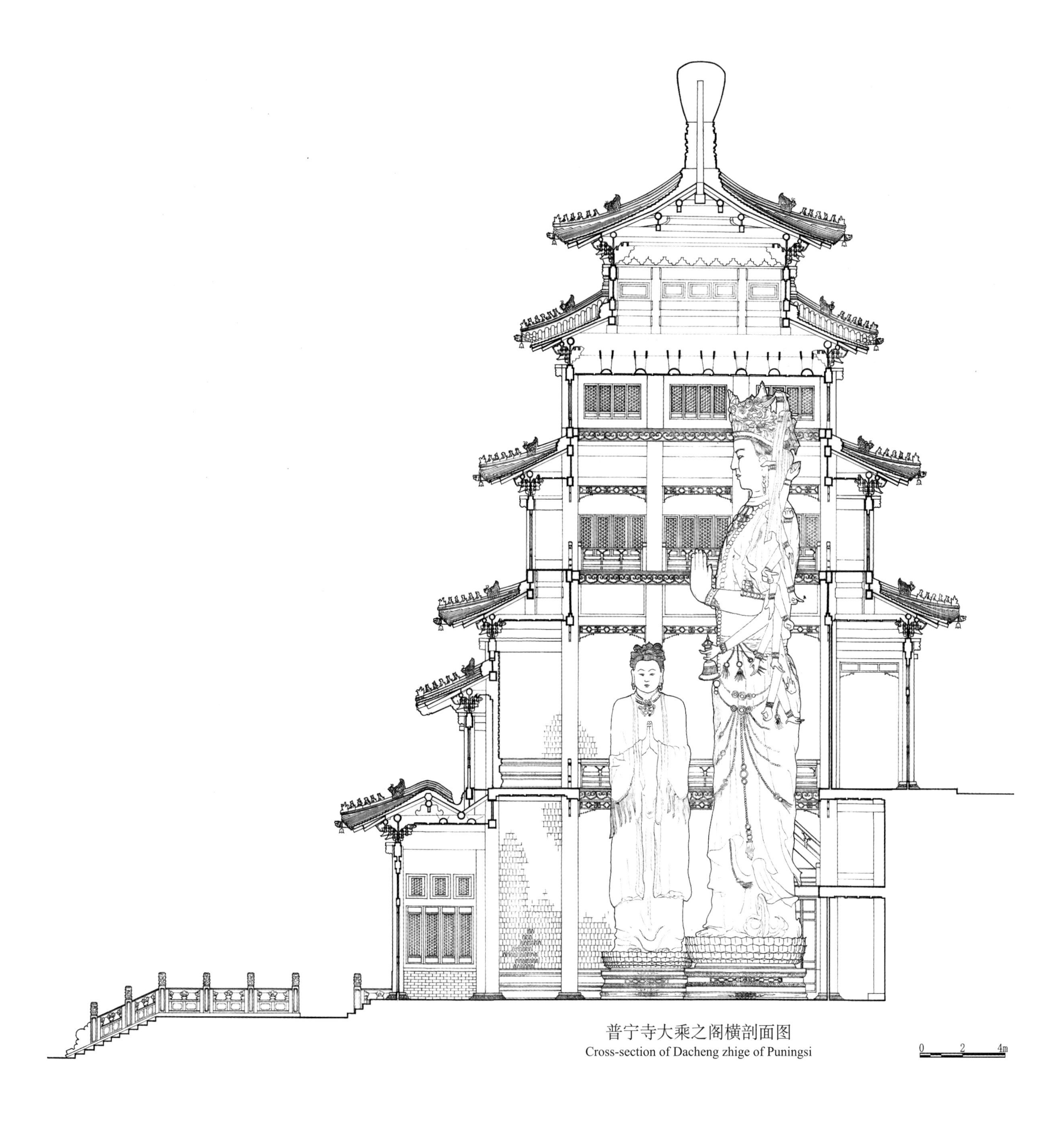

普宁寺大乘之阁横剖面图
Cross-section of Dacheng zhige of Puningsi

普宁寺大乘之阁纵剖面图
Longitudinal section of Dacheng zhige of Puningsi

0 2 4m

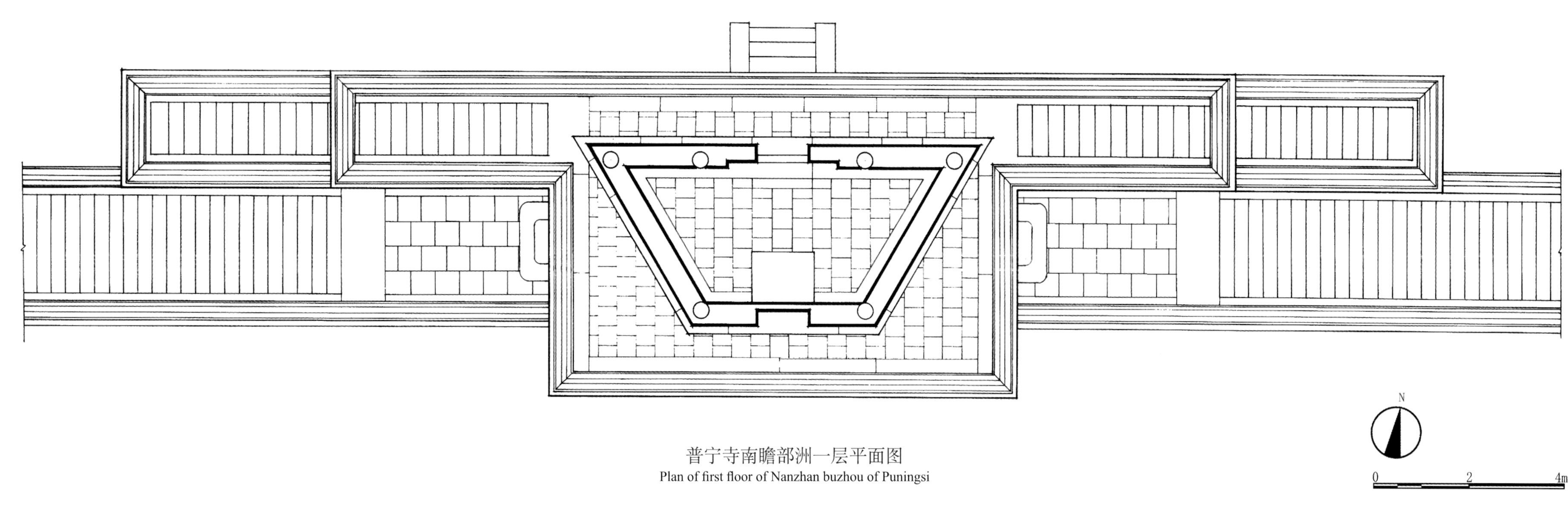

普宁寺南瞻部洲一层平面图

Plan of first floor of Nanzhan buzhou of Puningsi

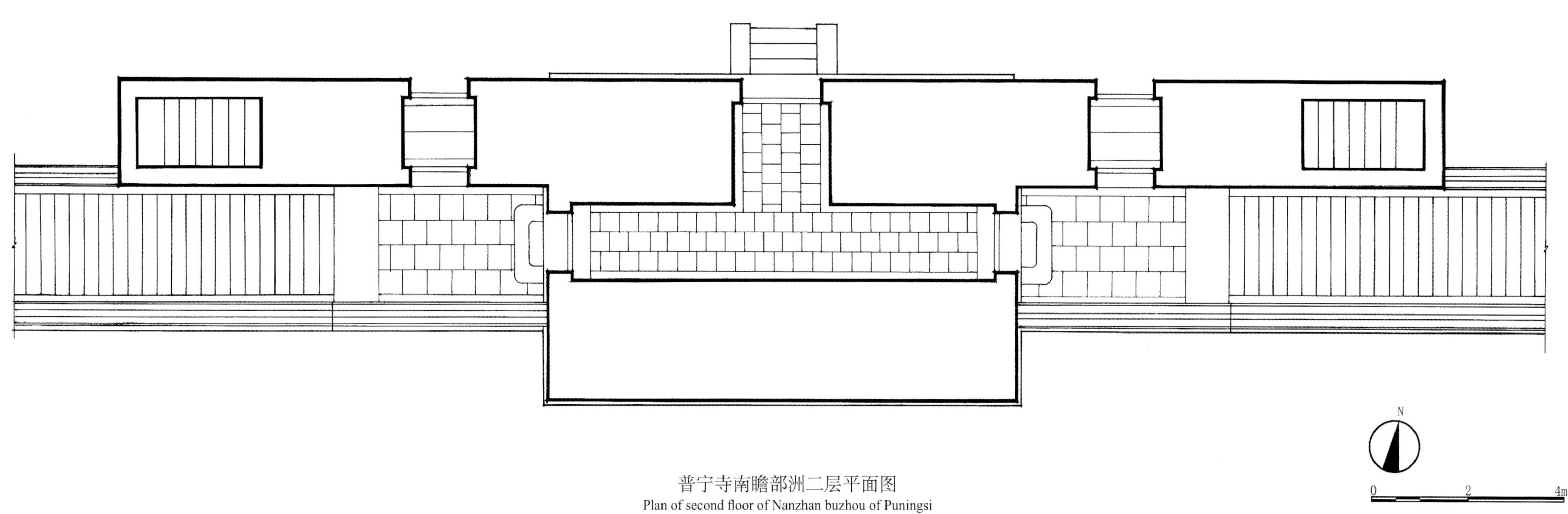

普宁寺南瞻部洲二层平面图

Plan of second floor of Nanzhan buzhou of Puningsi

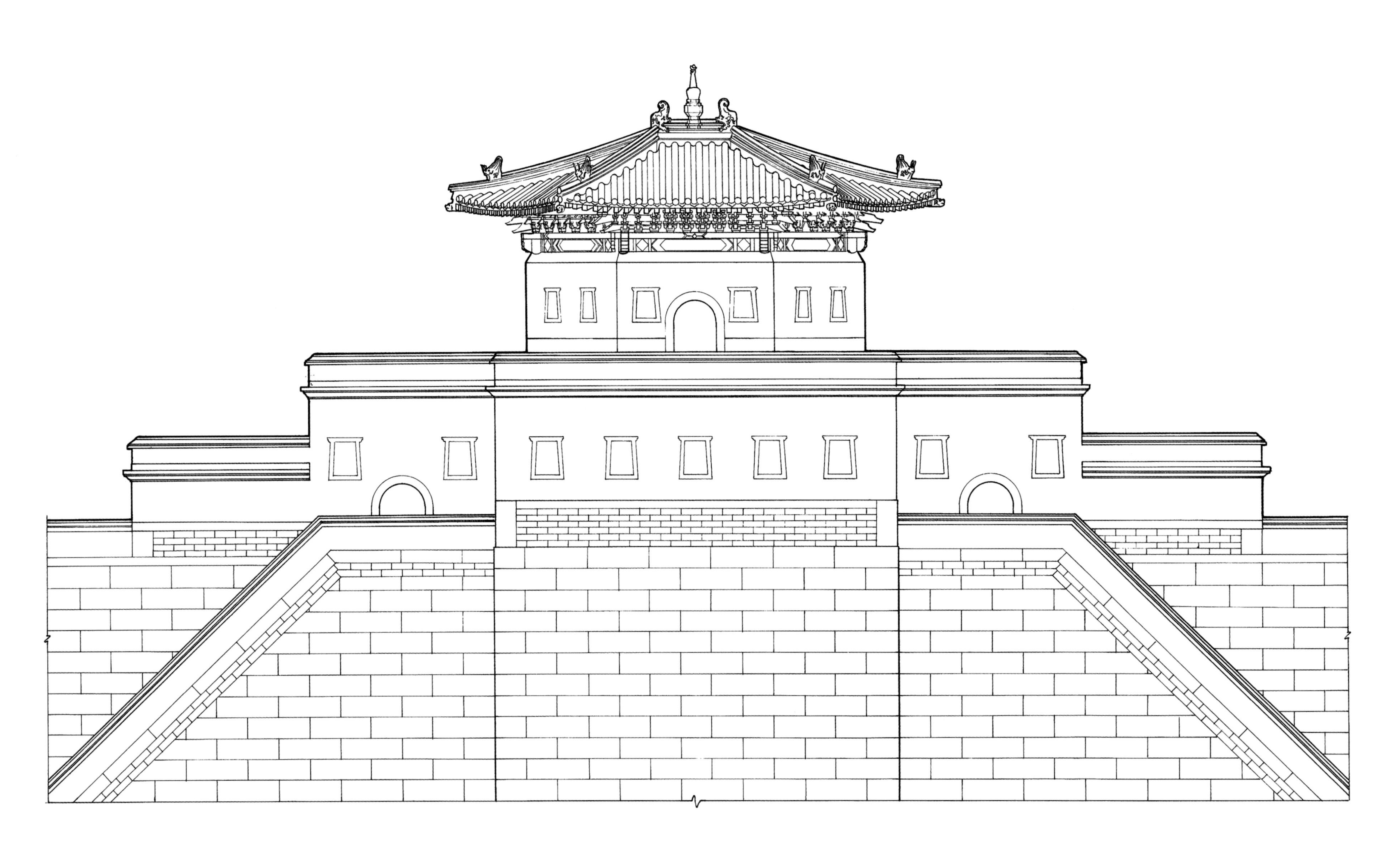

普宁寺南瞻部洲正立面图

Front elevation of Nanzhan buzhou of Puningsi

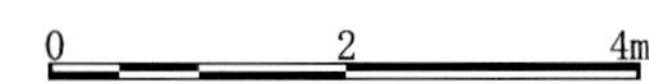

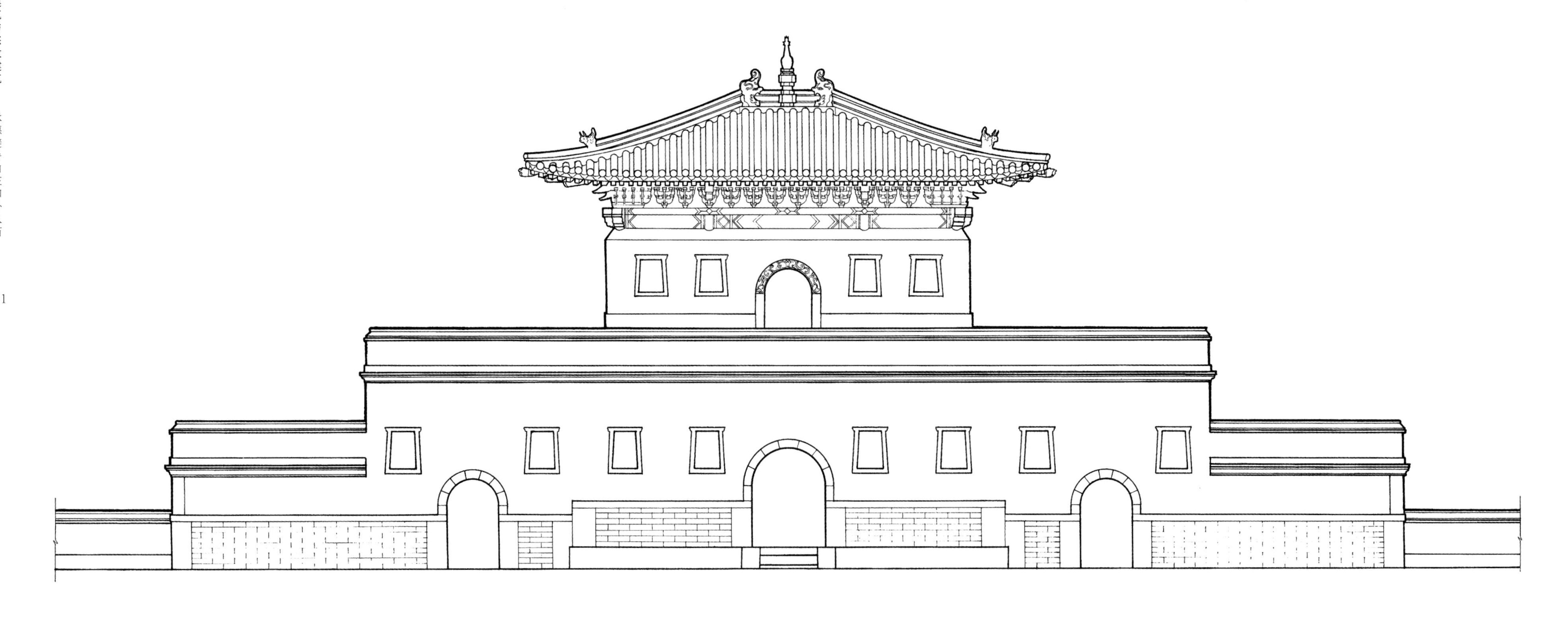

普宁寺南瞻部洲背立面图
Rear elevation of Nanzhan buzhou of Puningsi

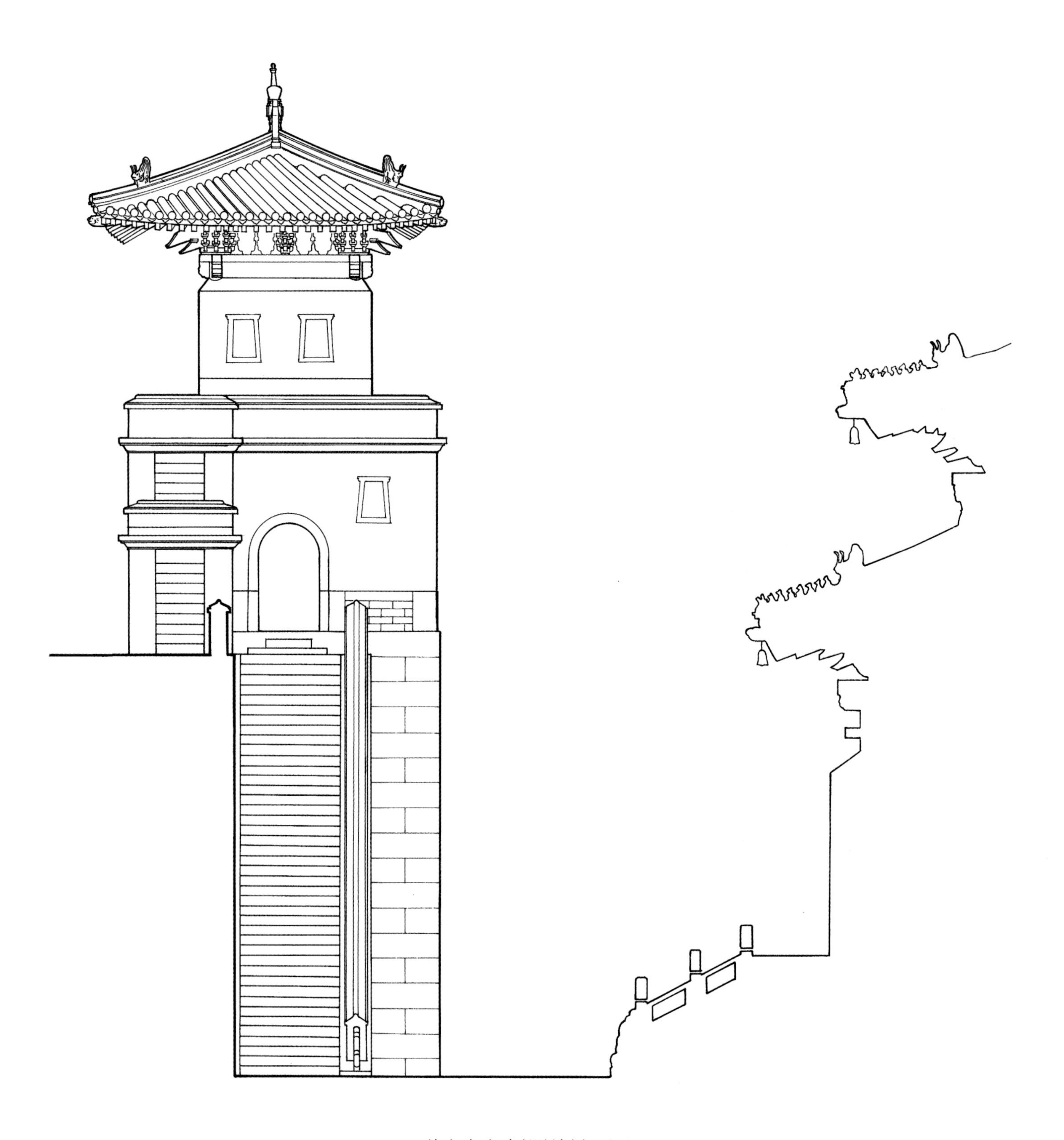

普宁寺南瞻部洲侧立面图
Side elevation of Nanzhan buzhou of Puningsi

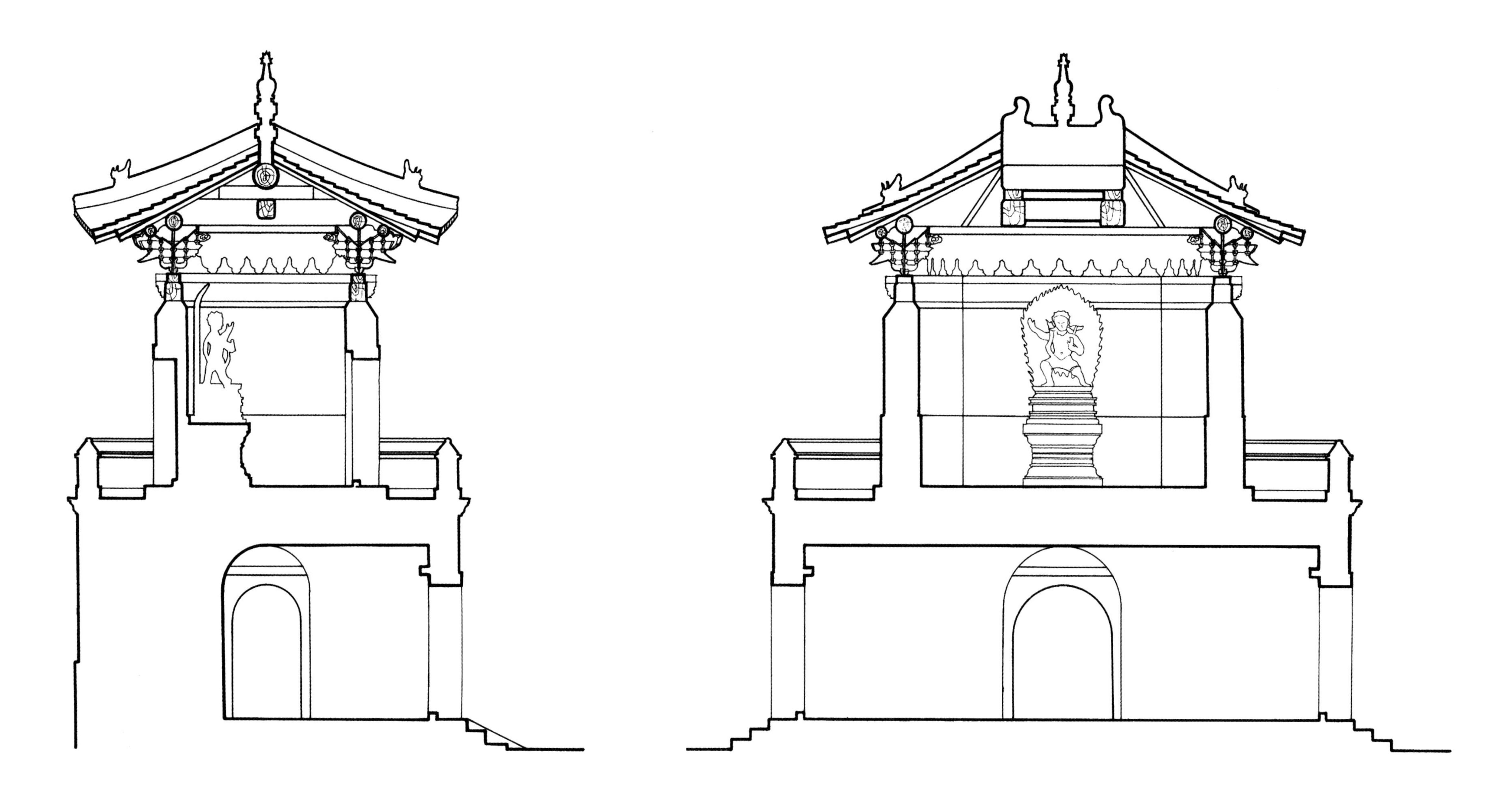

普宁寺南瞻部洲横剖面图
Cross-section of Nanzhan buzhou of Puningsi

普宁寺南瞻部洲纵剖面图
Longitudinal section of Nanzhan buzhou of Puningsi

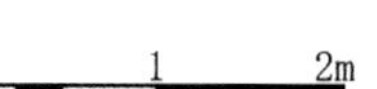

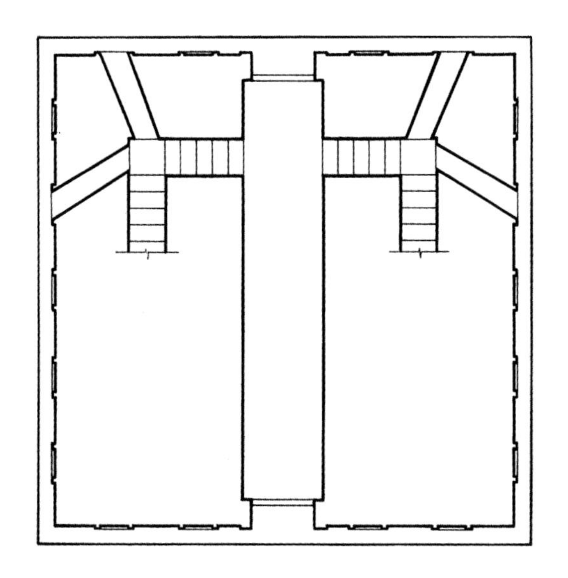

0 1 2m

普宁寺北俱卢洲一层平面图
Plan of first floor of Beiju luzhou of Puningsi

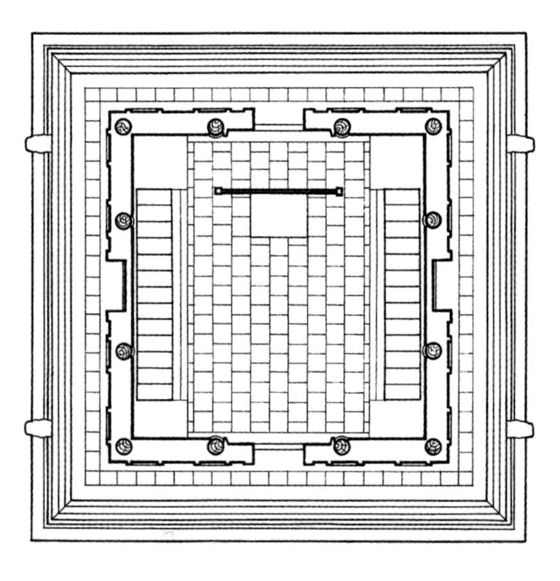

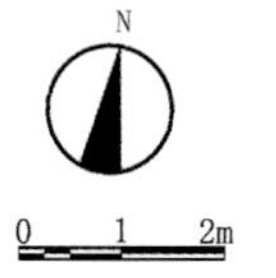

0 1 2m

普宁寺北俱卢洲二层平面图
Plan of second floor of Beiju luzhou of Puningsi

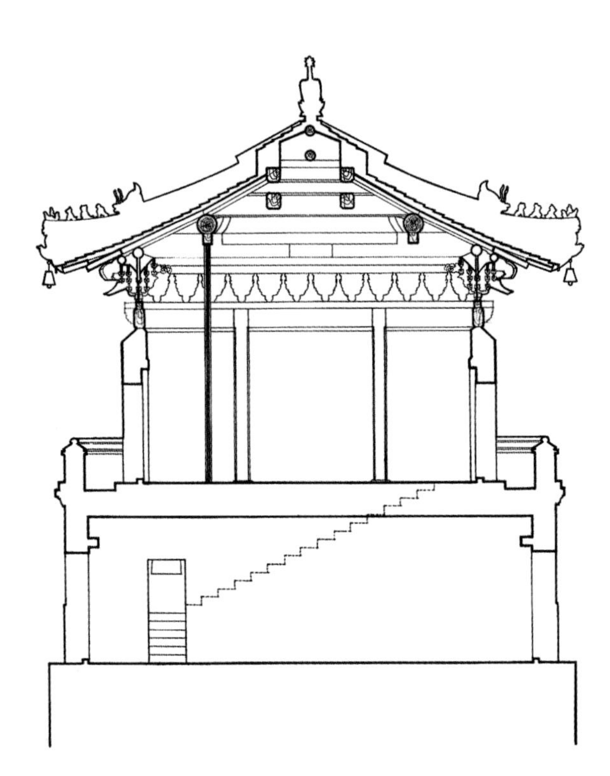

0 1 2m

普宁寺北俱卢洲横剖面图
Cross-section of Beiju luzhou of Puningsi

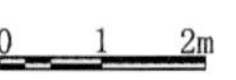

普宁寺北俱卢洲纵剖面图
Longitudinal section of Beiju luzhou of Puningsi

普宁寺北俱卢洲正立面图
Front elevation of Beiju luzhou of Puningsi

普宁寺北俱卢洲侧立面图

Side elevation of Beiju luzhou of Puningsi

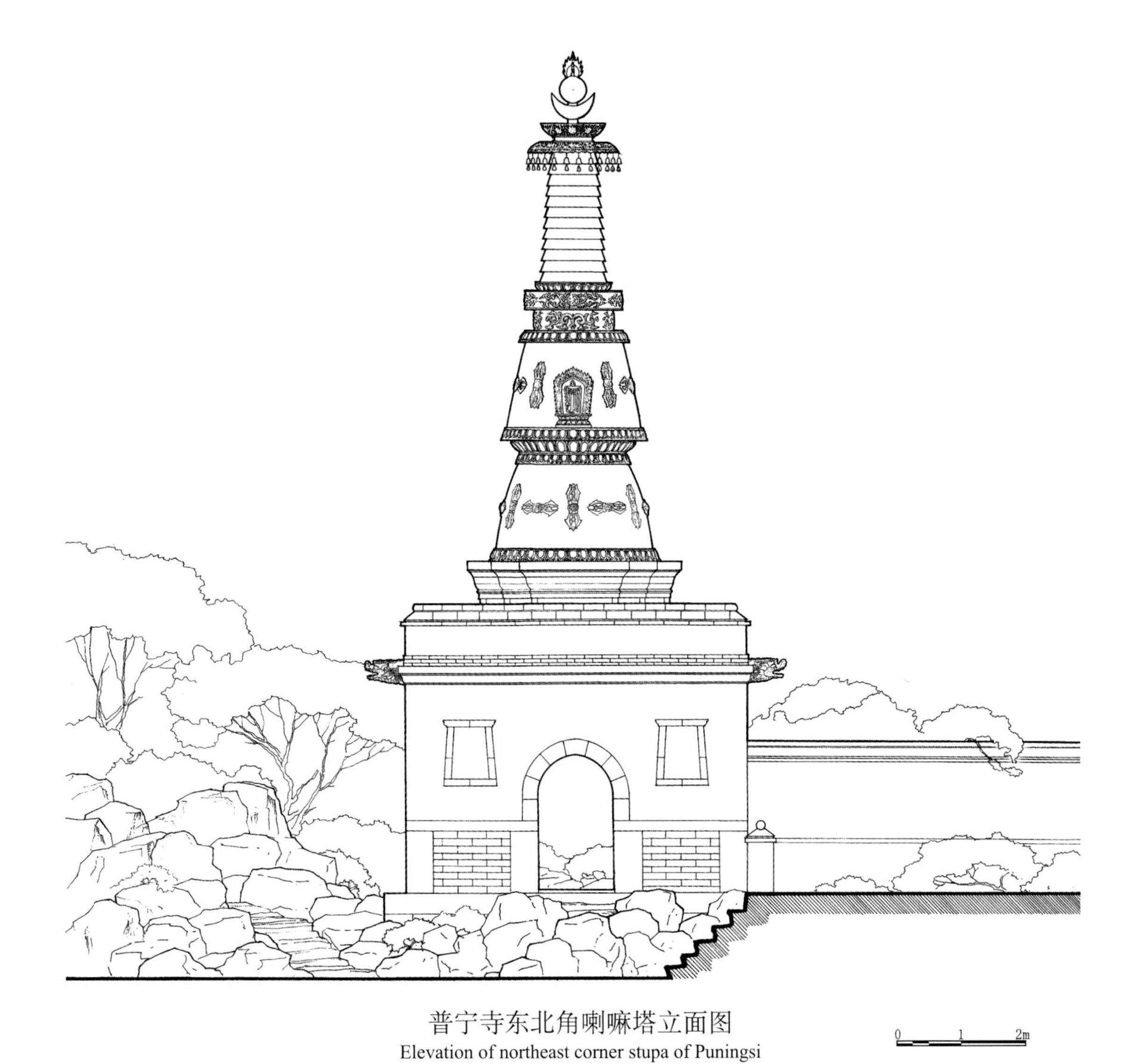

普宁寺东北角喇嘛塔立面图
Elevation of northeast corner stupa of Puningsi

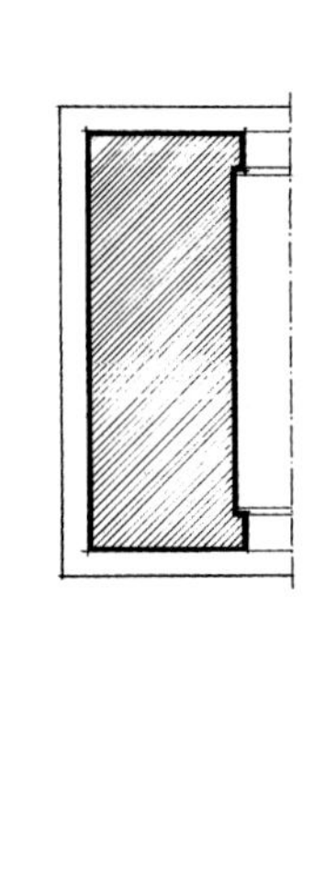

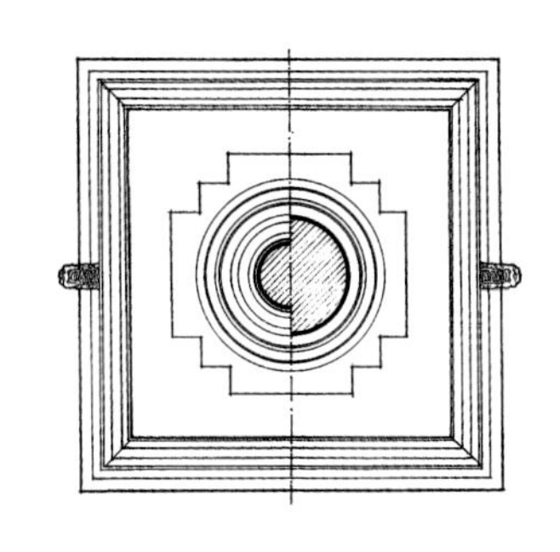

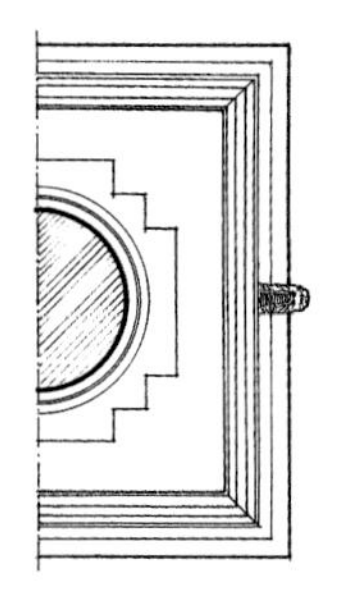

普宁寺东北角喇嘛塔平面图
Plan of northeast corner stupa of Puningsi

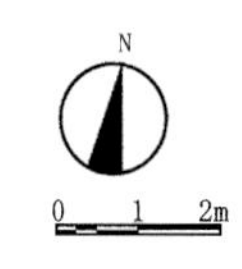

普宁寺东南角喇嘛塔立面图
Elevation of southeast corner stupa of Puningsi

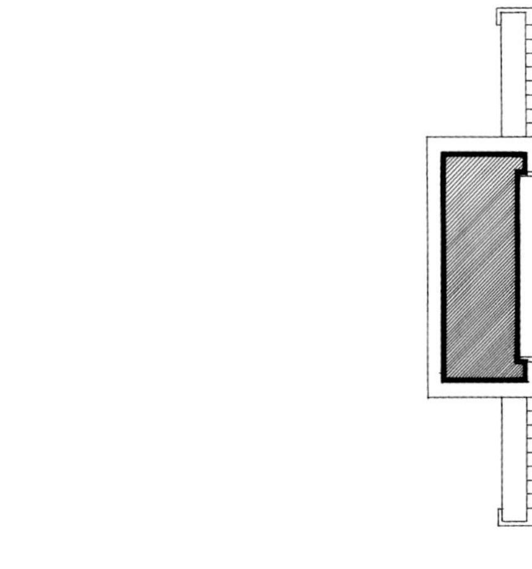

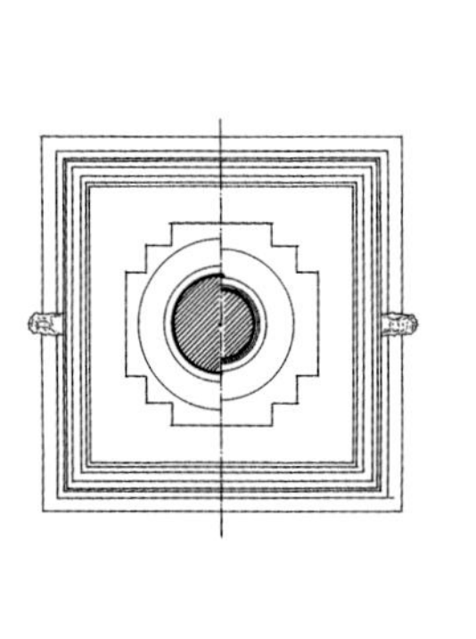

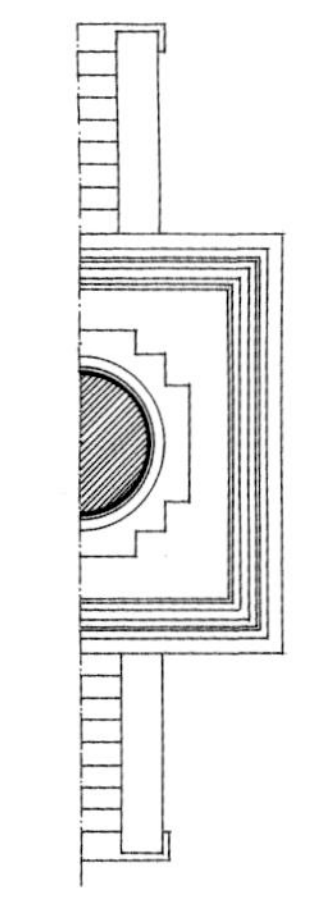

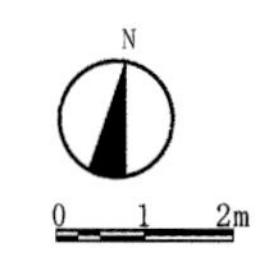

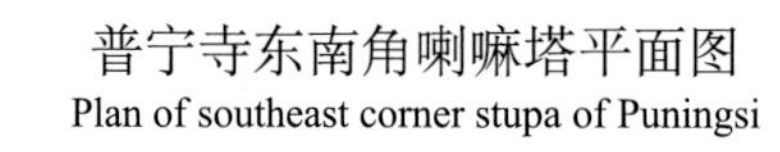

普宁寺东南角喇嘛塔平面图
Plan of southeast corner stupa of Puningsi

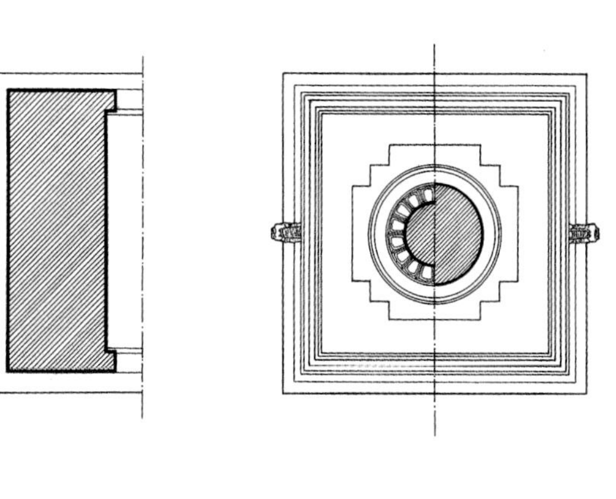

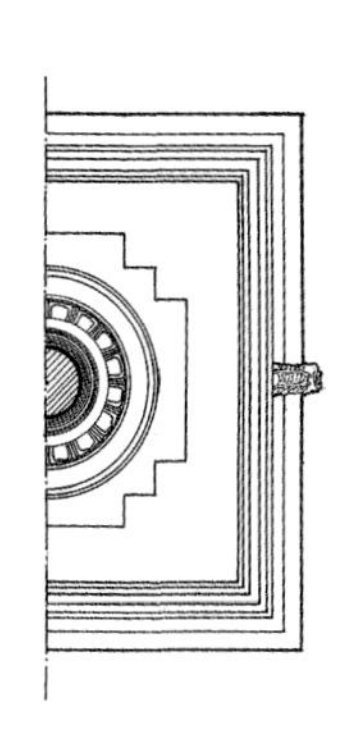

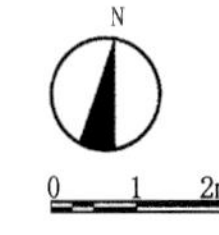

普宁寺西北角喇嘛塔平面图
Plan of northwest corner stupa of Puningsi

普宁寺西北角喇嘛塔立面图
Elevation of northwest corner stupa of Puningsi

0 1 2m

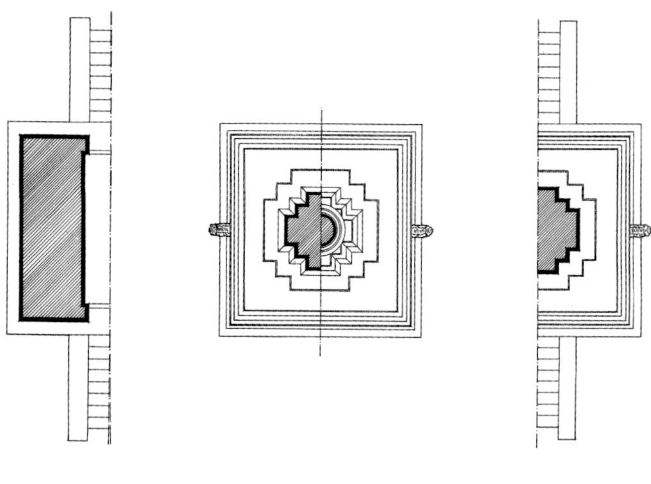

普宁寺西南角喇嘛塔平面图
Plan of southwest corner stupa of Puningsi

普宁寺西南角喇嘛塔立面图
Elevation of southwest corner stupa of Puningsi

普佑寺
Puyousi

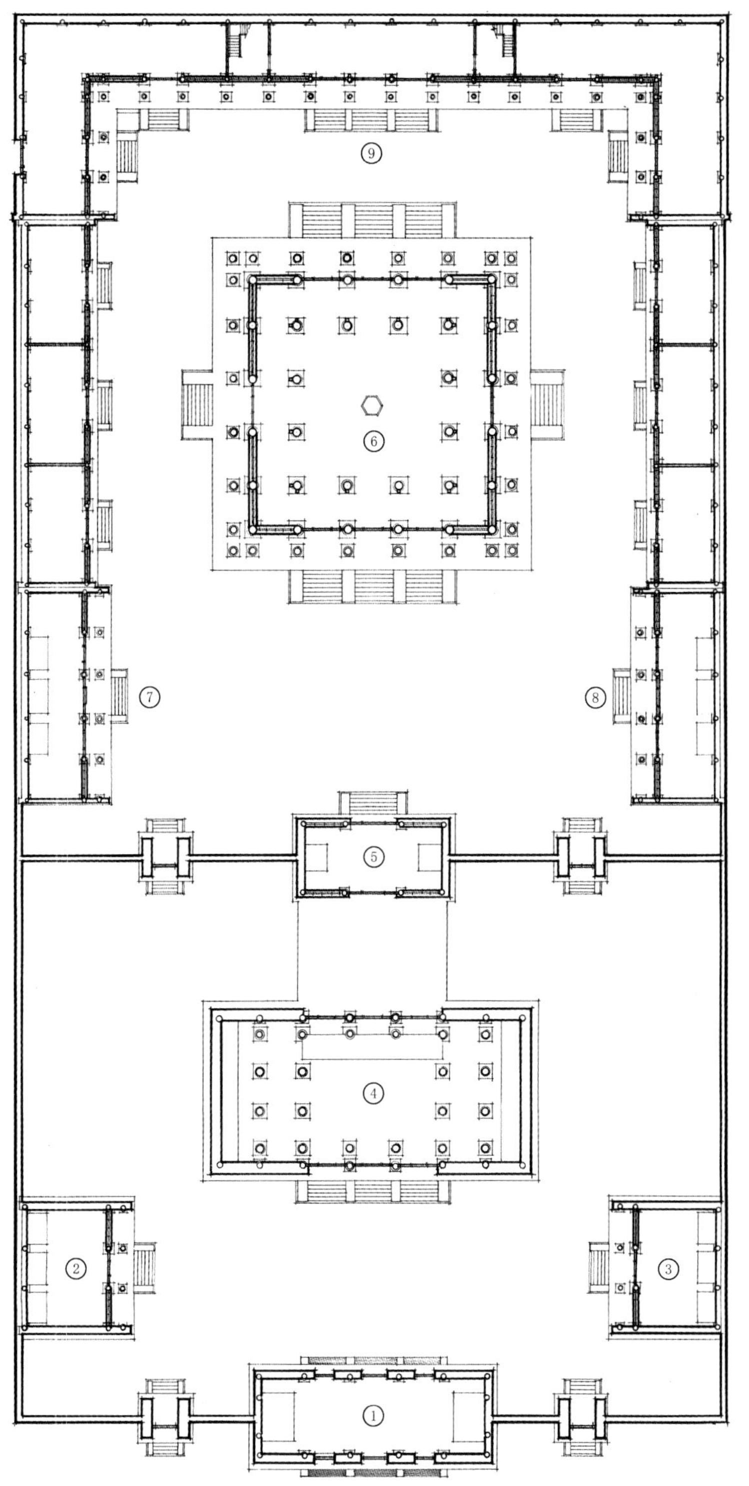

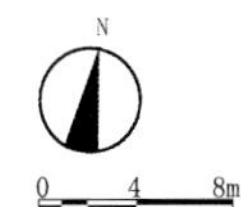

普佑寺总平面图
Site plan of Puyousi

1 山门 Shanmen (Gate Hall)
2 西配殿 West Side Hall
3 东配殿 East Side Hall
4 大方广殿 Dafangguangdian
5 天王殿 Tianwangdian (Hall of Heavenly King)
6 法轮殿 Falundian
7 西配殿 West Side Hall
8 东配殿 East Side Hall
9 经楼 Jinglou (Depository of Buddhist Sutras)

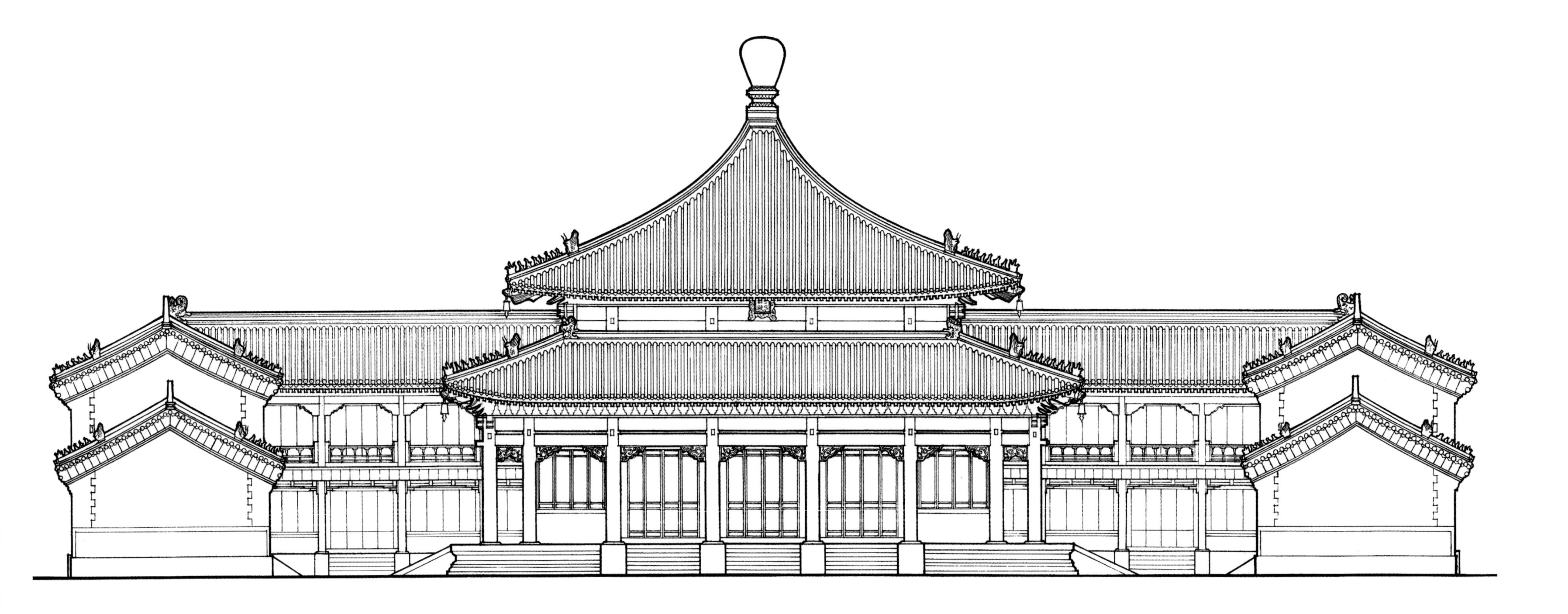

普佑寺法轮殿群楼正立面图
Front elevation of qunlou of Falundian of Puyousi

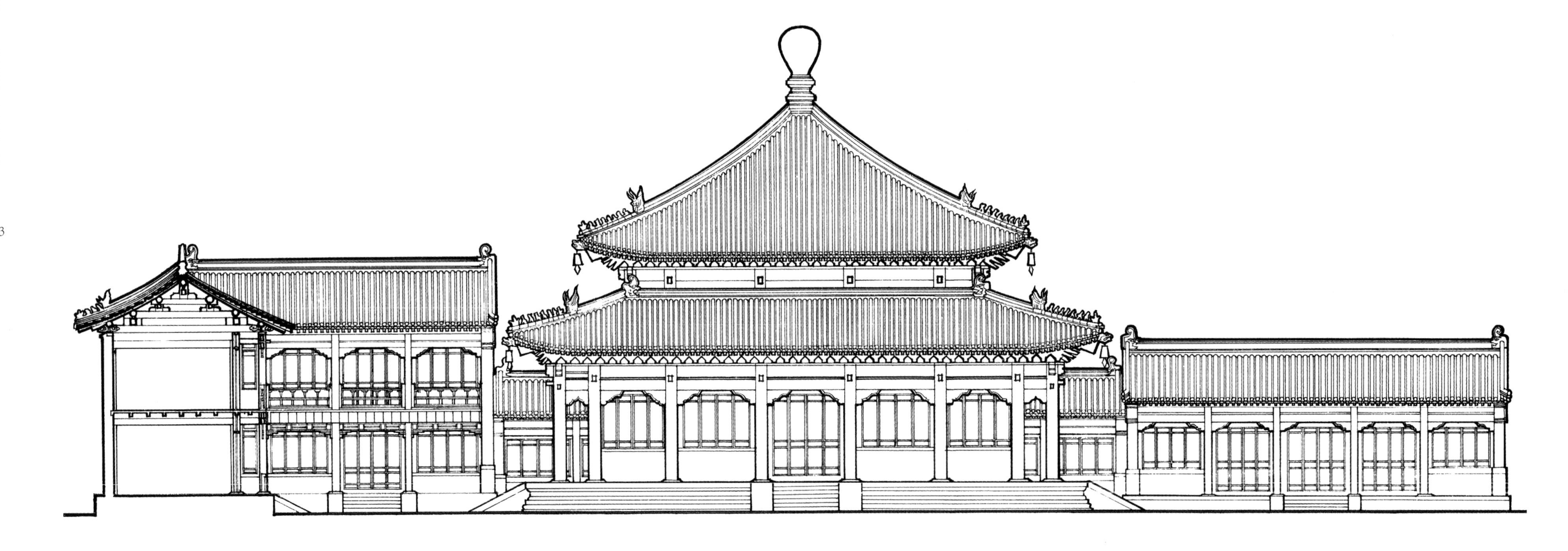

普佑寺法轮殿群楼侧立面图
Side elevation of qunlou of Falundian of Puyousi

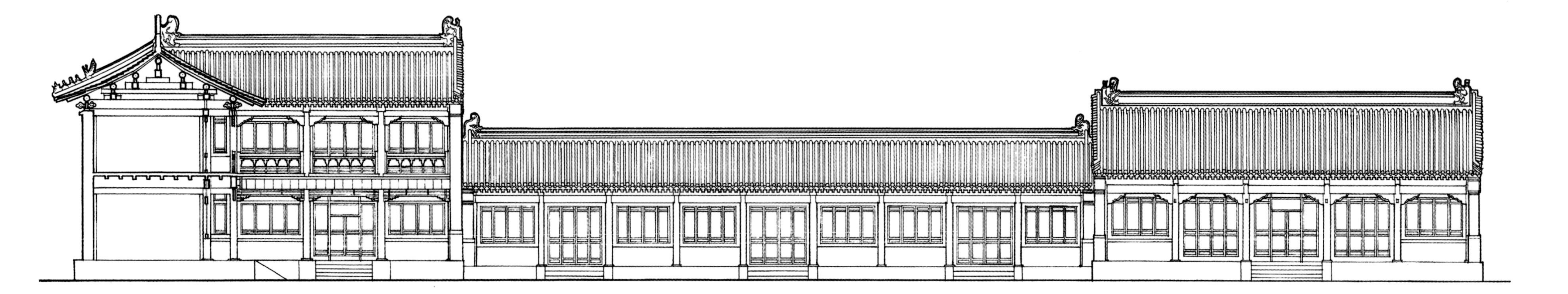

普佑寺群楼东立面图
East elevation of qunlou of Puyousi

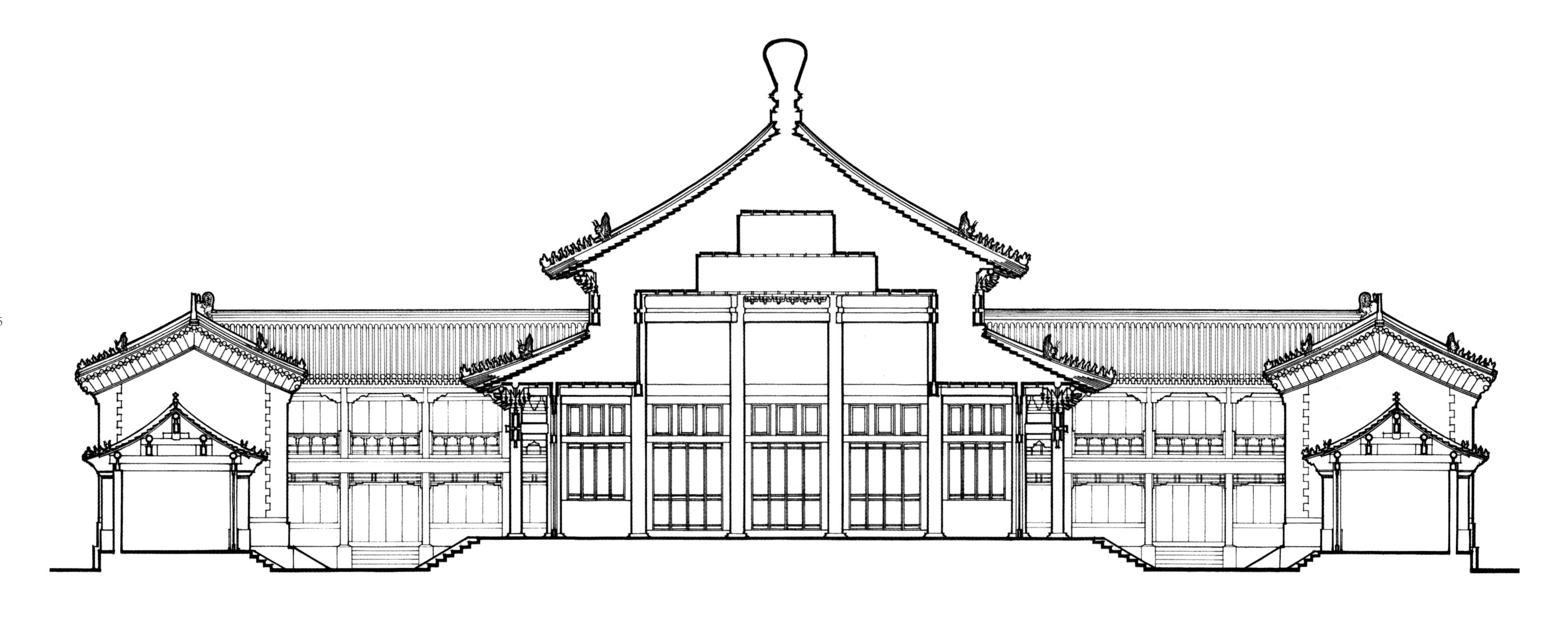

普佑寺法轮殿群楼剖面图
Section of qunlou of Falundian of Puyousi

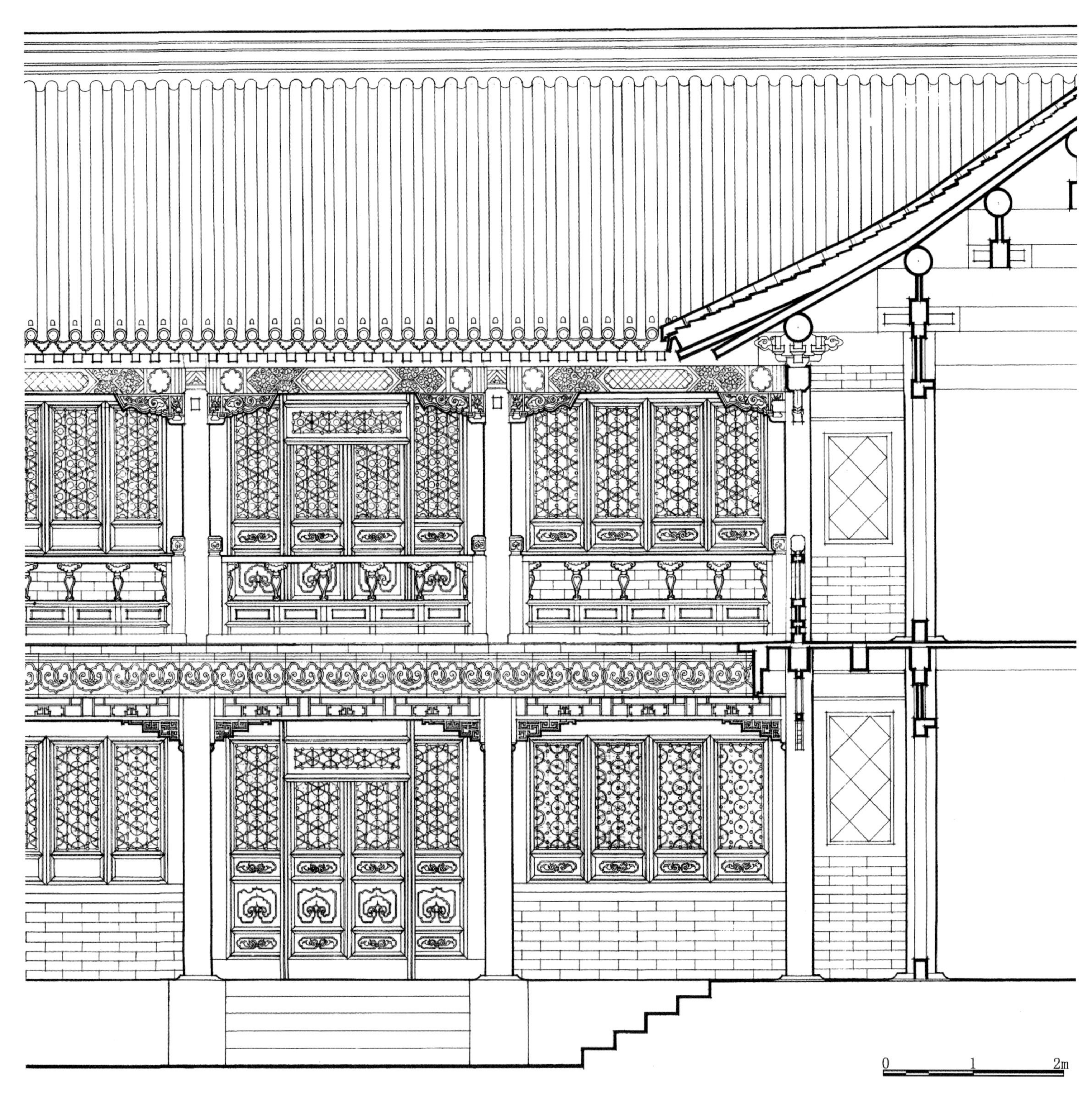

普佑寺群楼外檐大样图
Outer eaves of qunlou of Puyousi

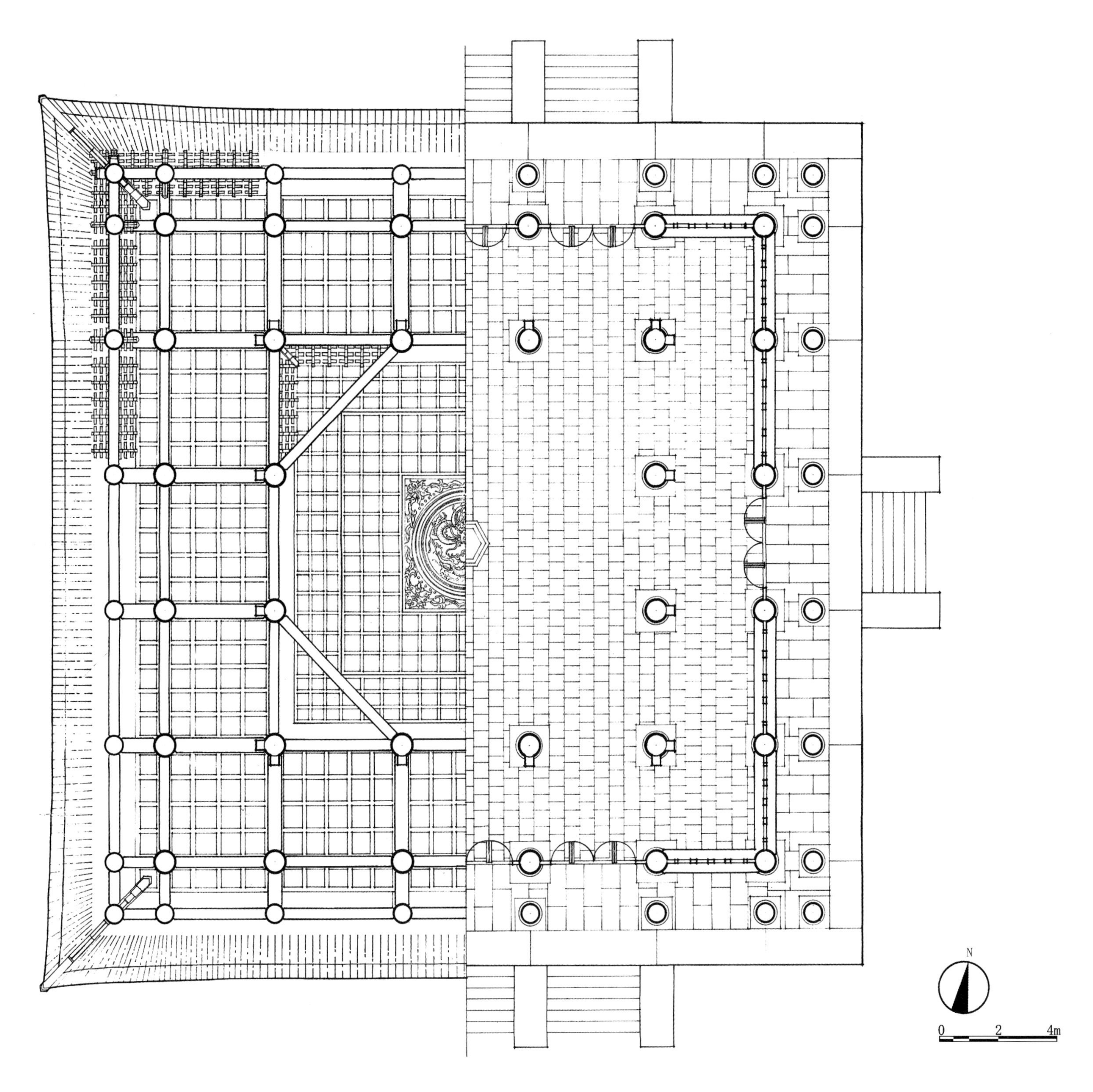

普佑寺法轮殿平面图、天花仰视图
Plan and reflected ceiling plan of Falundian of Puyousi

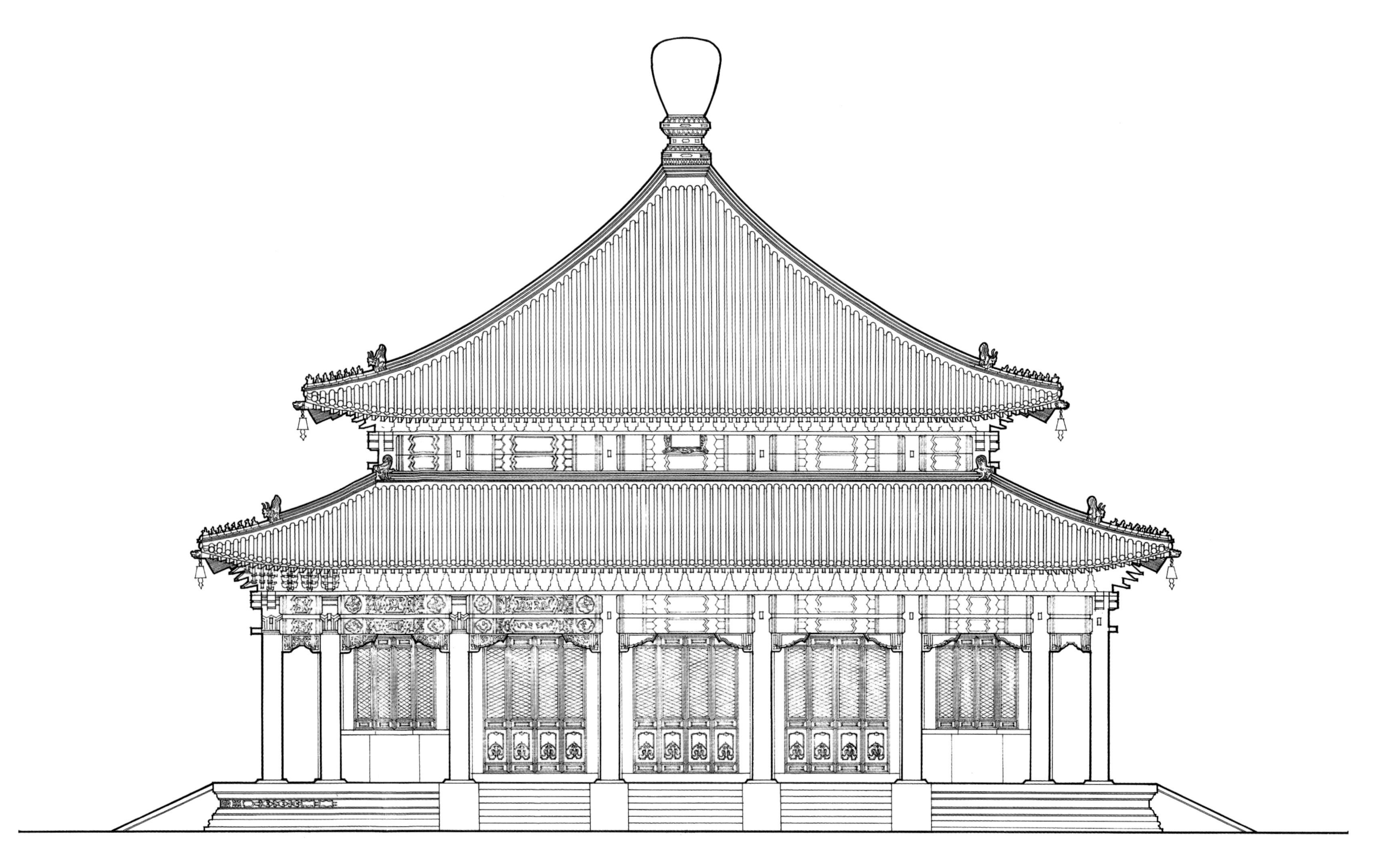

普佑寺法轮殿南立面图
South elevation of Falundian of Puyousi

安远庙
Anyuanmiao

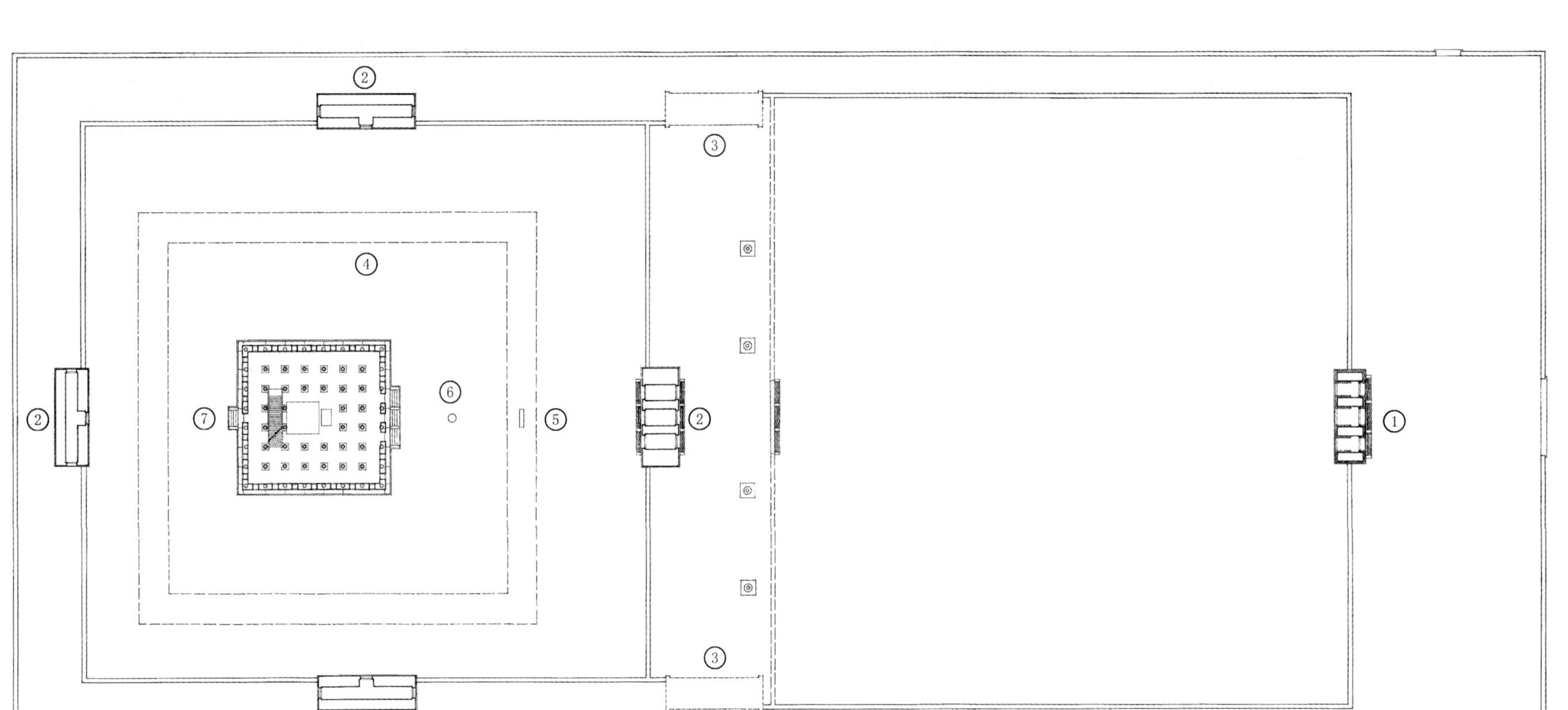

安远庙总平面图
Site plan of Anyuanmiao

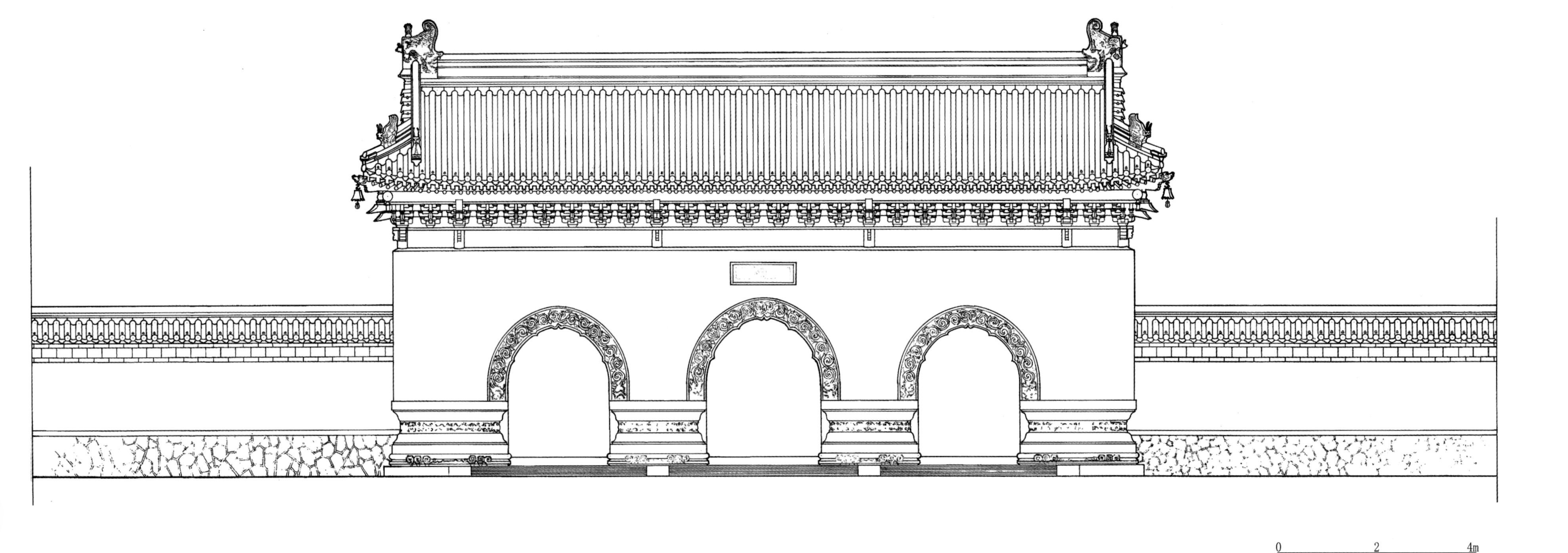

安远庙山门正立面图
Front elevation of shanmen of Anyuanmiao

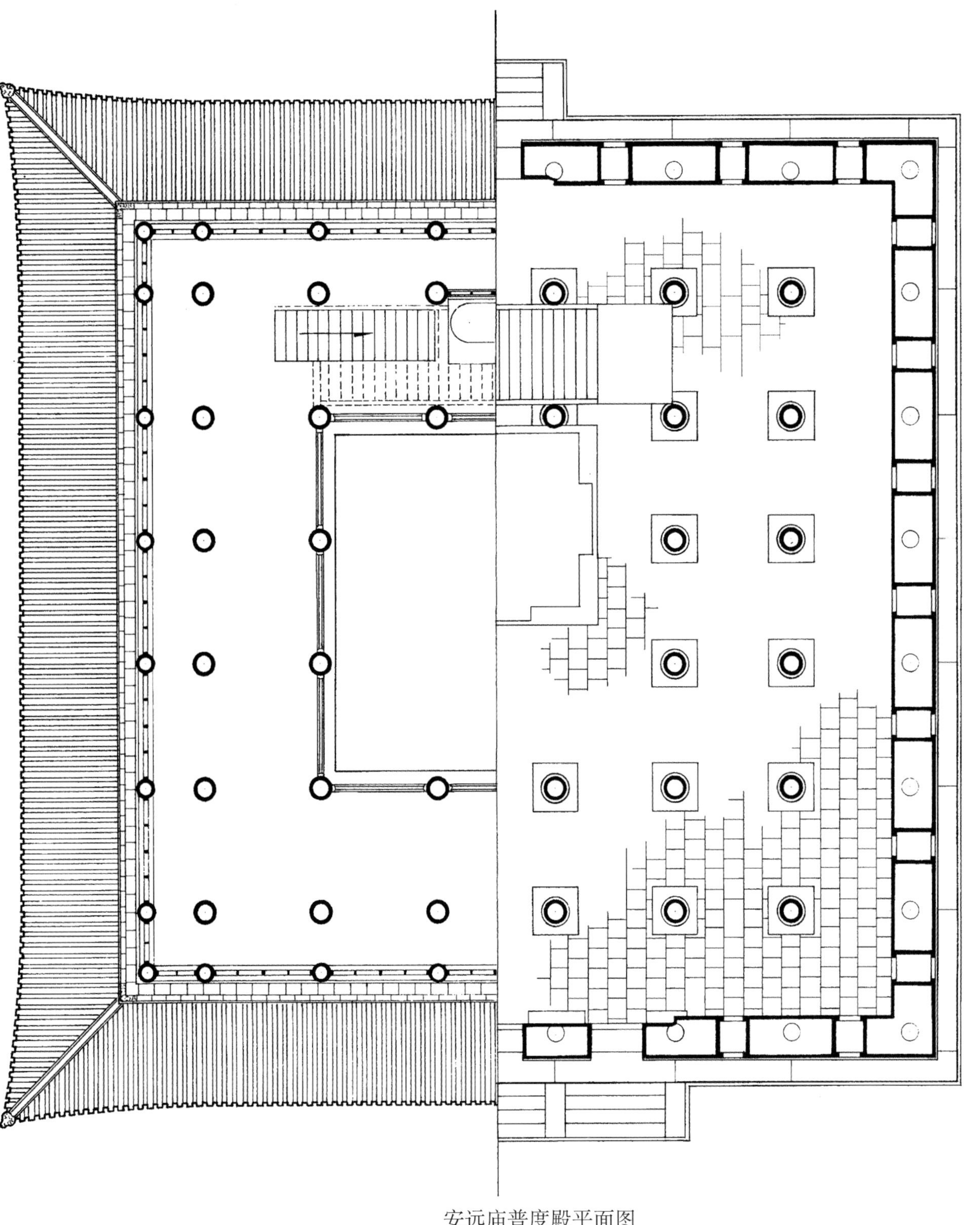

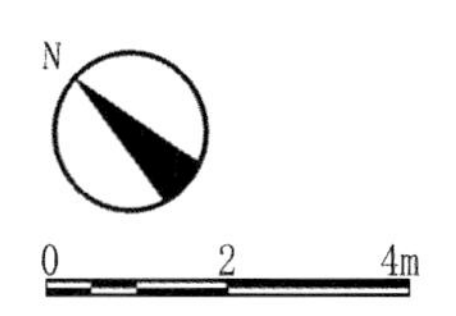

安远庙普度殿平面图
Plan of Pududian of Anyuanmiao

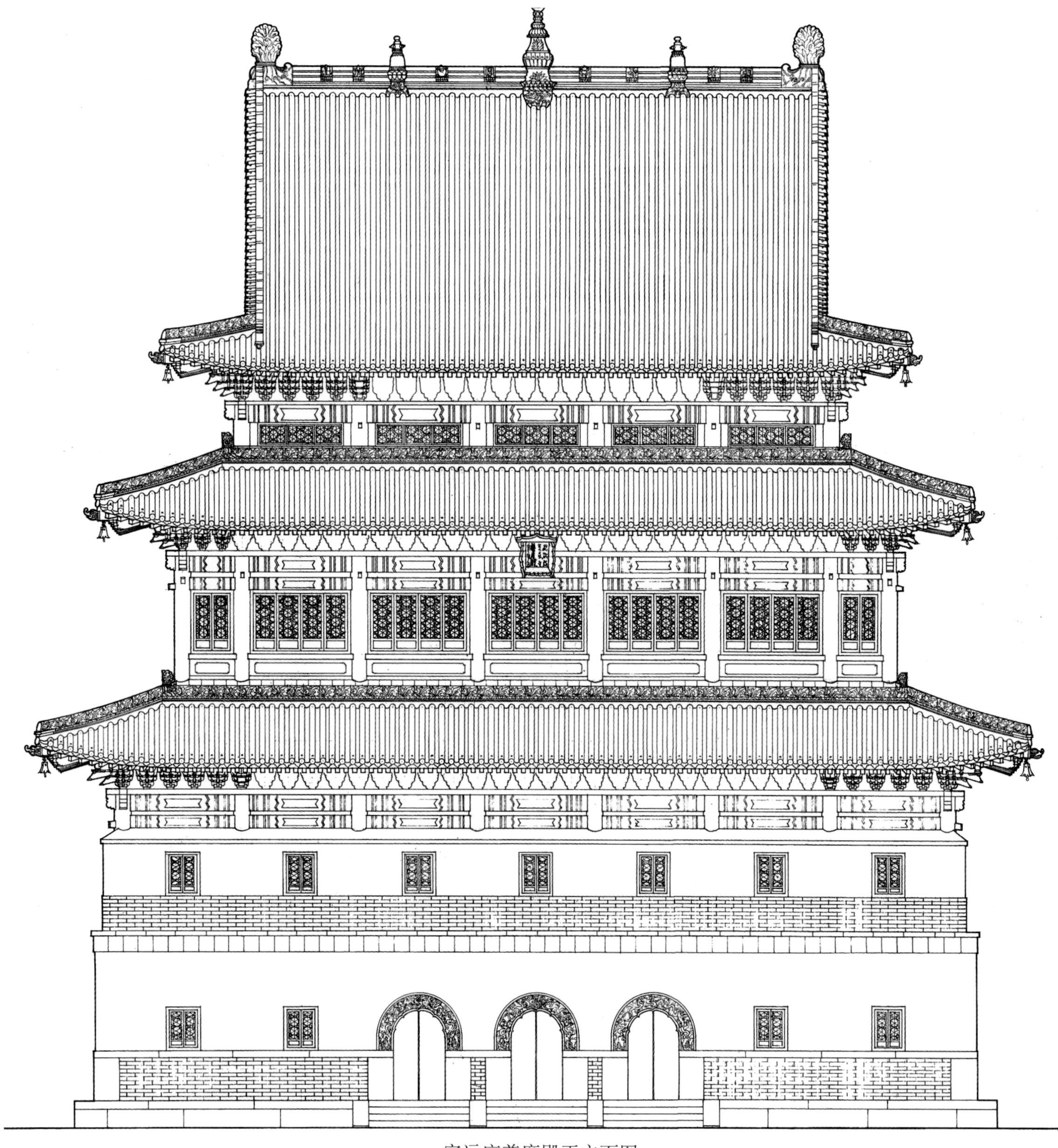

安远庙普度殿正立面图

Front elevation of Pududian of Anyuanmiao

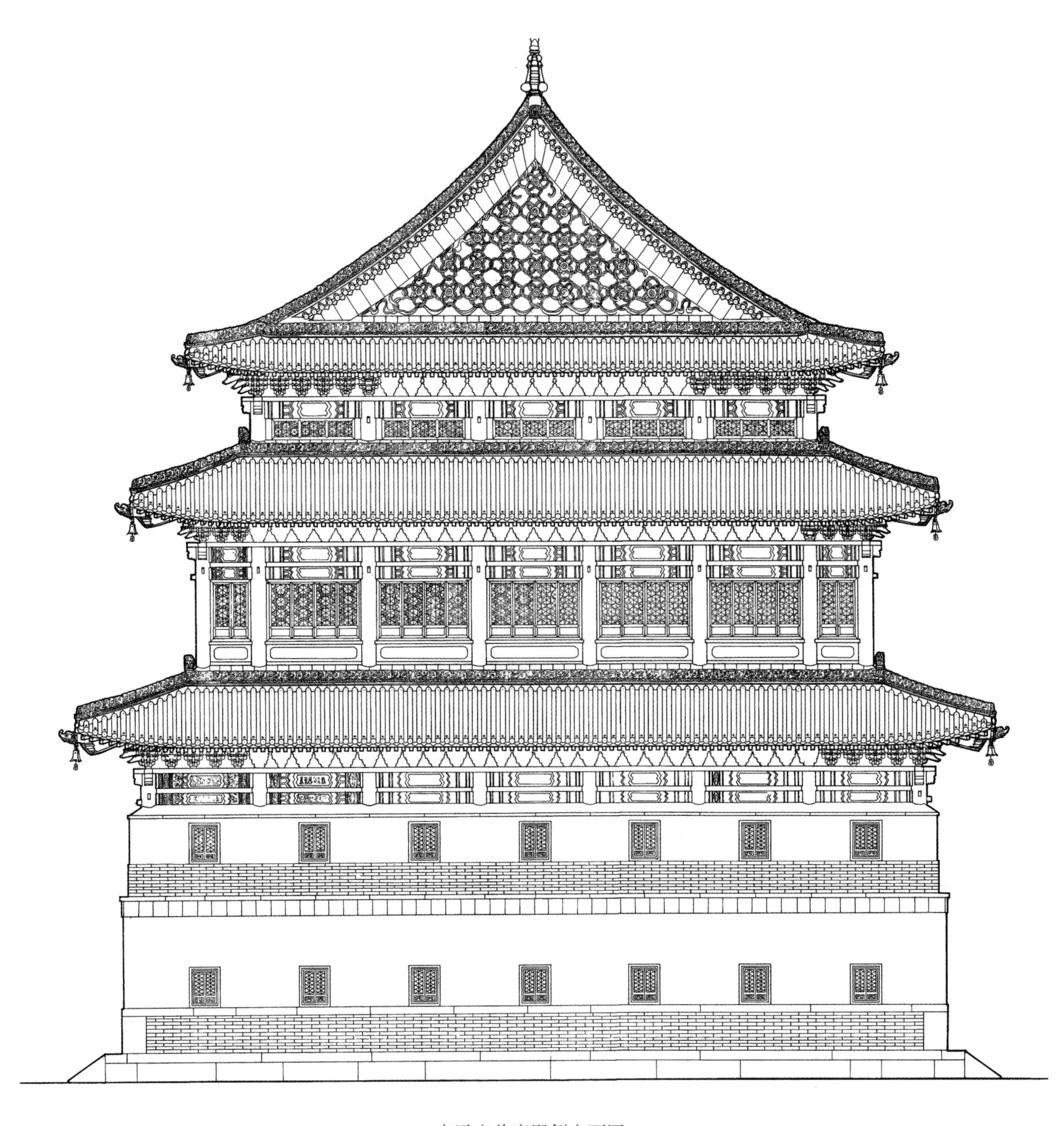

安远庙普度殿侧立面图

Side elevation of Pududian of Anyuanmiao

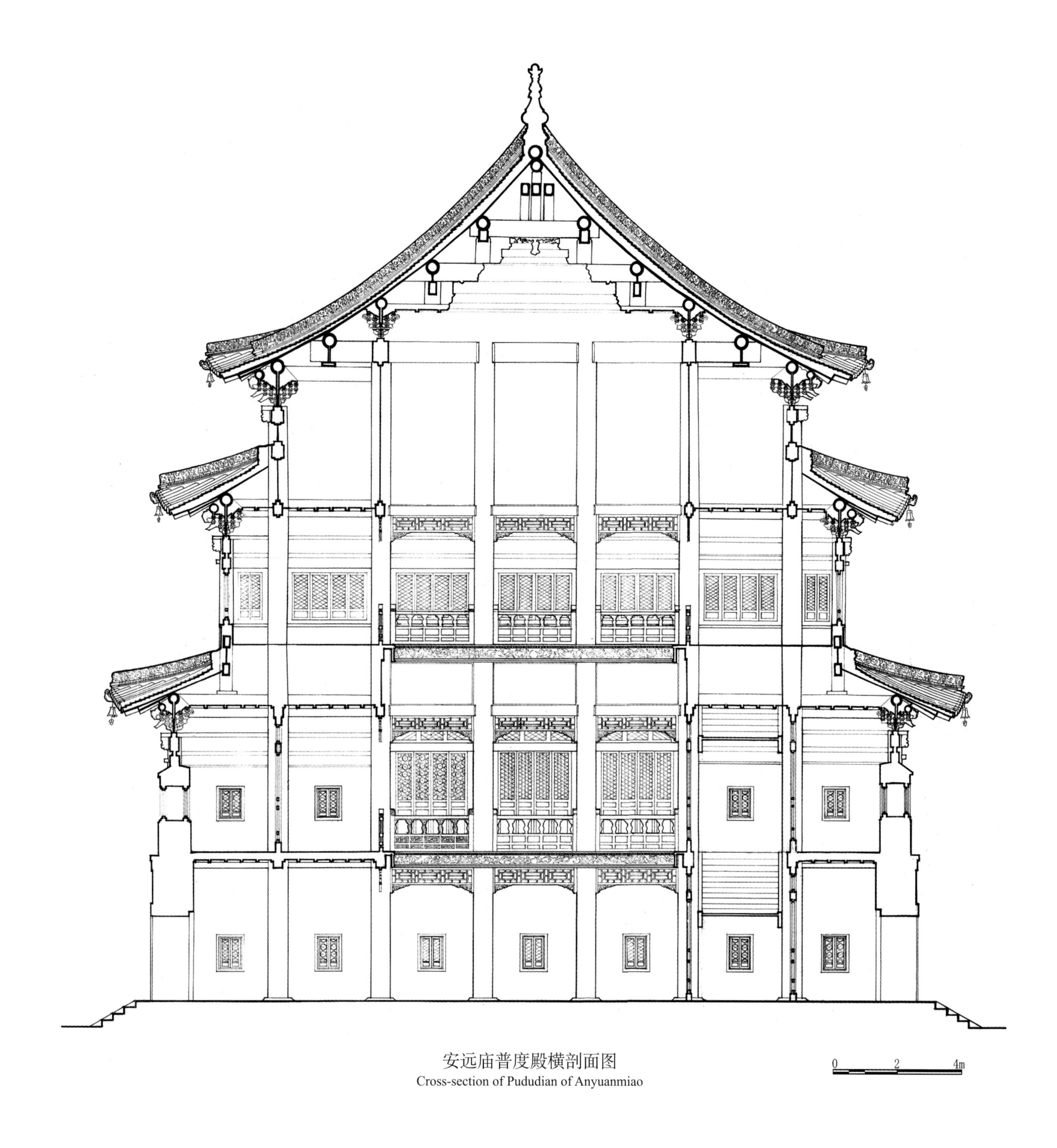

安远庙普度殿横剖面图
Cross-section of Pududian of Anyuanmiao

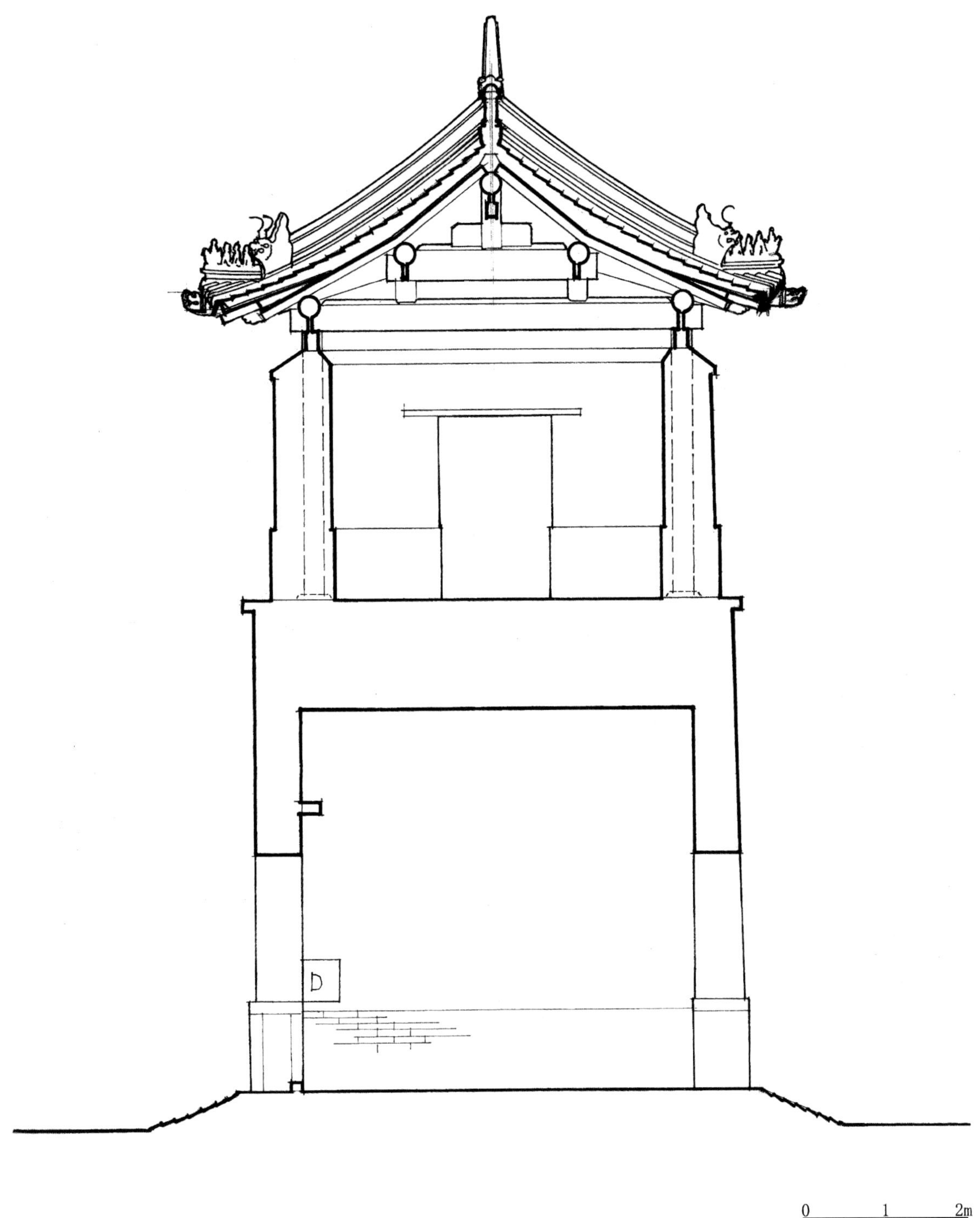

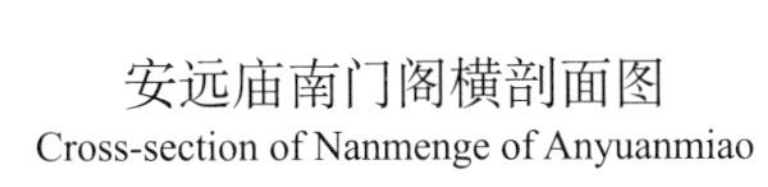

安远庙南门阁横剖面图
Cross-section of Nanmenge of Anyuanmiao

安远庙山门侧立面图
Side elevation of shanmen of Anyuanmiao

普乐寺
Pulesi

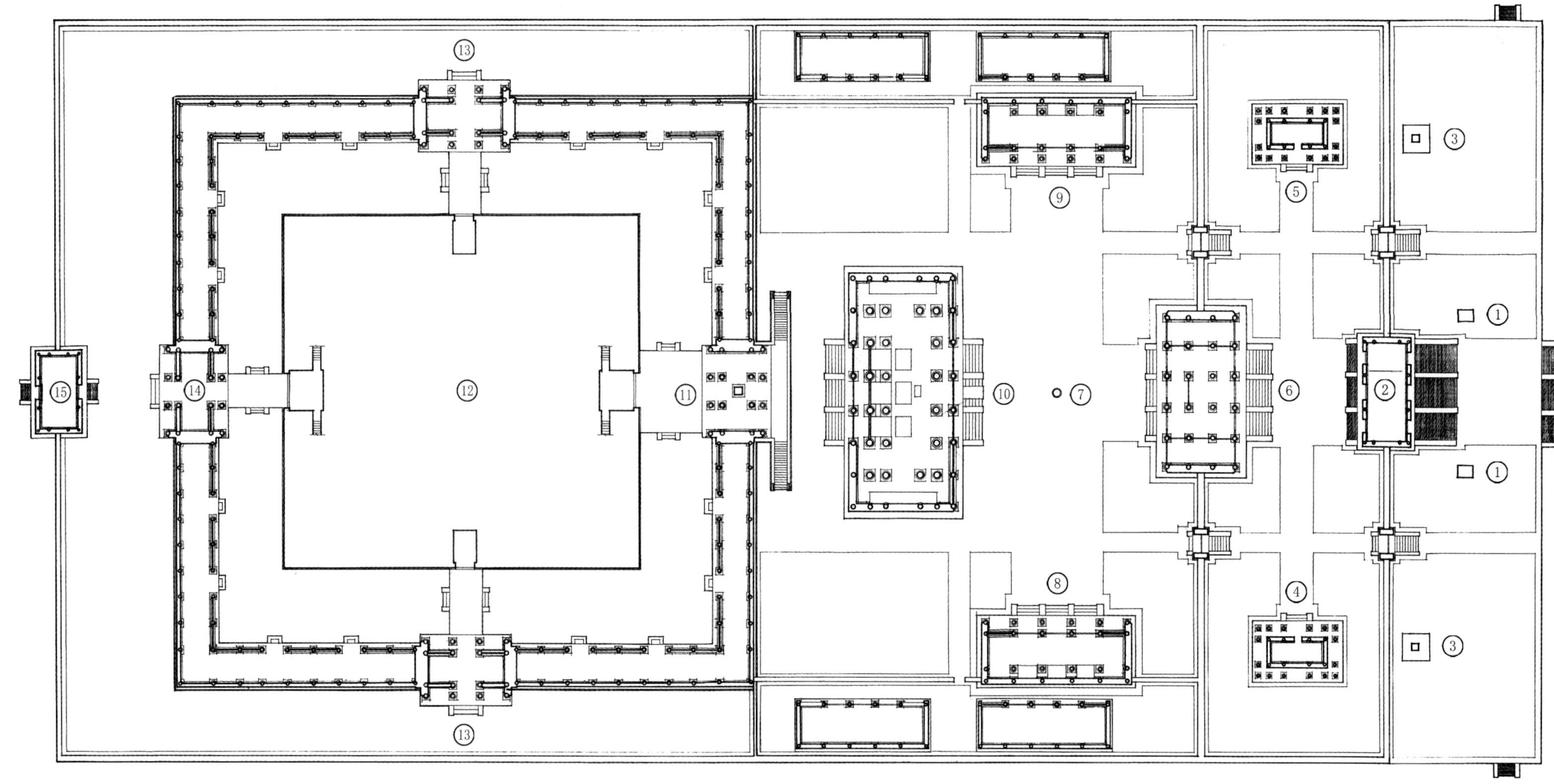

1 石狮 Stone Lions
2 山门 Shanmen (Gate Hall)
3 幢杆支石 Supporting Stones of Flagpole
4 鼓楼 Gulou (Drum-tower)
5 钟楼 Zhonglou (Bell-tower)
6 天王殿 Tianwangdian (Hall of Heavenly King)
7 铁香炉 Incense Burner
8 胜因殿 Shengyindian
9 慧心殿 Huixindian
10 宗印殿 Zongyindian
11 前门 Front Gate
12 阇城 Ducheng
13 侧门 Side Gate
14 后门 Back Gate
15 通梵门 Tongfan Gate

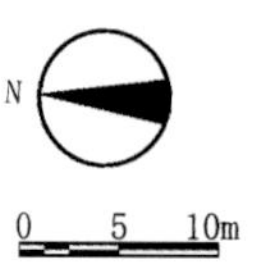

普乐寺总平面图
Site plan of Pulesi

普乐寺山门正立面图
Front elevation of shanmen of Pulesi

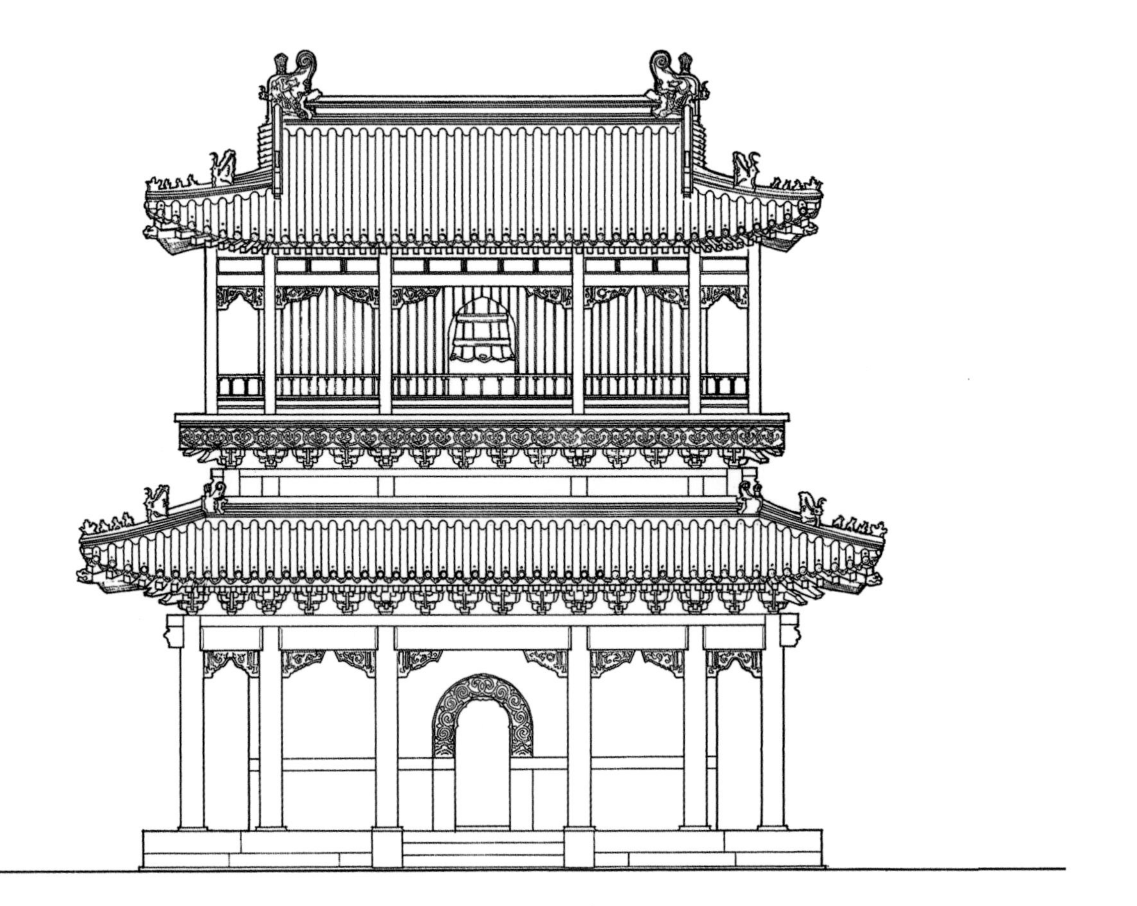

普乐寺钟楼立面图
Elevation of zhonglou of Pulesi

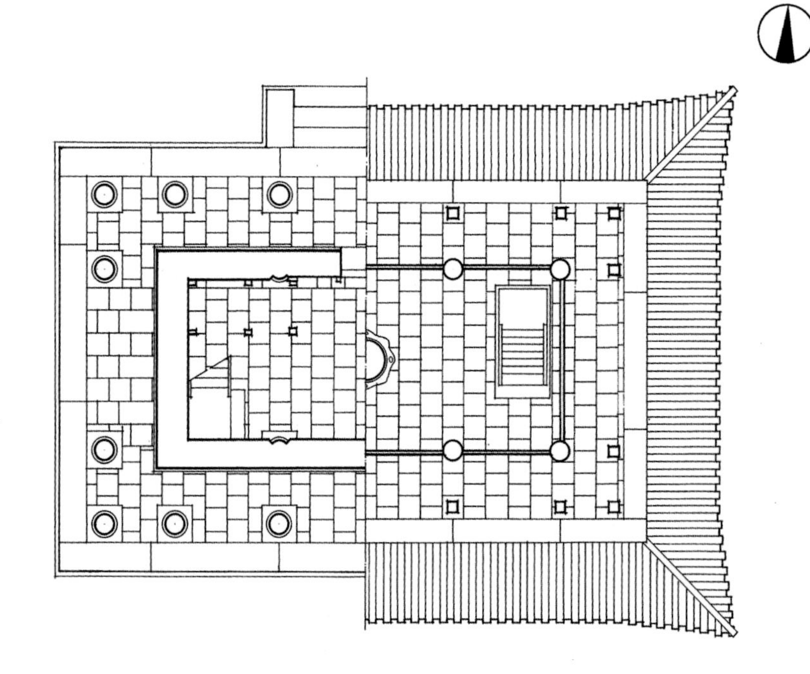

普乐寺钟楼平面图
Plan of zhonglou of Pulesi

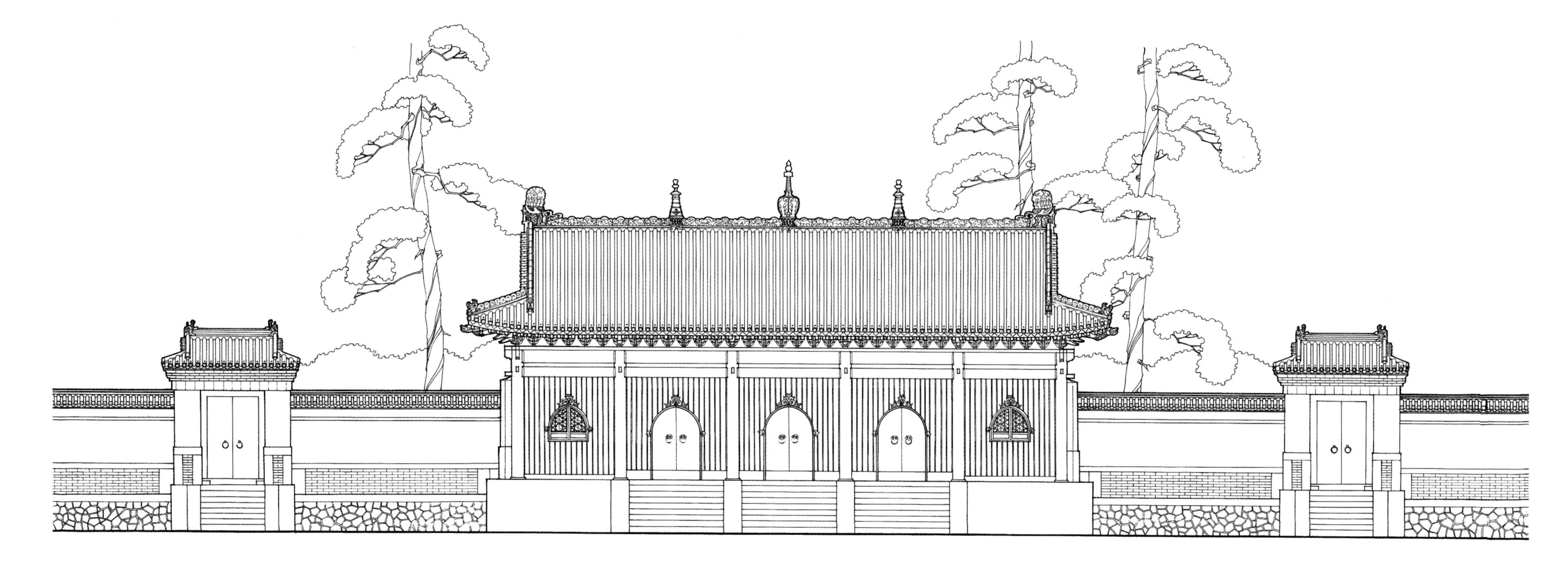

普乐寺天王殿正立面图
Front elevation of Tianwangdian of Pulesi

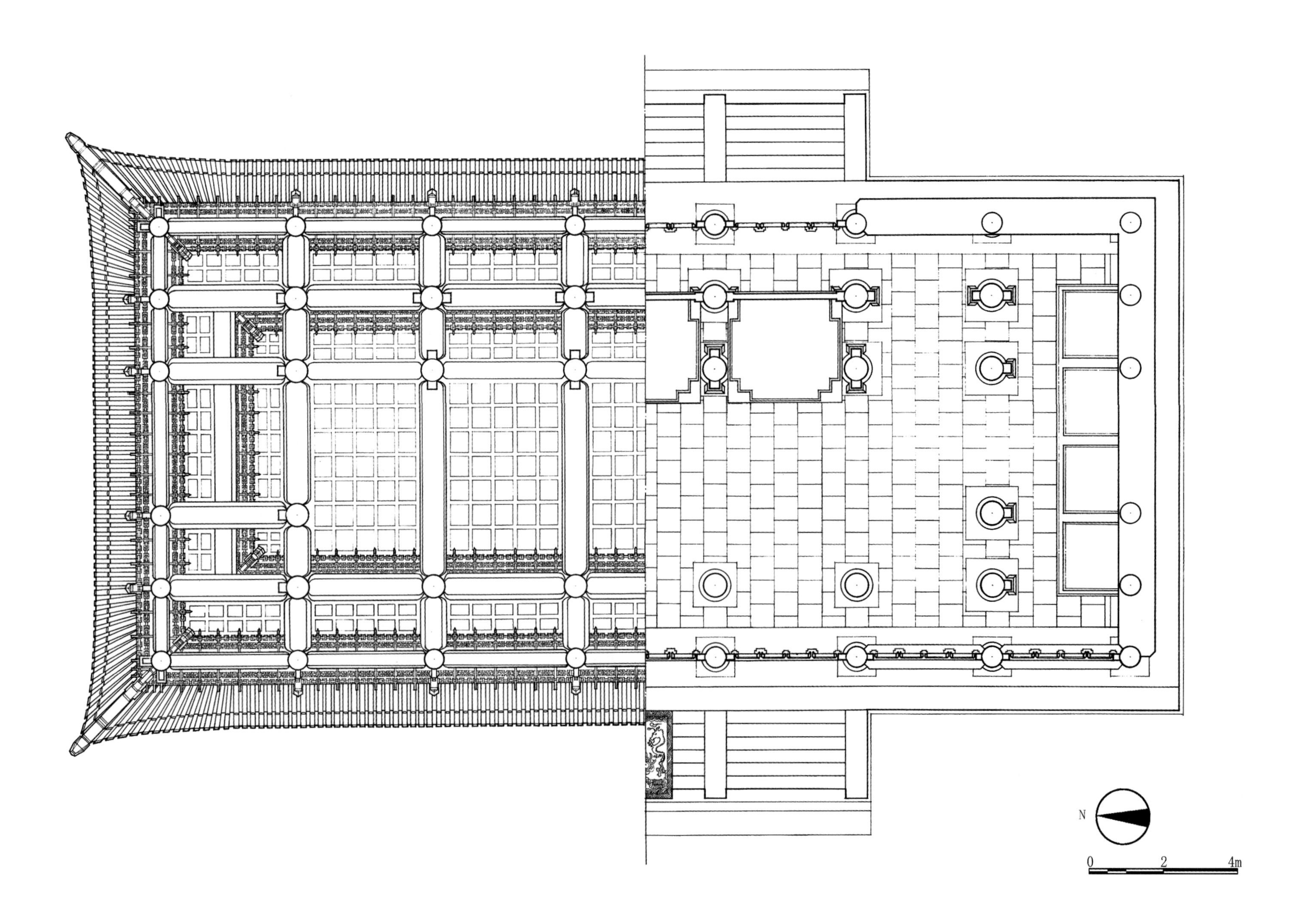

普乐寺宗印殿平面图、天花仰视图
Plan and reflected ceiling plan of Zongyindian of Pulesi

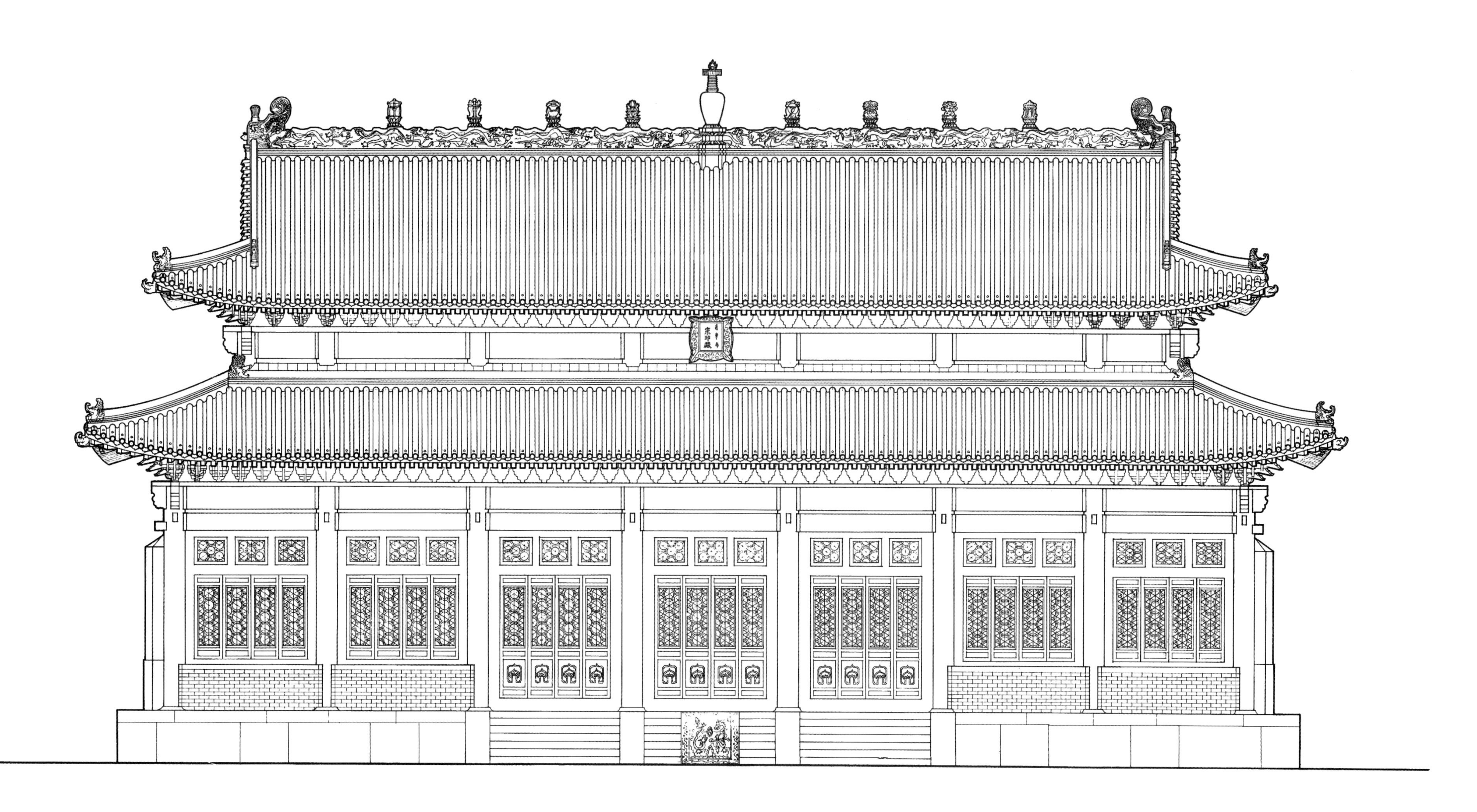

普乐寺宗印殿正立面图
Front elevation of Zongyindian of Pulesi

普乐寺宗印殿纵剖面图

Longitudinal section of Zongyindian of Pulesi

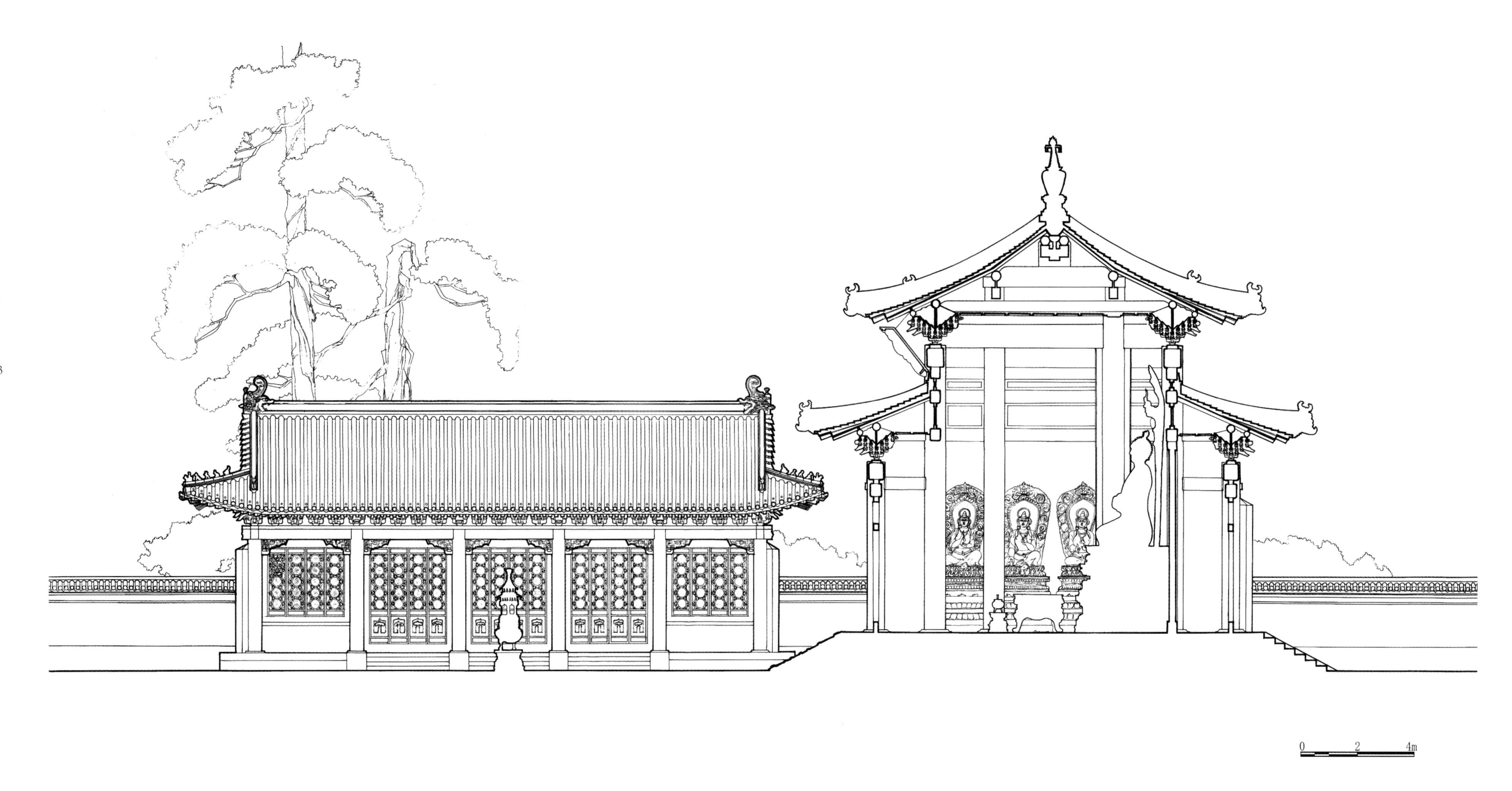

普乐寺宗印殿横剖面图、配殿正立面图
Cross-section of Zongyindian and front elevation of peidian of Pulesi

普乐寺中轴线前半部剖面图
Section along central-axis of front half of Pulesi

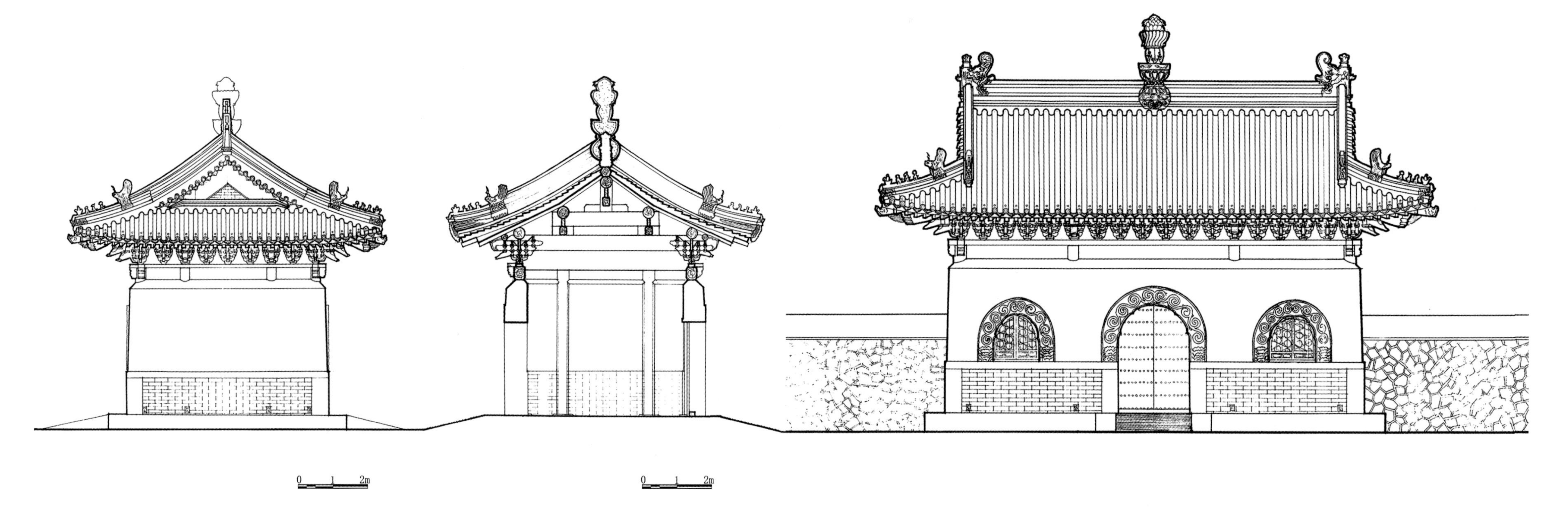

普乐寺后山门侧立面图
Side elevation of rear shanmen of Pulesi

普乐寺后山门横剖面图
Cross-section of rear shanmen of Pulesi

普乐寺后山门正立面图
Front elevation of rear shanmen of Pulesi

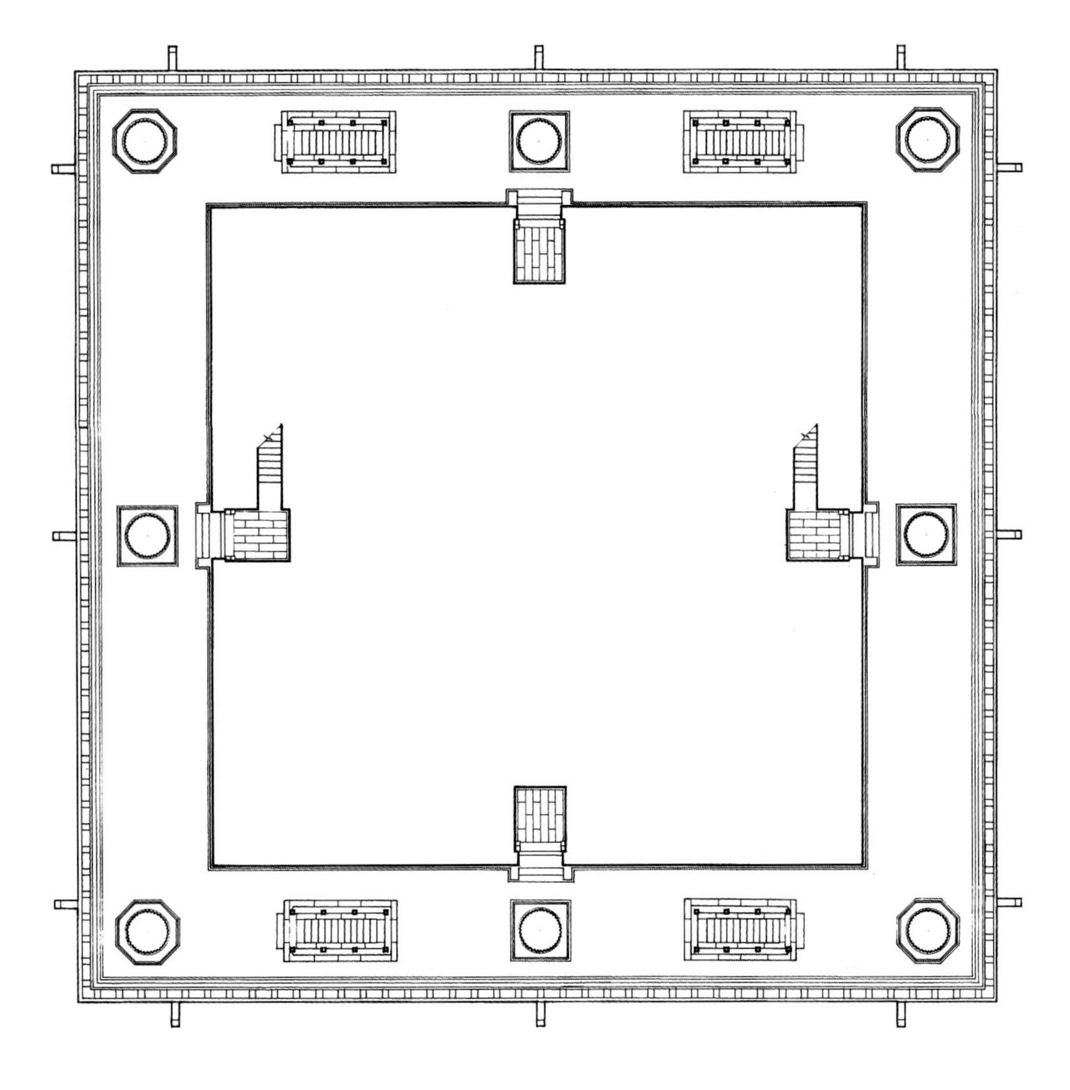

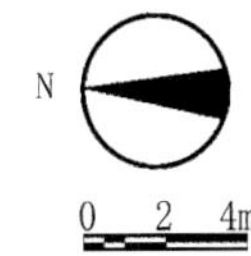

普乐寺阇城二层平面图
Plan of second floor of Ducheng of Pulesi

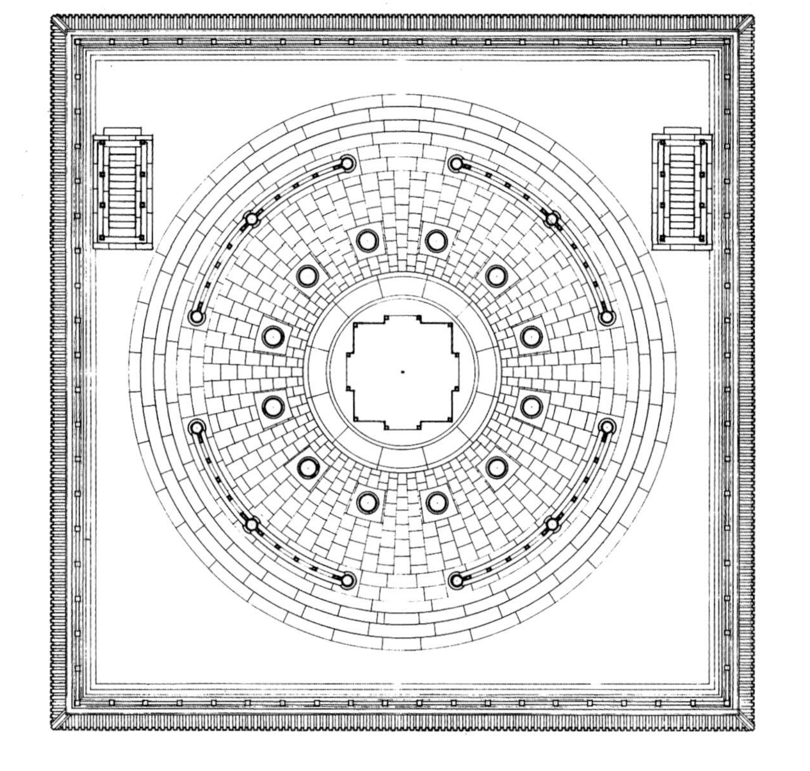

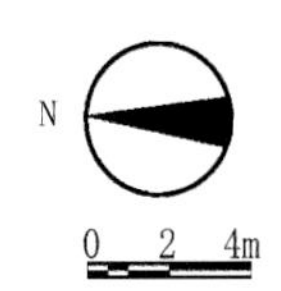

普乐寺阇城三层平面图
Plan of third floor of Ducheng of Pulesi

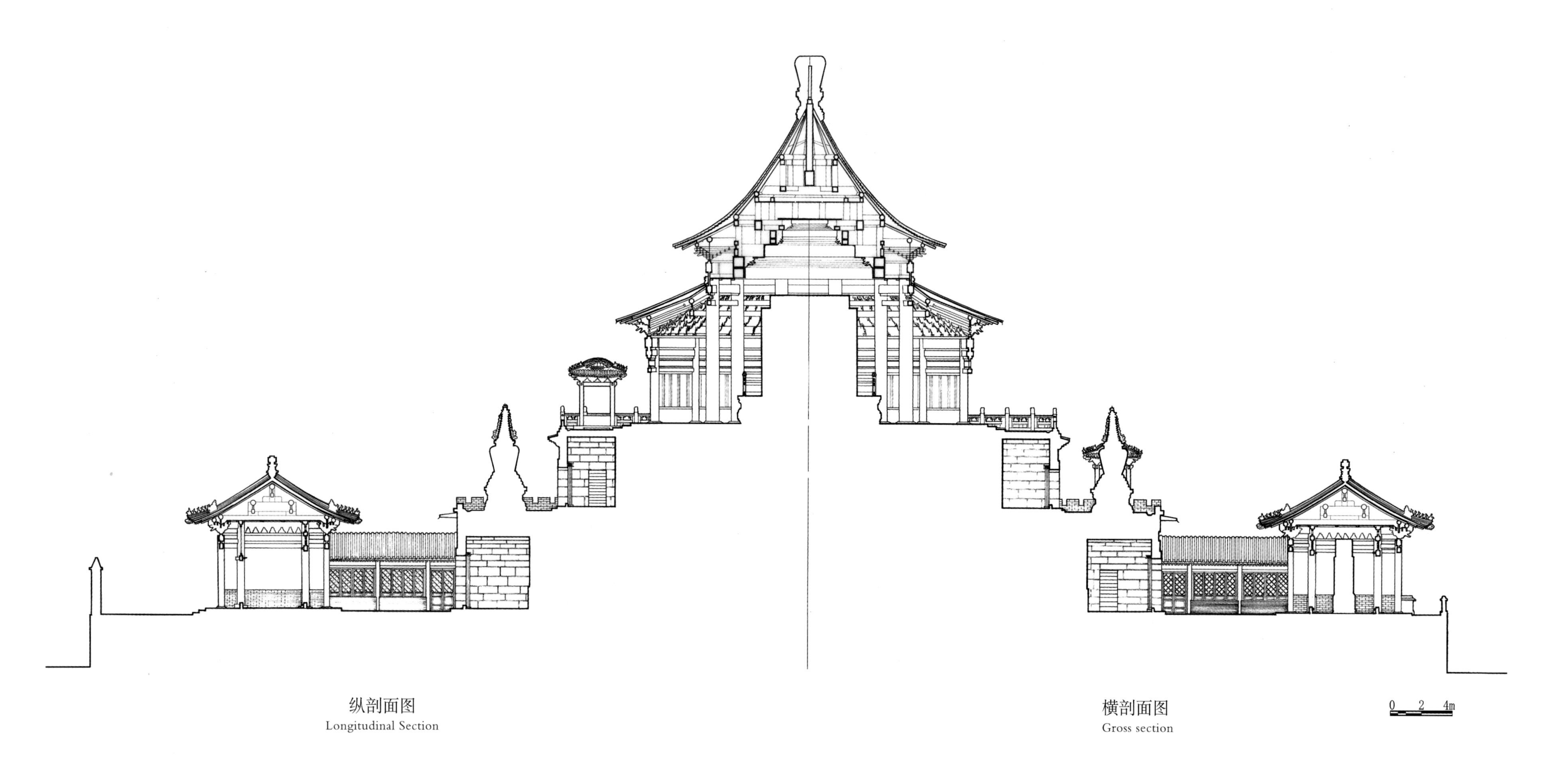

纵剖面图
Longitudinal Section

横剖面图
Gross section

普乐寺阇城剖面图
Section of Ducheng of Pulesi

普乐寺旭光阁东立面图
East elevation of Xuguangge of Pulesi

普陀宗乘之庙
Putuo zongcheng Temple

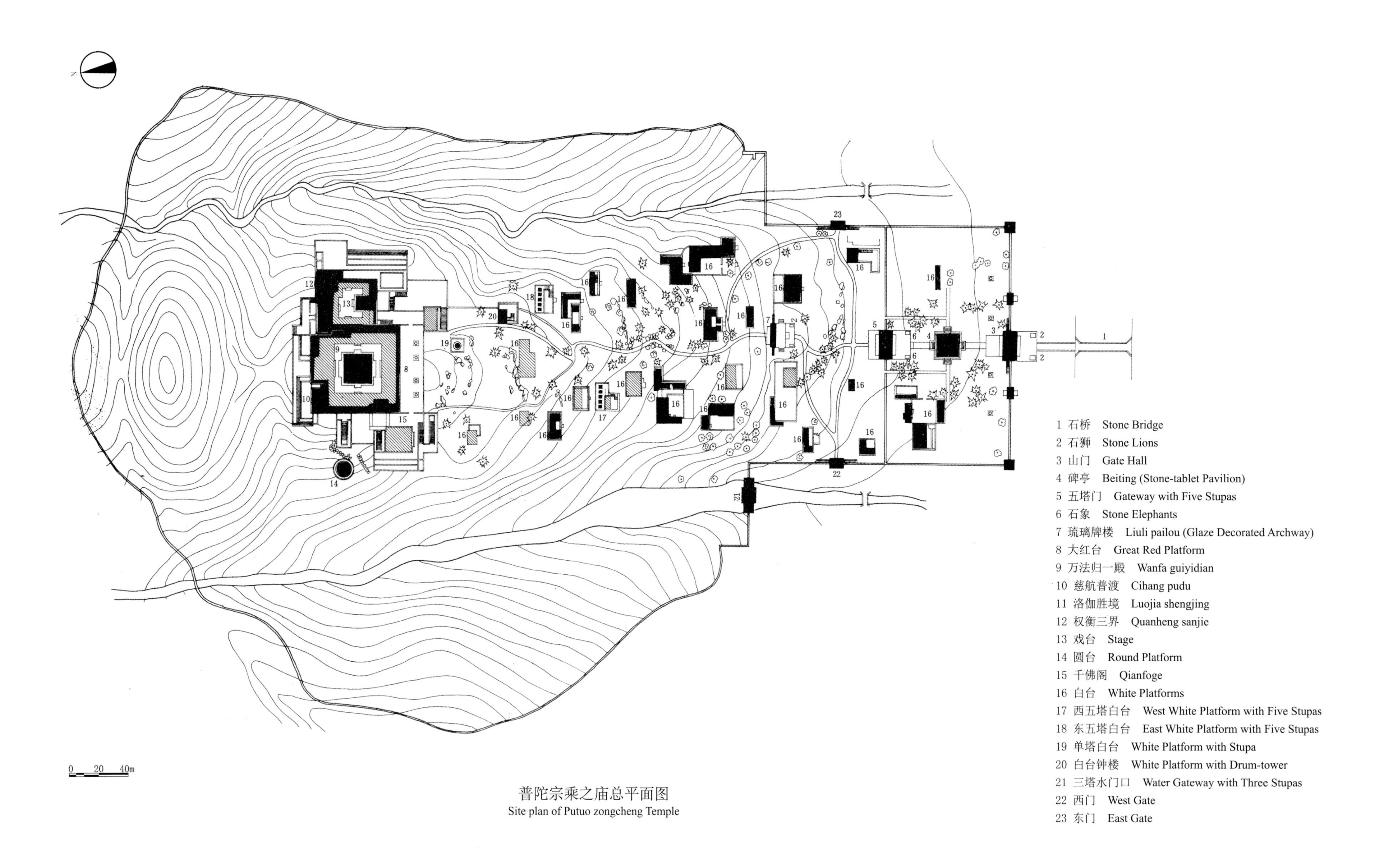

普陀宗乘之庙总平面图
Site plan of Putuo zongcheng Temple

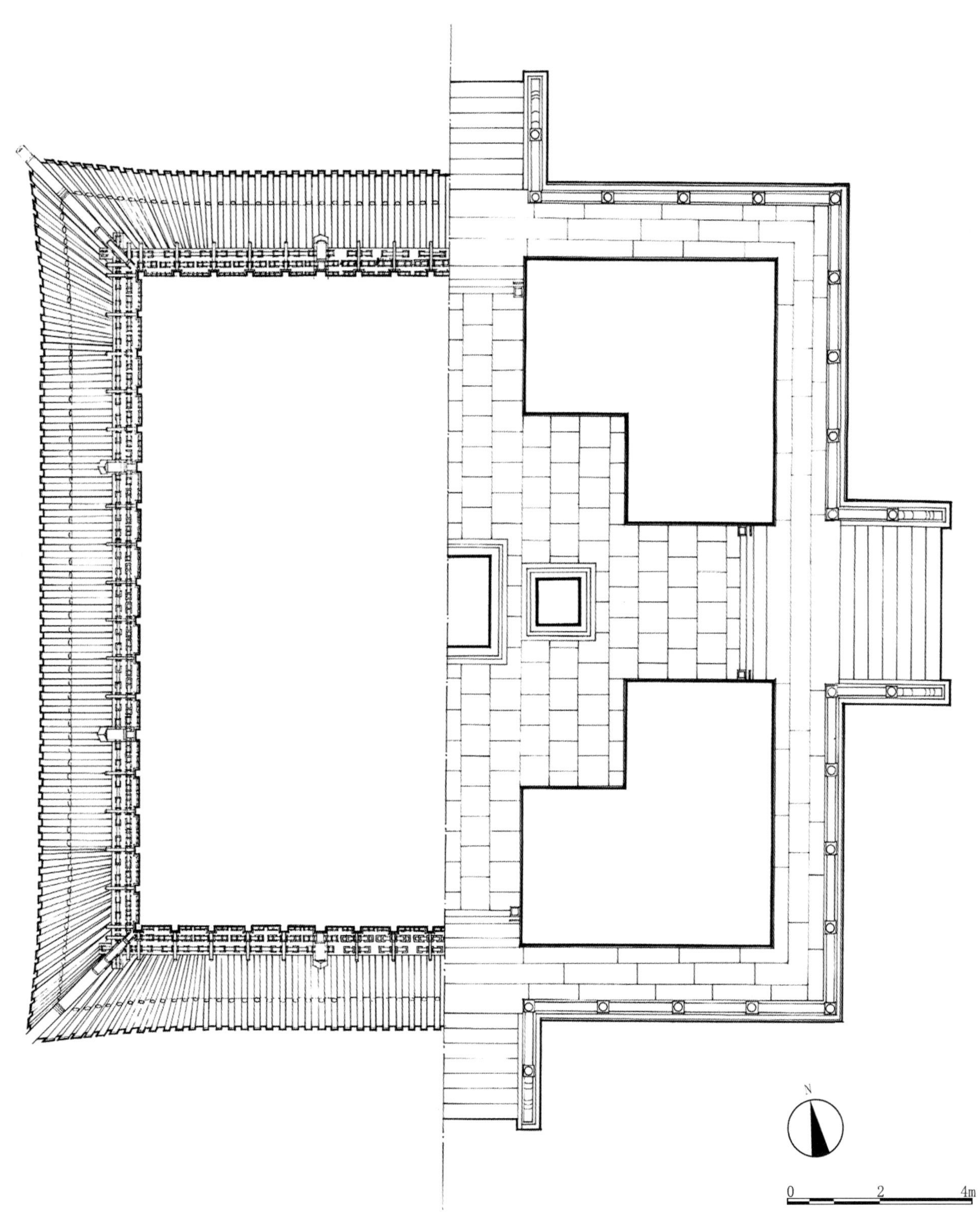

普陀宗乘之庙碑亭平面图、仰视图

Plan and plan as seen from below of beiting of Putuo zongcheng Temple

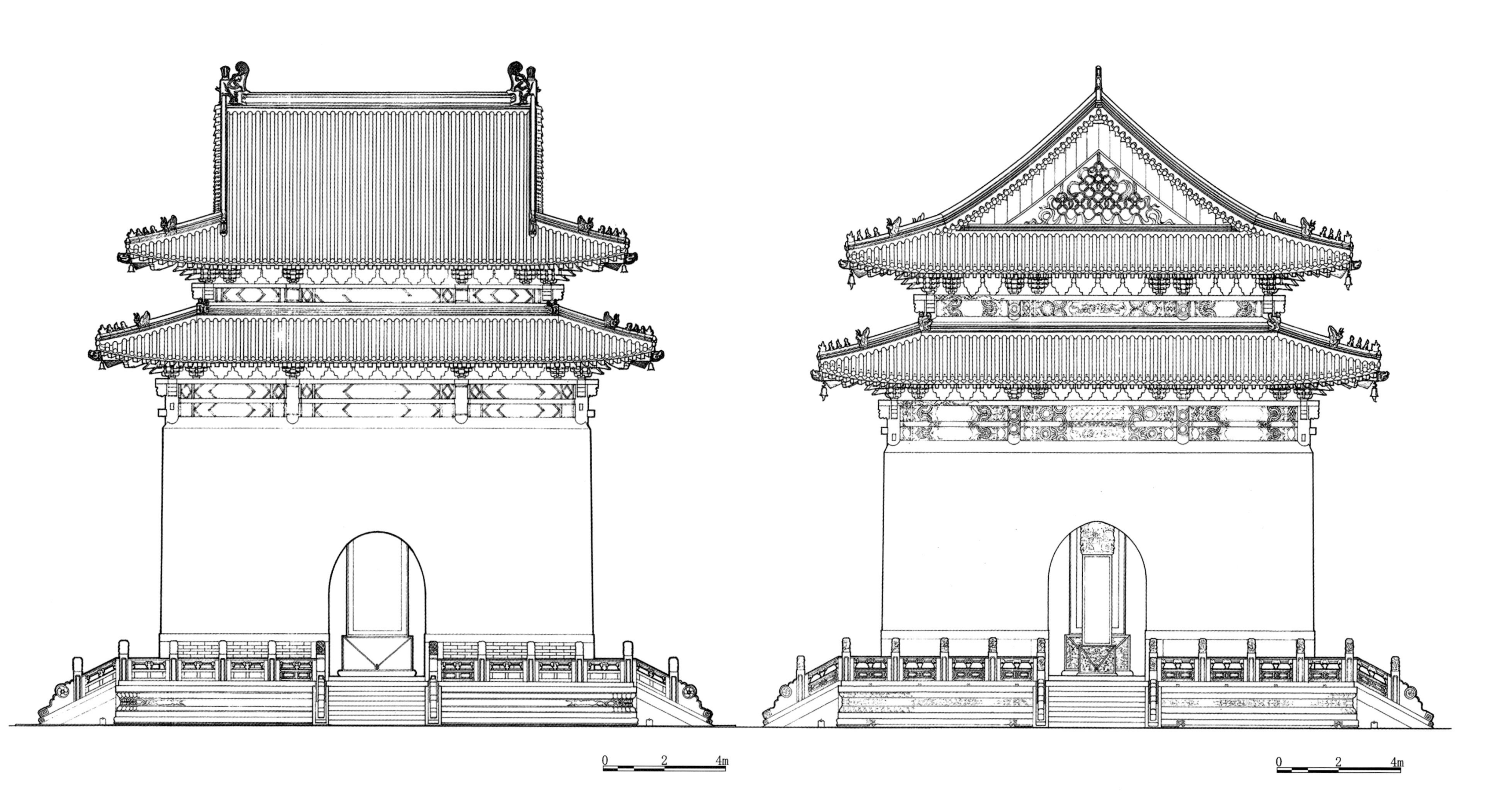

普陀宗乘之庙碑亭正立面图
Front elevation of beiting of Putuo zongcheng Temple

普陀宗乘之庙碑亭侧立面图
Side elevation of beiting of Putuo zongcheng Temple

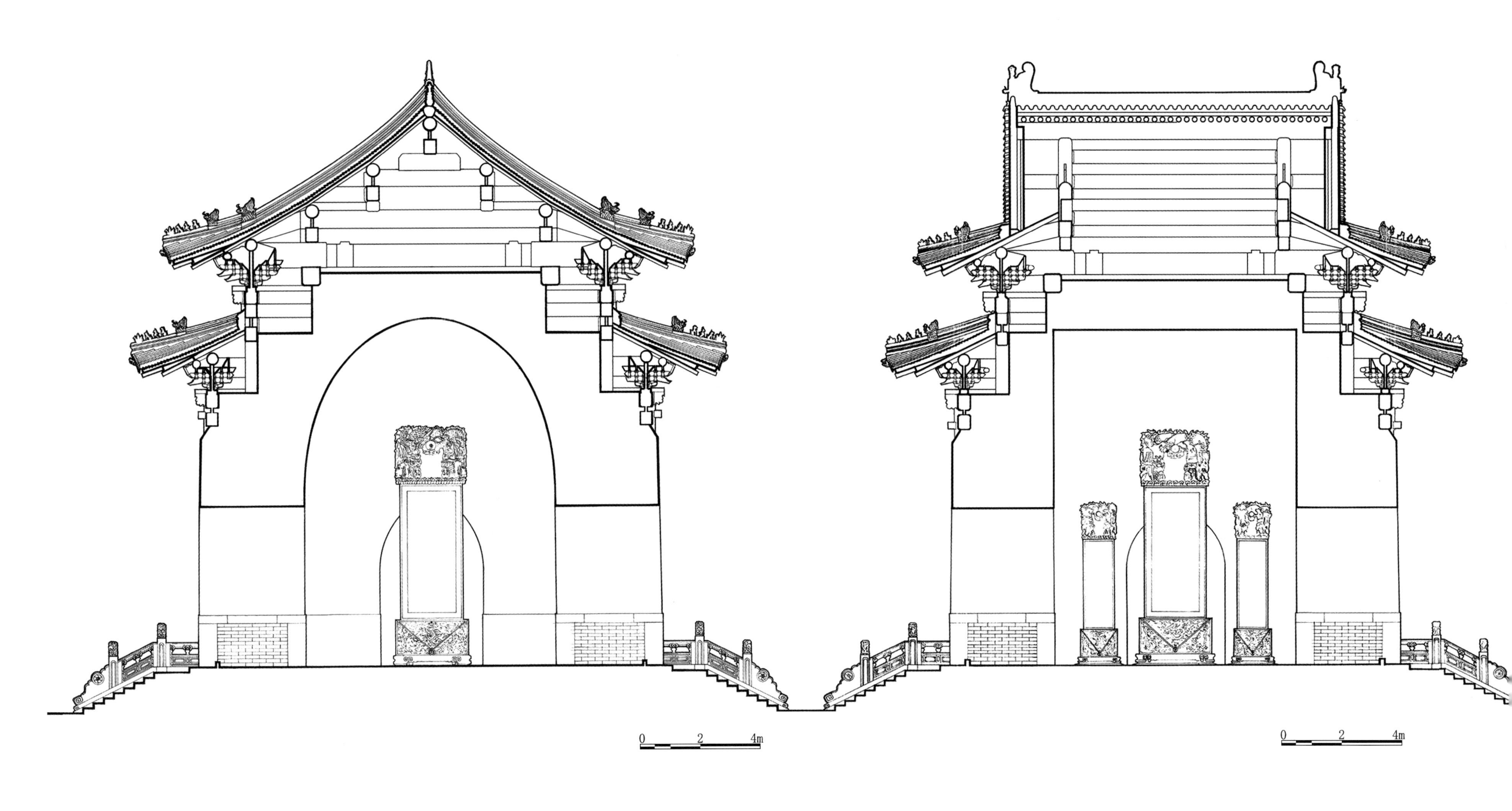

普陀宗乘之庙碑亭横剖面图
Cross-section of beiting of Putuo zongcheng Temple

普陀宗乘之庙碑亭纵剖面图
Longitudinal section of beiting of Putuo zongcheng Temple

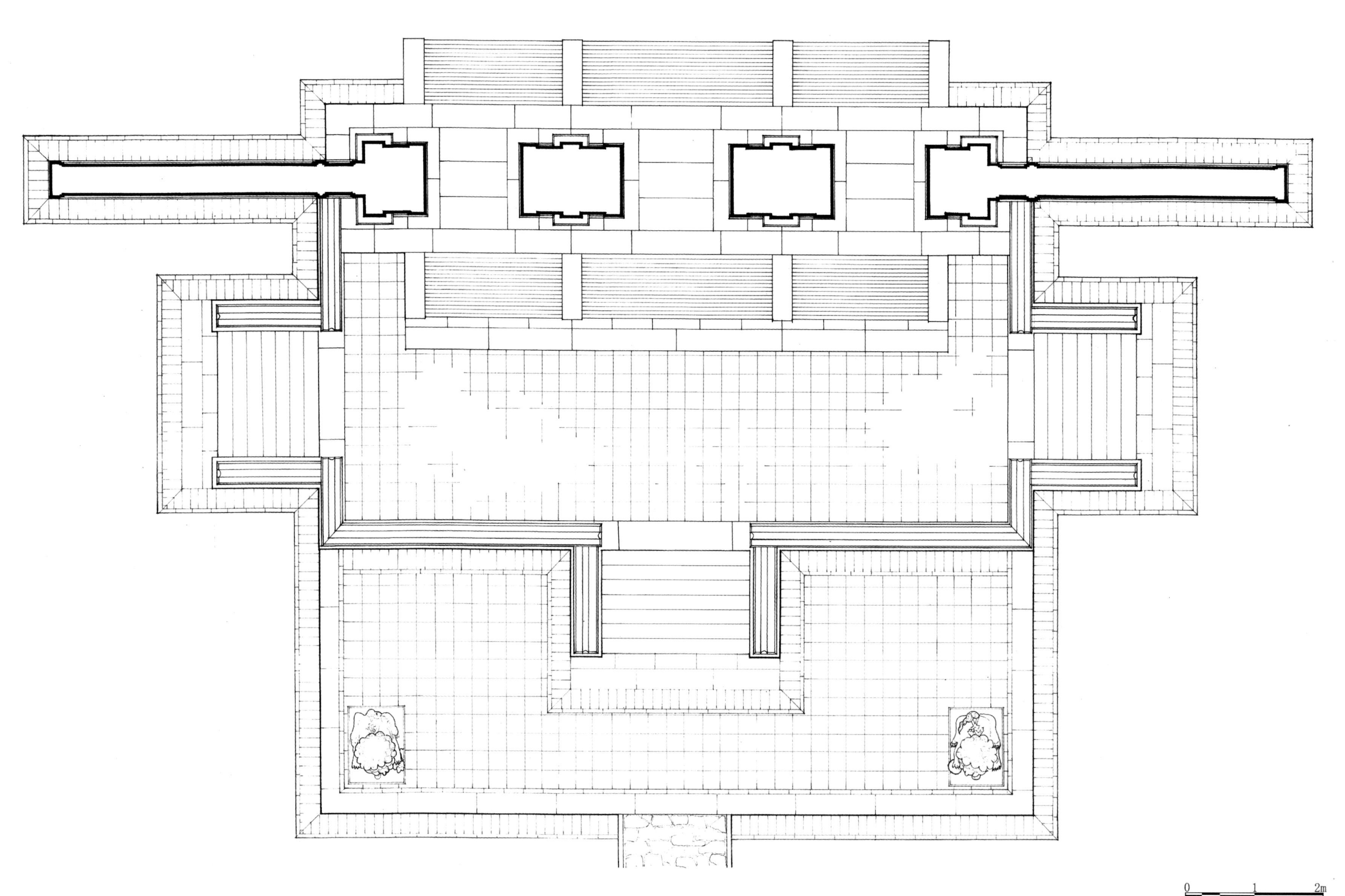

普陀宗乘之庙琉璃牌楼平面图
Plan of Liuli pailou of Putuo zongcheng Temple

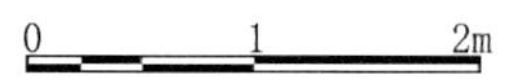

普陀宗乘之庙琉璃牌楼正立面图

Front elevation of Liuli pailou of Putuo zongcheng Temple

普陀宗乘之庙琉璃牌楼剖面图
Section of Liuli pailou of Putuo zongcheng Temple

普陀宗乘之庙琉璃牌楼侧立面图
Side elevation of Liuli pailou of Putuo zongcheng Temple

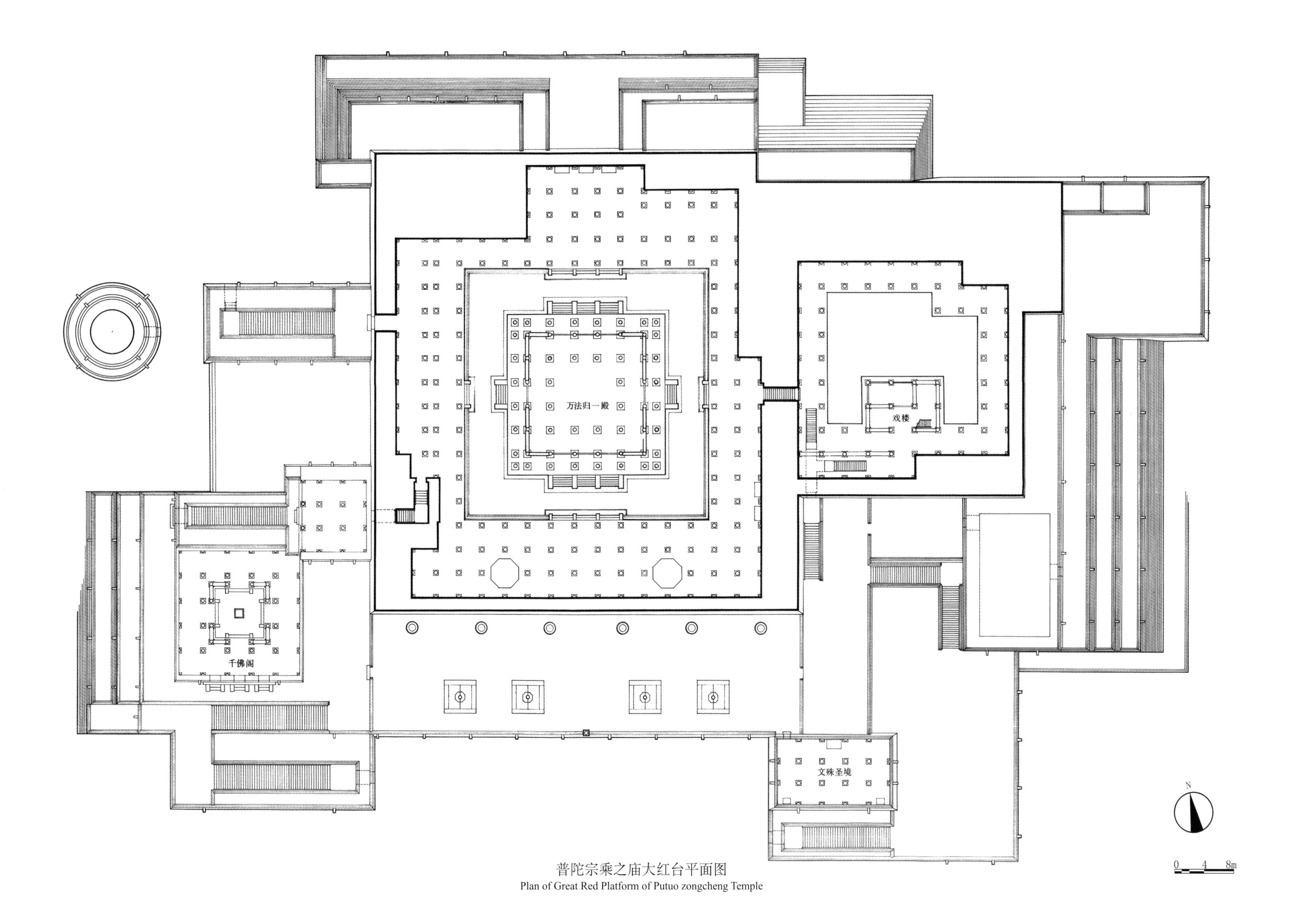

普陀宗乘之庙大红台平面图
Plan of Great Red Platform of Putuo zongcheng Temple

普陀宗乘之庙大红台南立面图
South elevation of Great Red Platform of Putuo zongcheng Temple

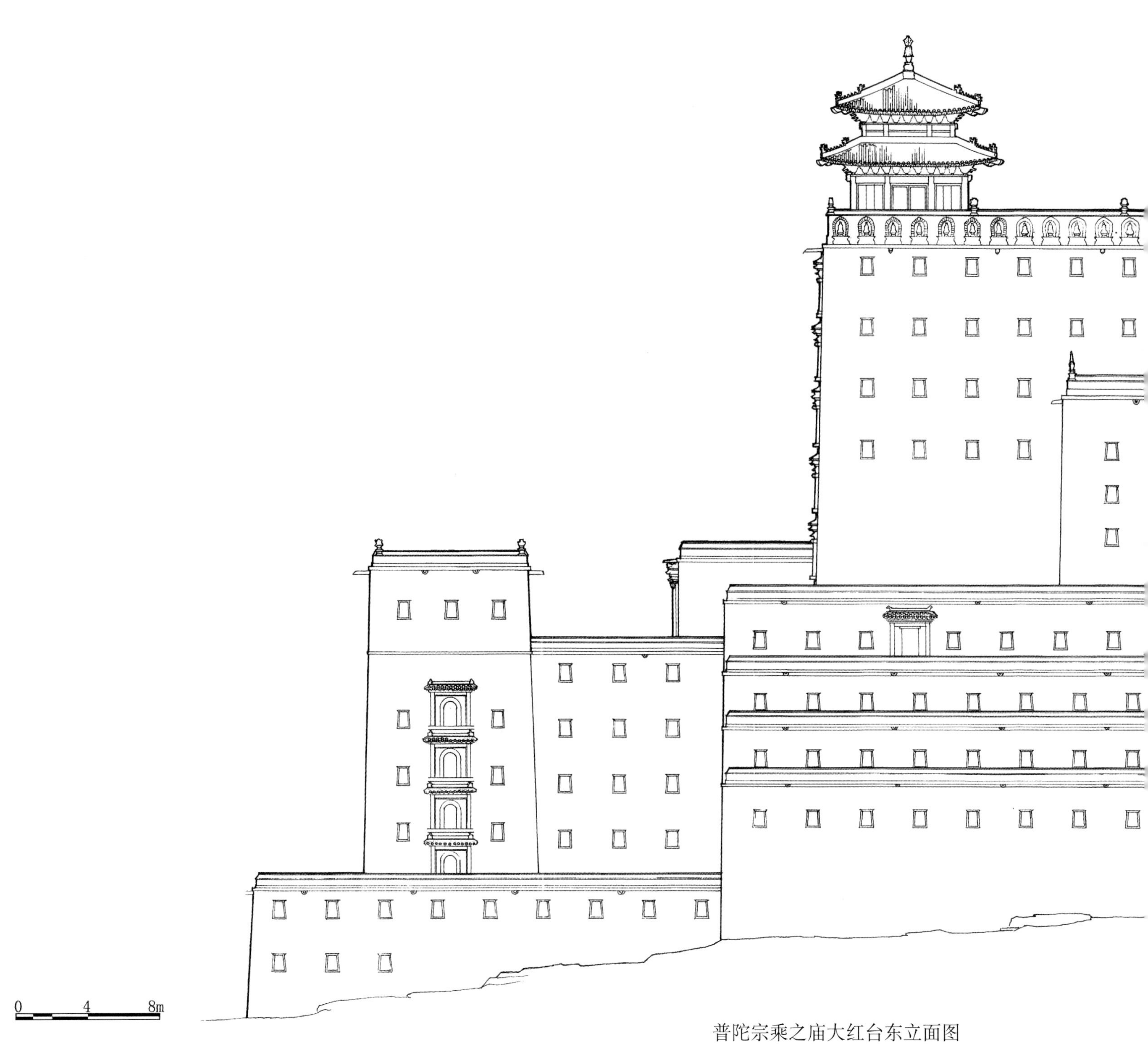

普陀宗乘之庙大红台东立面图
East elevation of Great Red Platform of Putuo zongcheng Temple

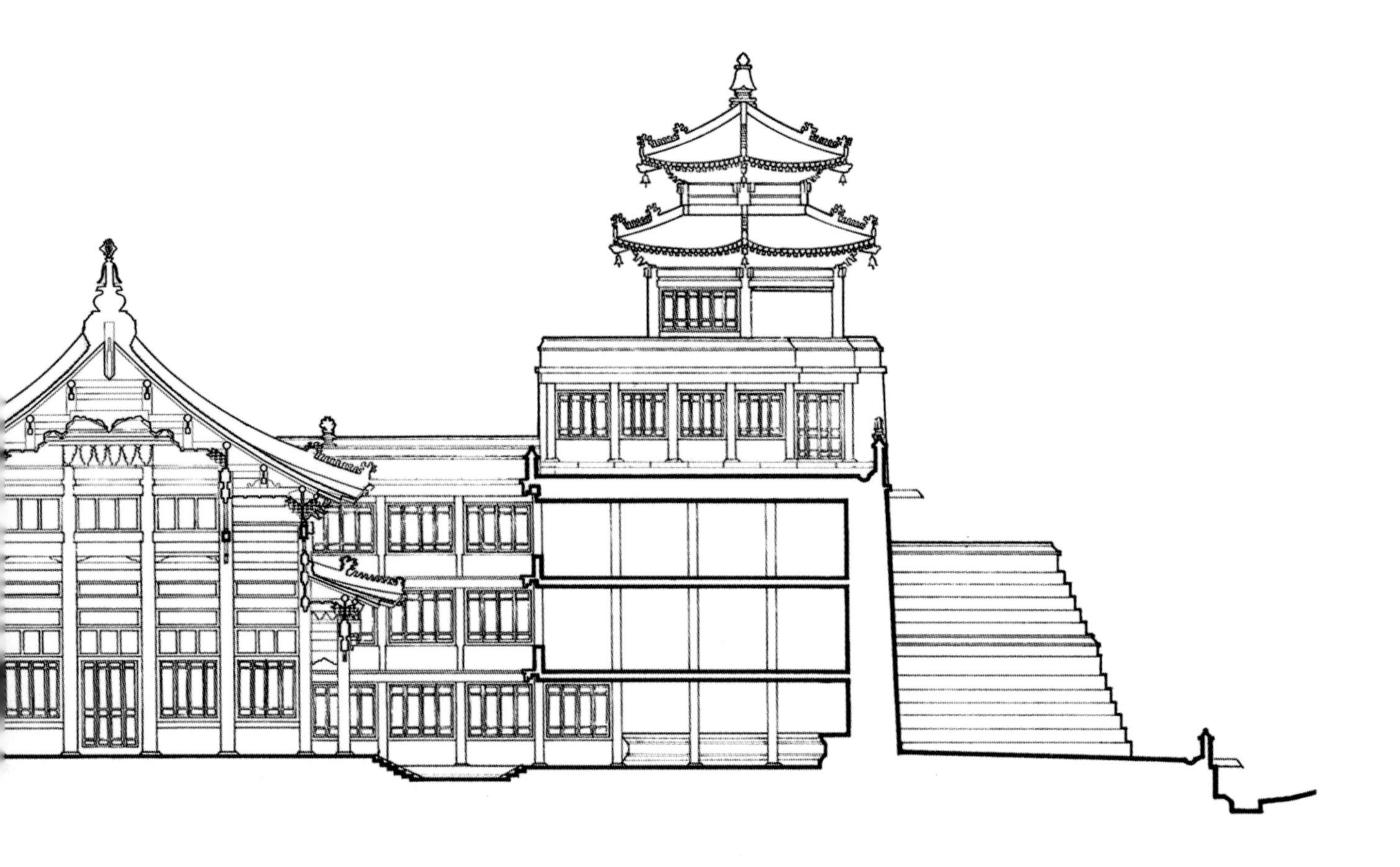

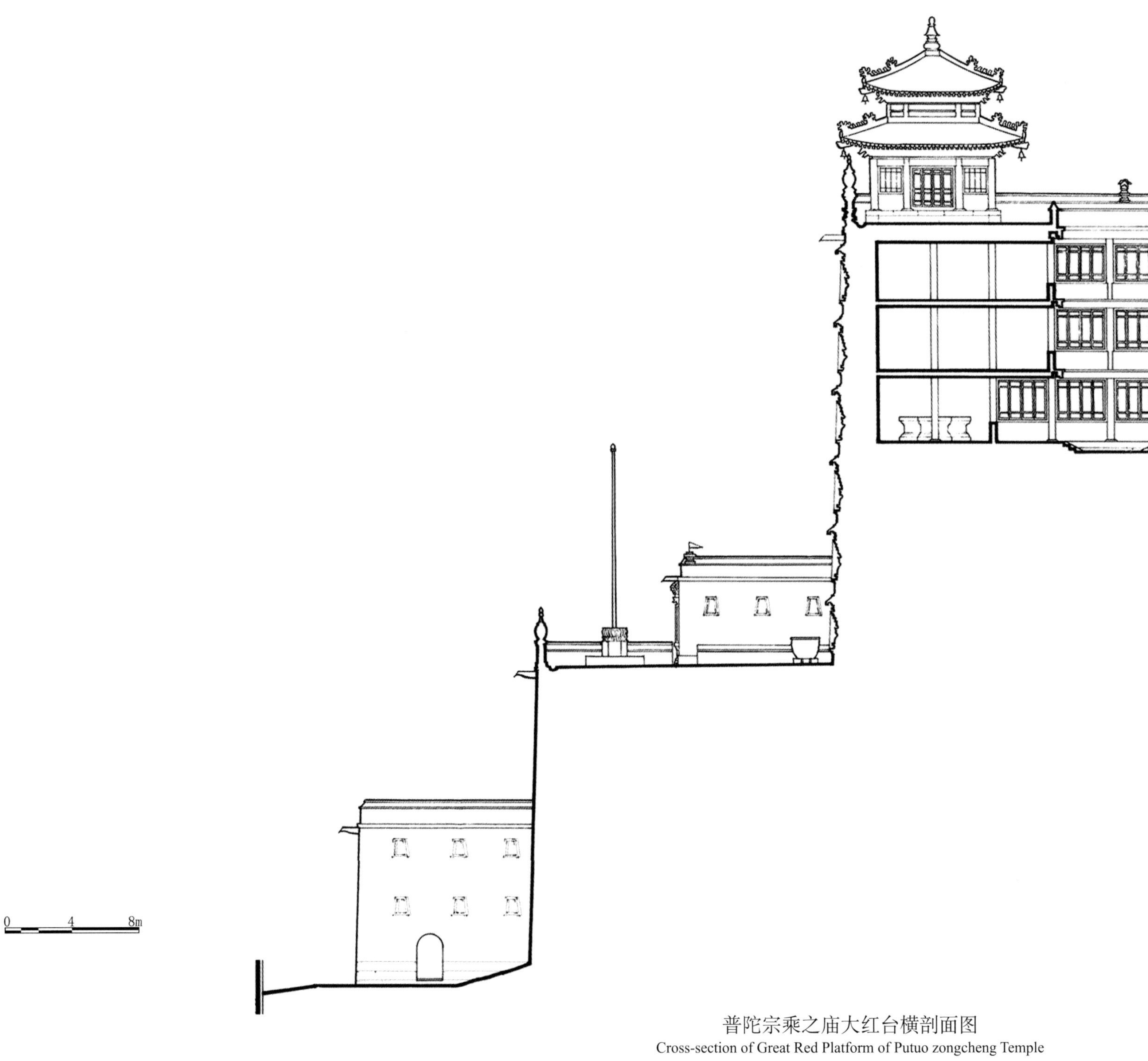

普陀宗乘之庙大红台横剖面图
Cross-section of Great Red Platform of Putuo zongcheng Temple

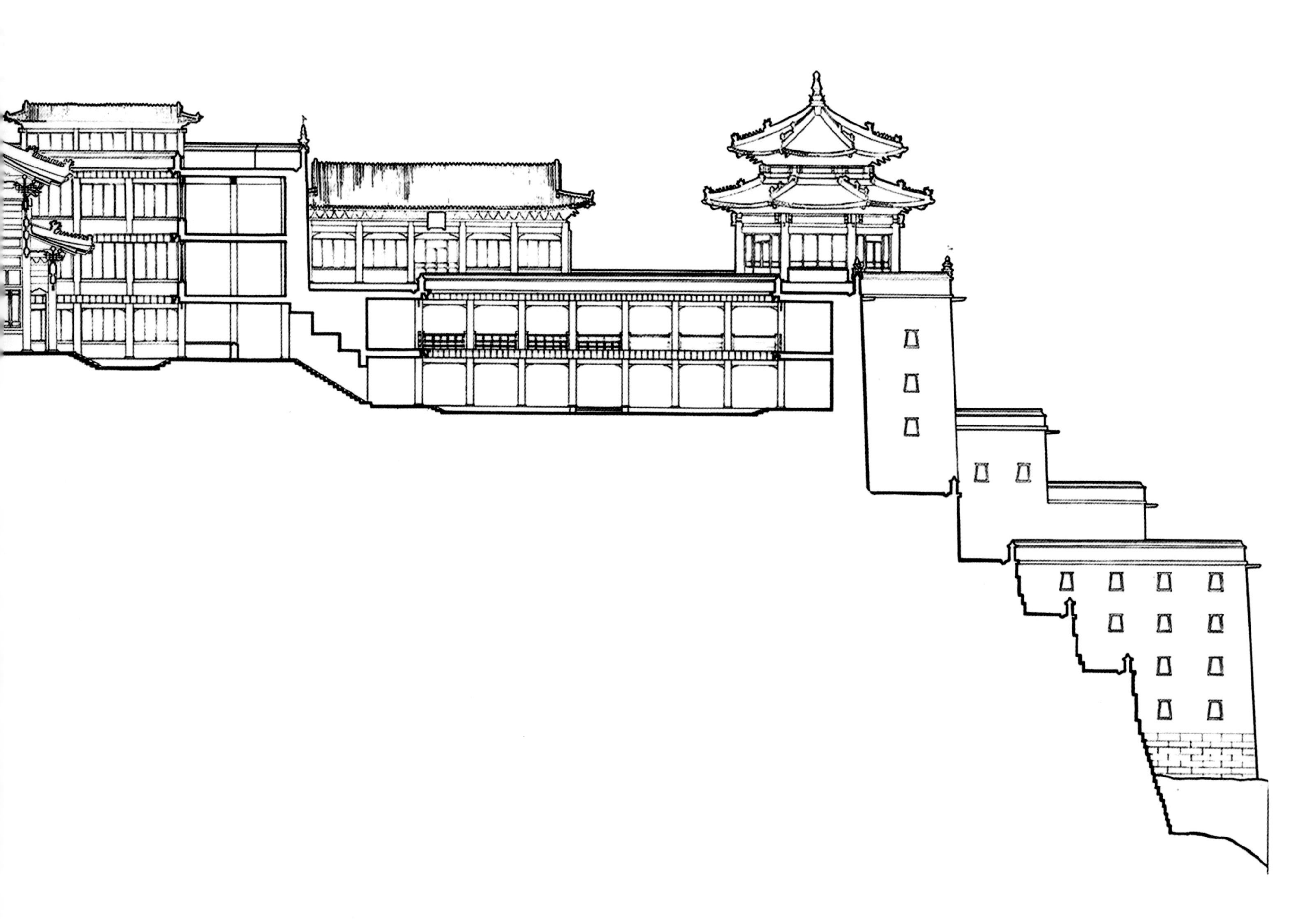

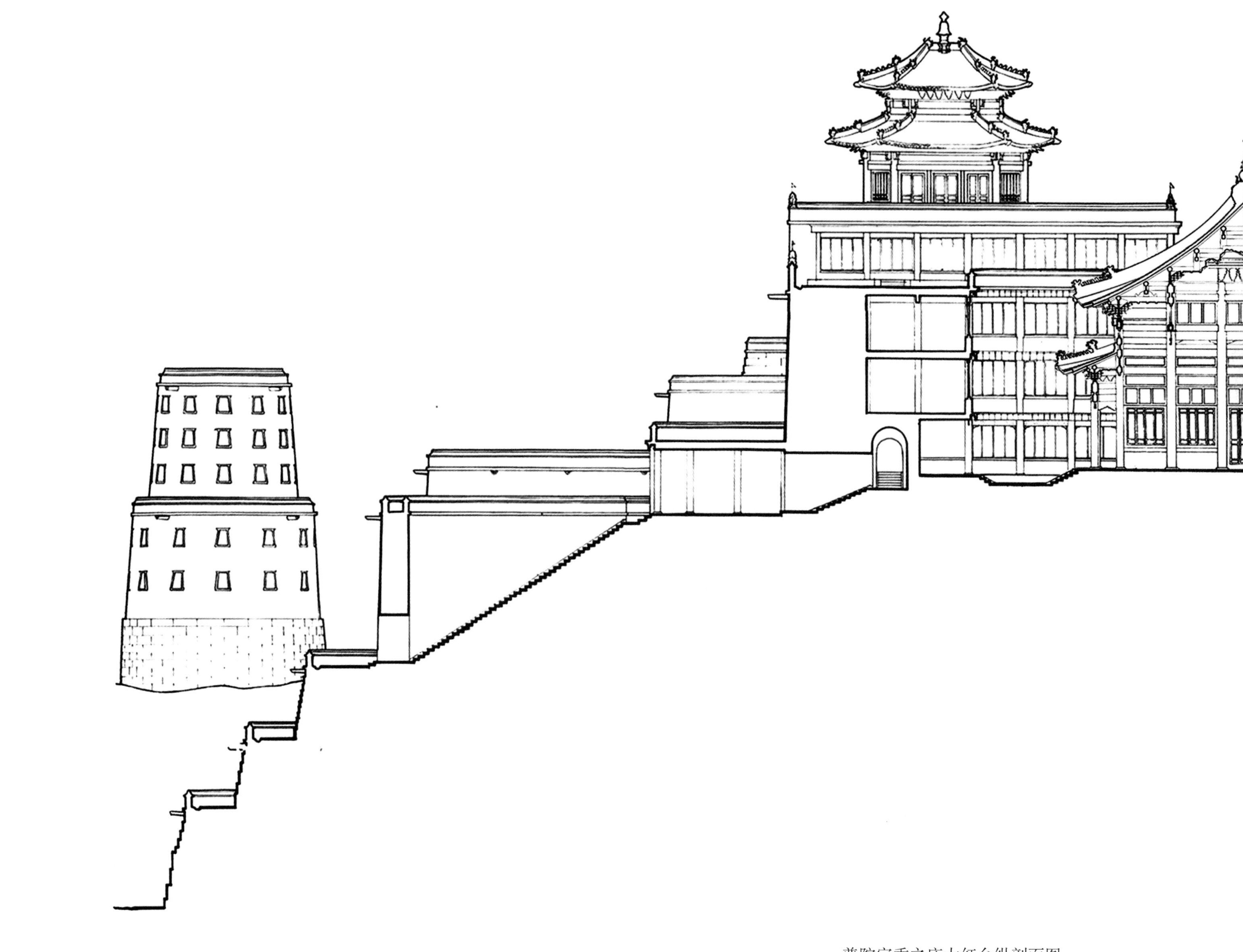

普陀宗乘之庙大红台纵剖面图
Longitudinal section of Great Red Platform of Putuo zongcheng Temple

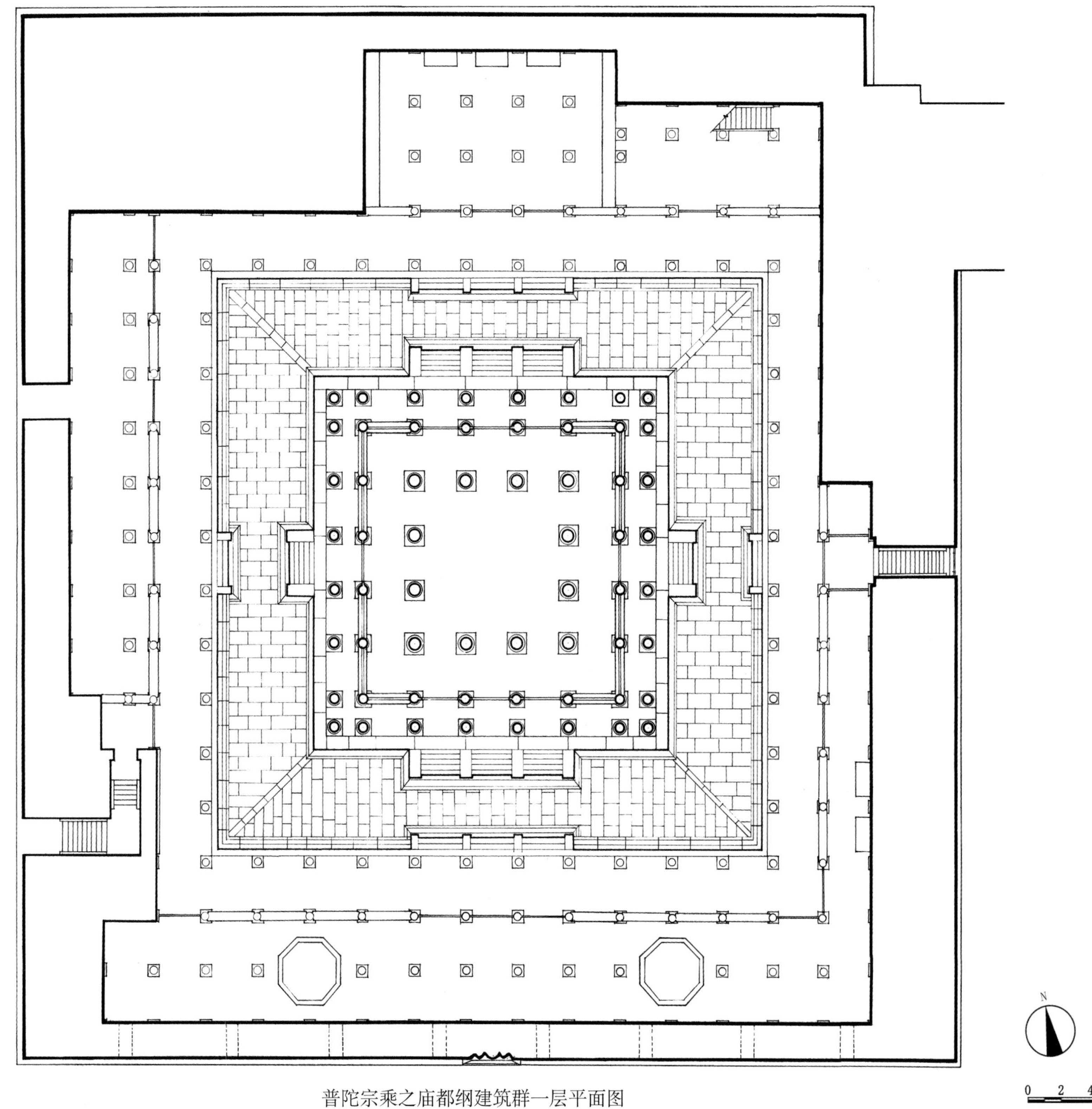

普陀宗乘之庙都纲建筑群一层平面图
Plan of first floor of buildings around Dugang of Putuo zongcheng Temple

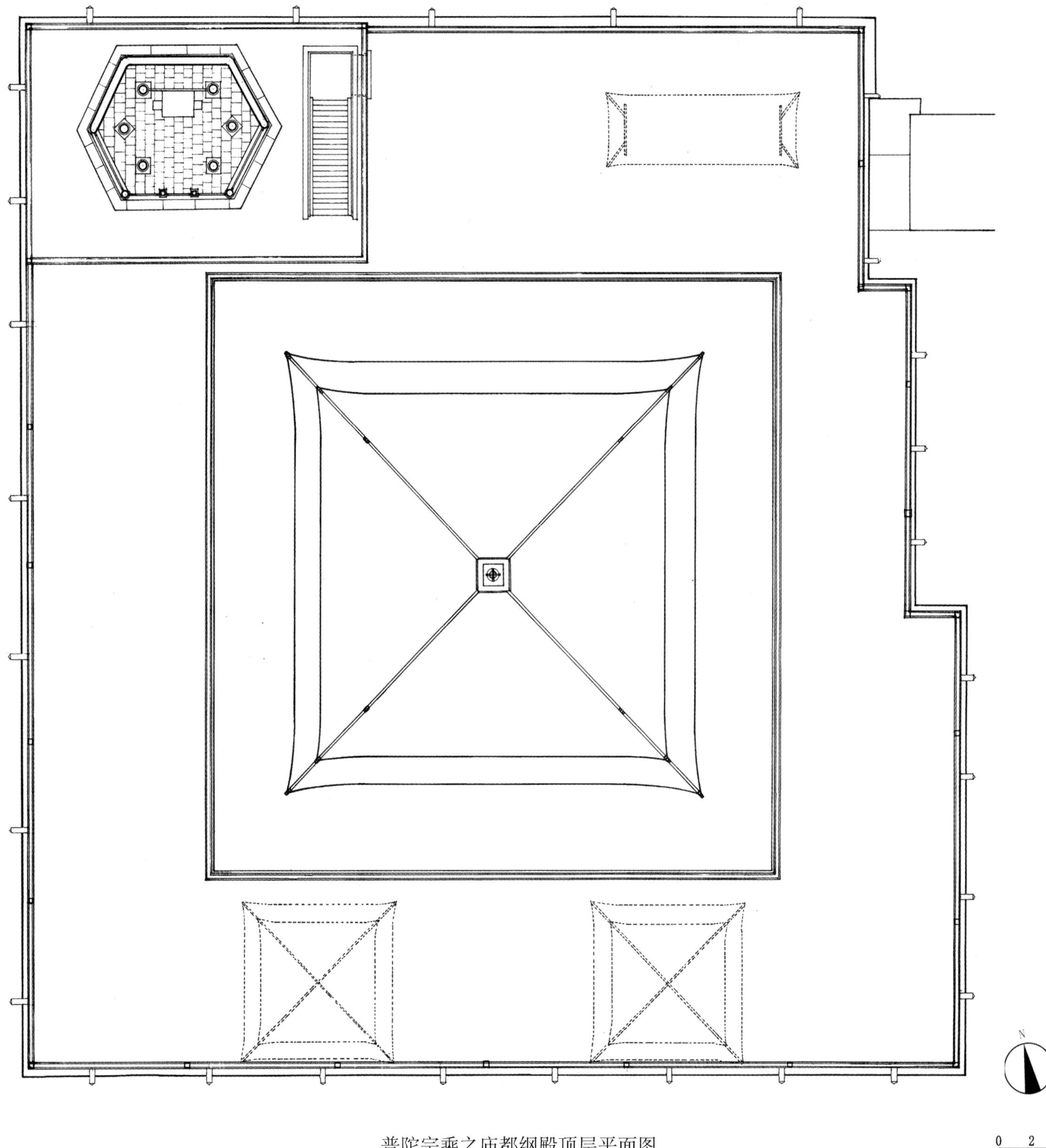

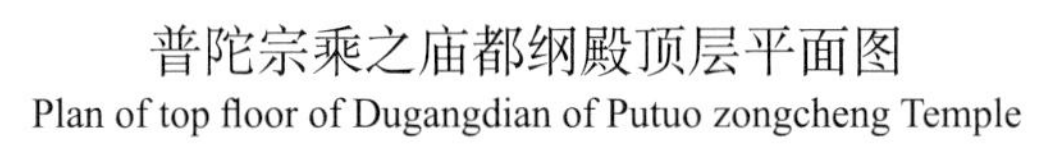
普陀宗乘之庙都纲殿顶层平面图
Plan of top floor of Dugangdian of Putuo zongcheng Temple

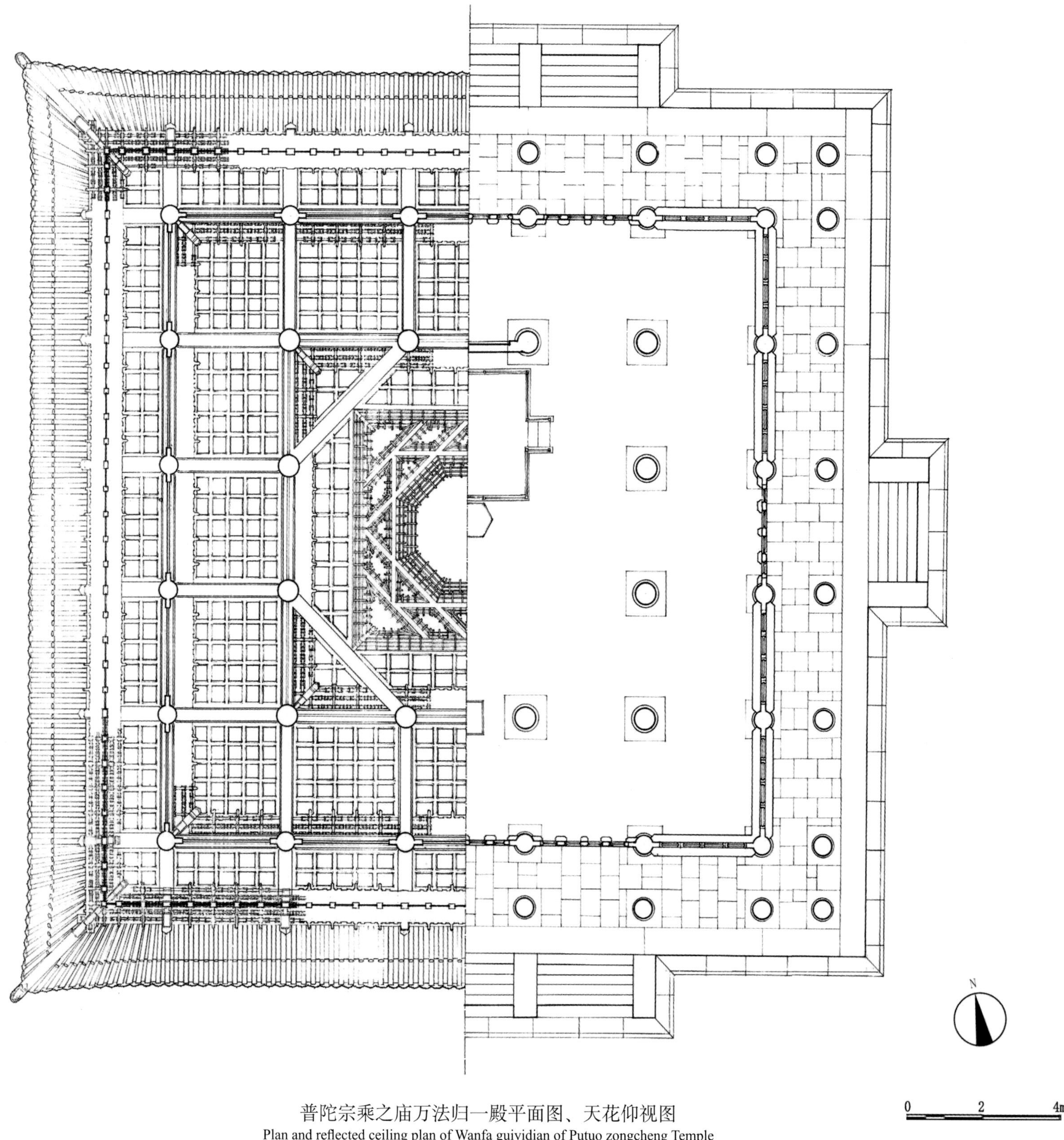

普陀宗乘之庙万法归一殿平面图、天花仰视图

Plan and reflected ceiling plan of Wanfa guiyidian of Putuo zongcheng Temple

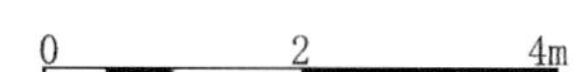

普陀宗乘之庙万法归一殿南立面图

South elevation of Wanfa guiyidian of Putuo zongcheng Temple

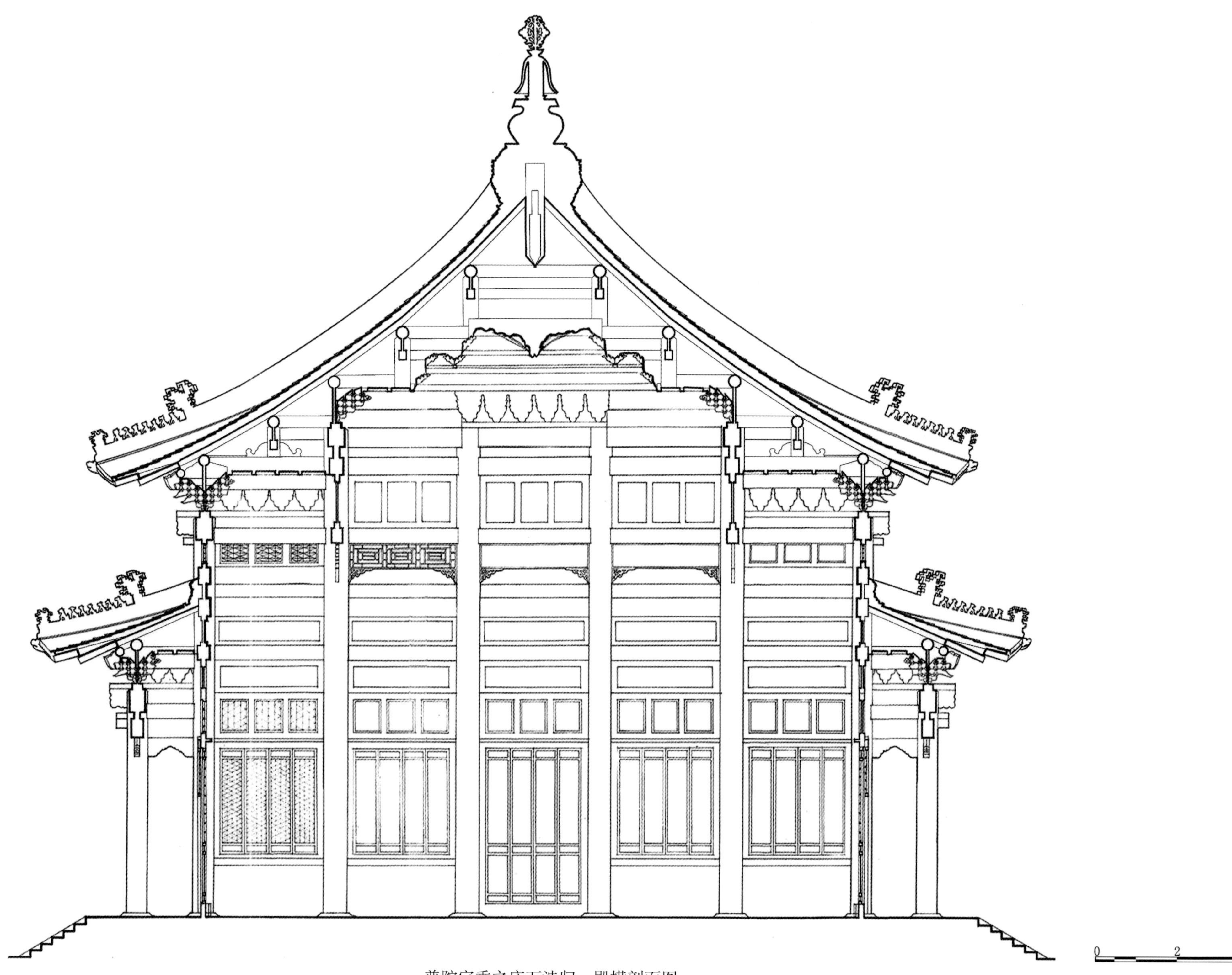

普陀宗乘之庙万法归一殿横剖面图
Cross-section of Wanfa guiyidian of Putuo zongcheng Temple

普陀宗乘之庙慈航普渡南立面图
South elevation of Cihang pudu of Putuo zongcheng Temple

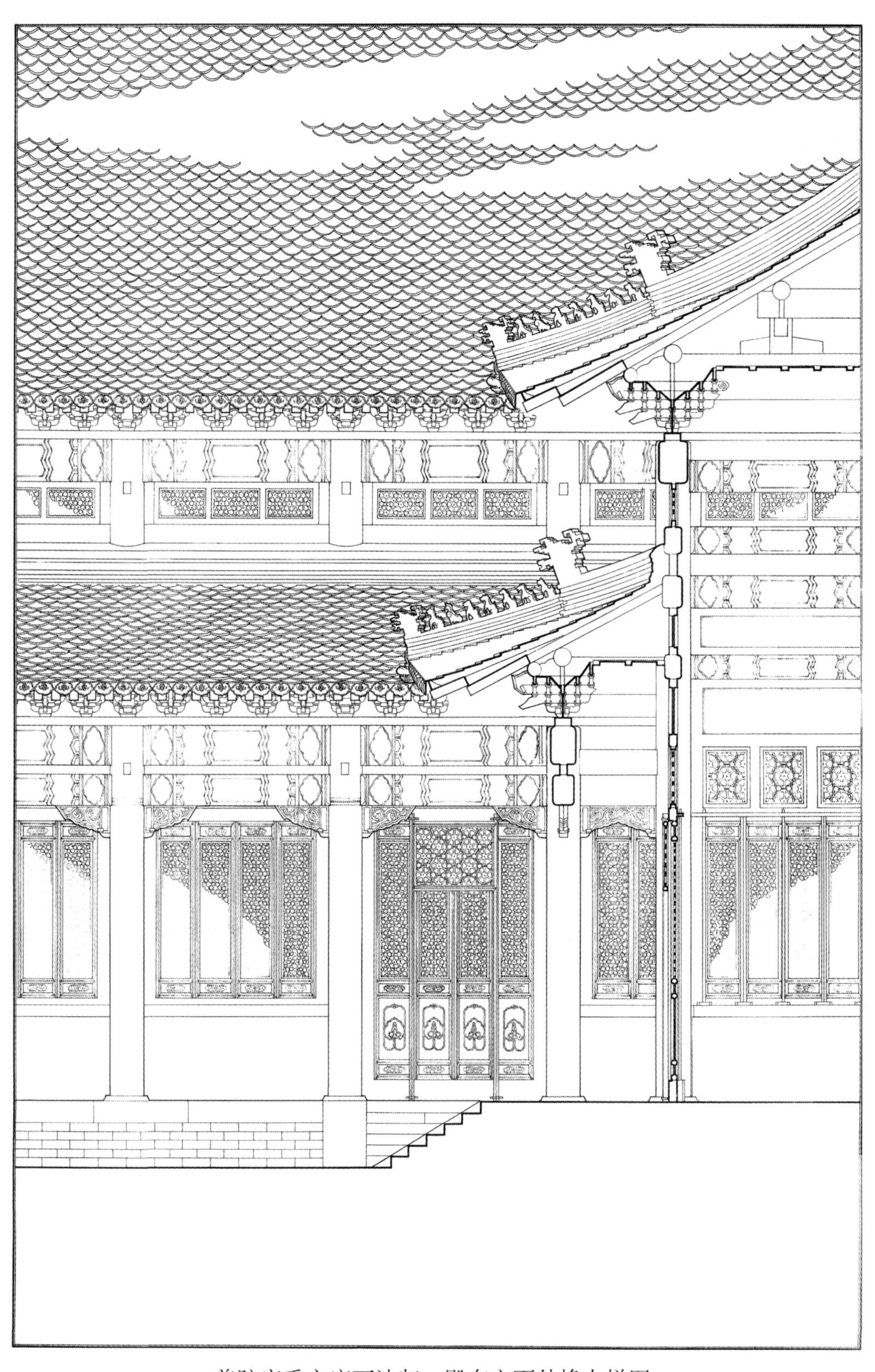

普陀宗乘之庙万法归一殿东立面外檐大样图
East elevation of outer eaves of Wanfa guiyidian of Putuo zongcheng Temple

普陀宗乘之庙慈航普渡东立面图
East elevation of Cihang pudu of Putuo zongcheng Temple

普陀宗乘之庙慈航普渡剖面图
Section of Cihang pudu of Putuo zongcheng Temple

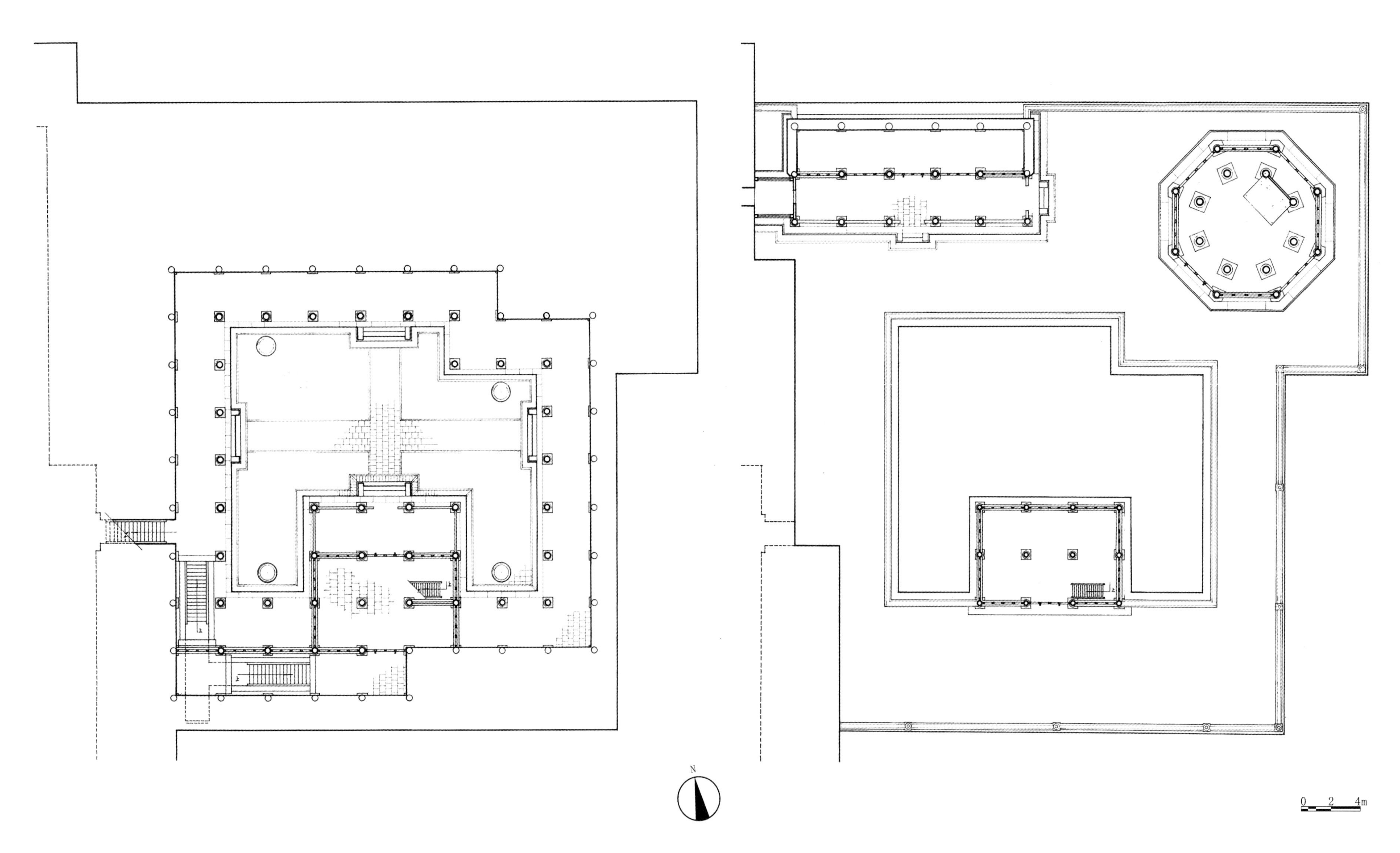

普陀宗乘之庙洛伽胜境组群平面图

Plan of building cluster around Luojia shengjing of Putuo zongcheng Temple

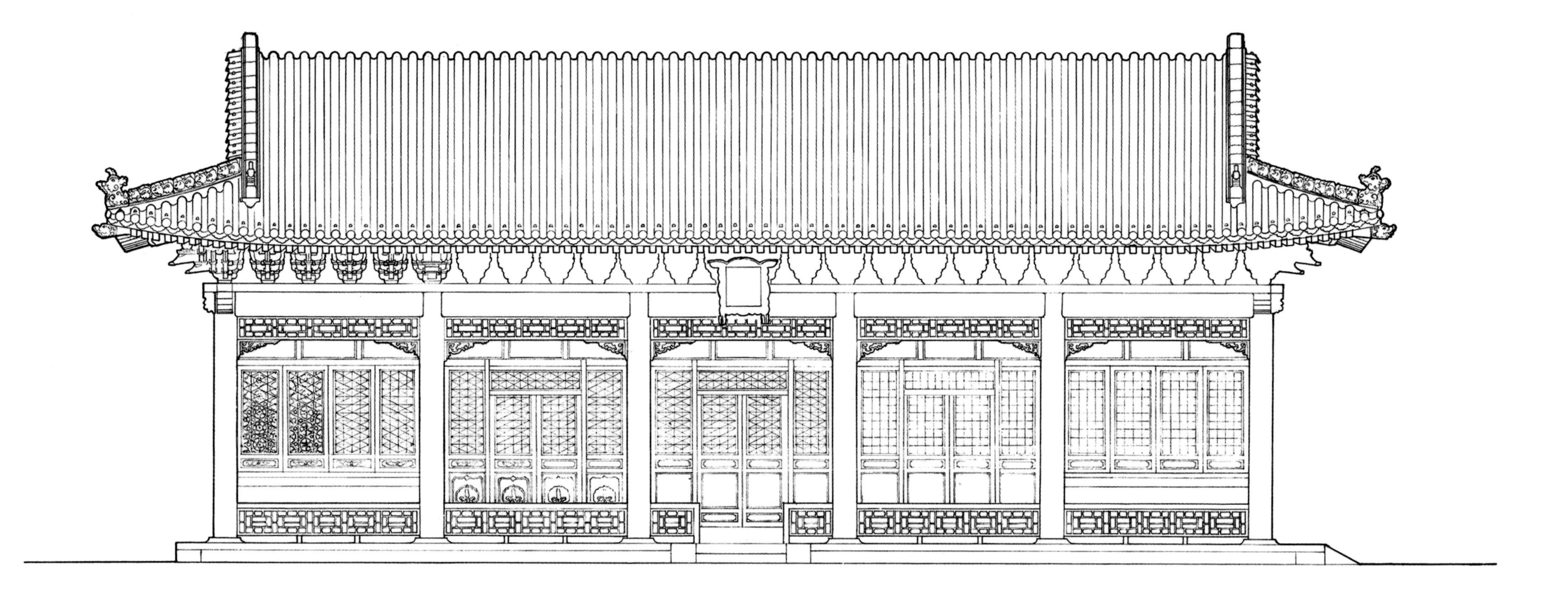

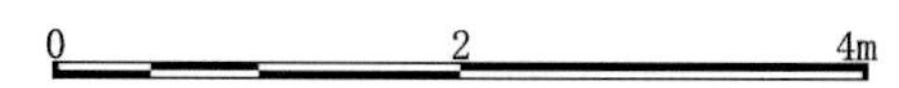

普陀宗乘之庙洛伽胜境南立面图
South elevation of Luojia shengjing of Putuo zongcheng Temple

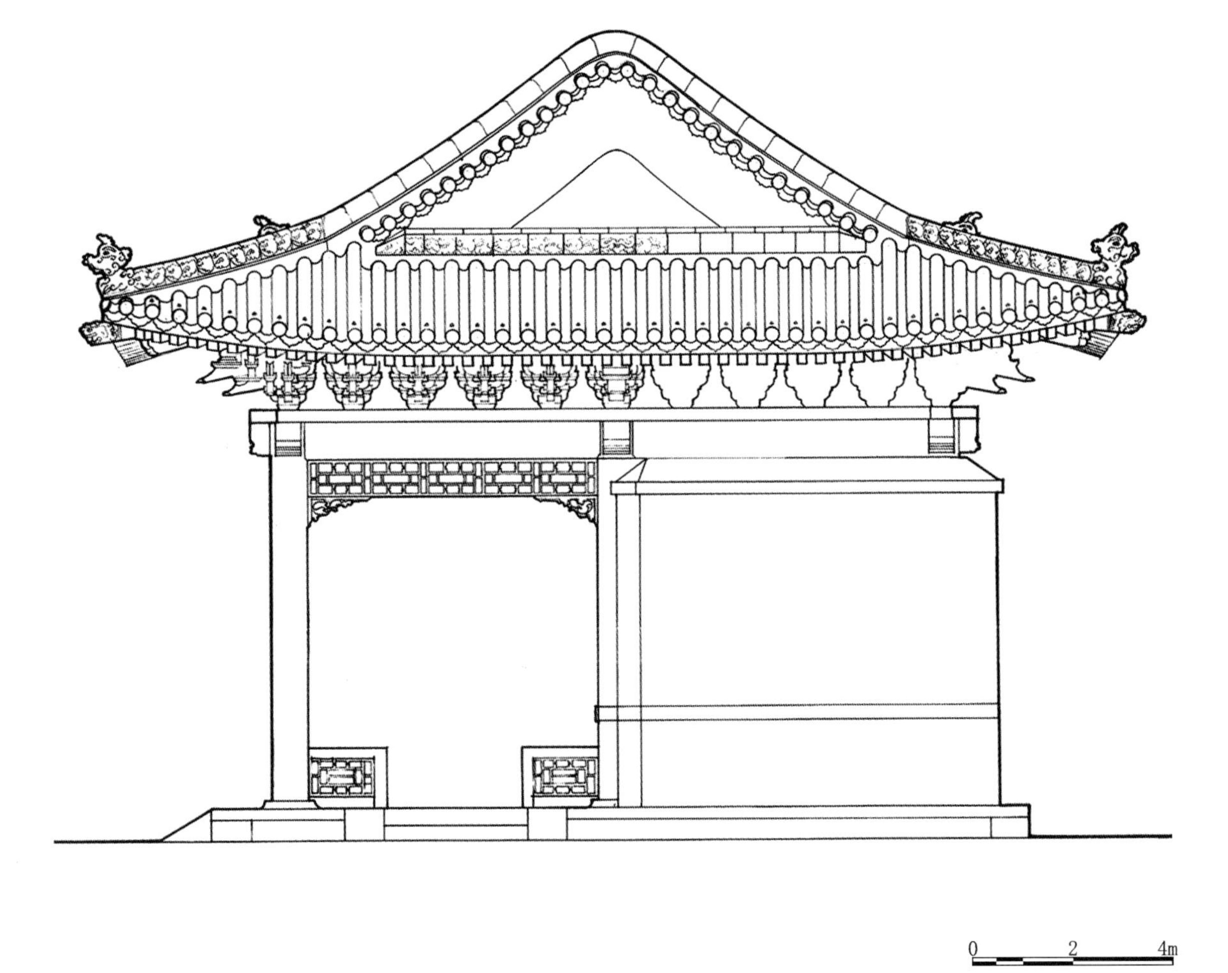

普陀宗乘之庙洛伽胜境侧立面图
Side elevation of Luojia shengjing of Putuo zongcheng Temple

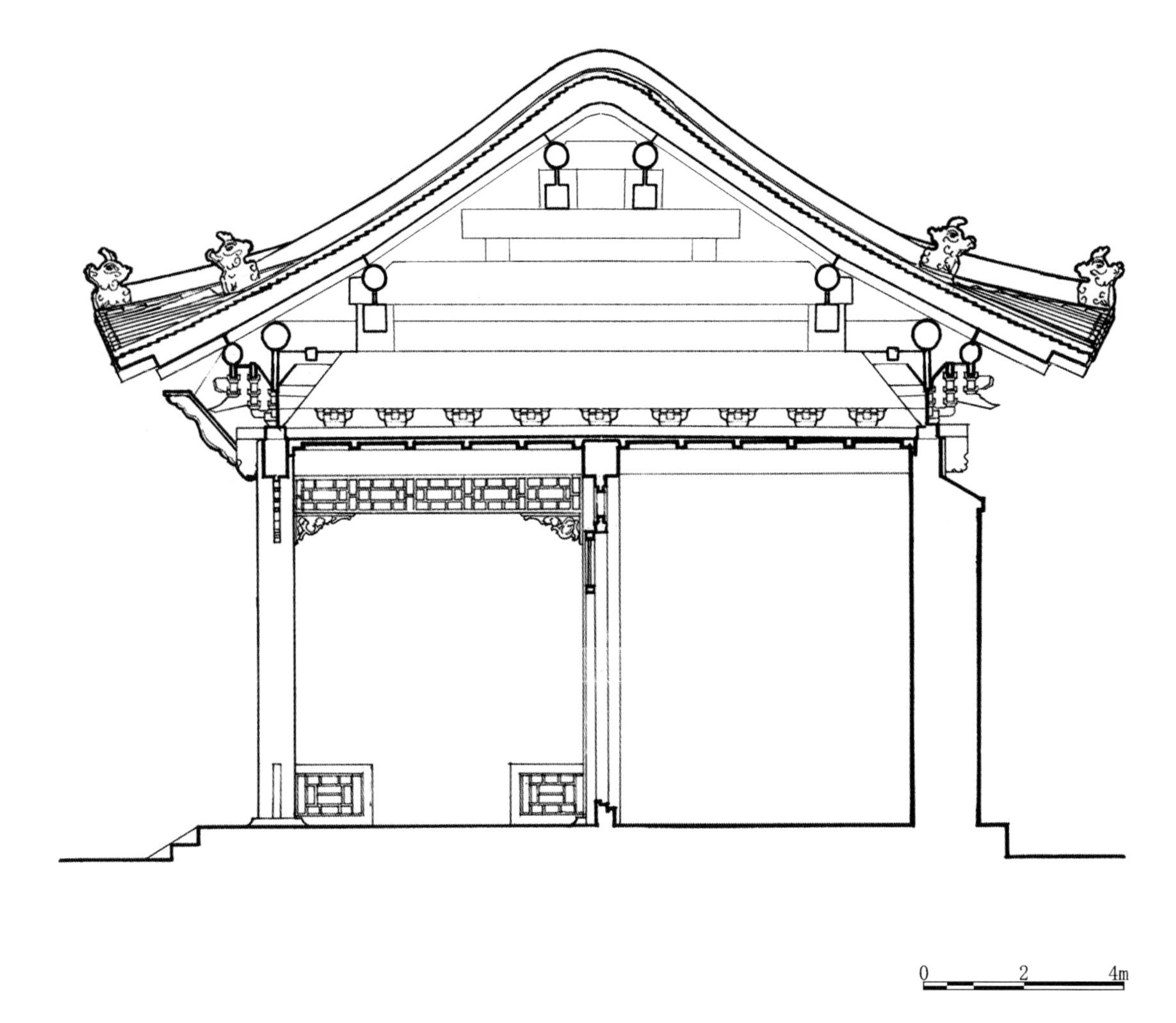

普陀宗乘之庙洛伽胜境横剖面图
Cross-section of Luojia shengjing of Putuo zongcheng Temple

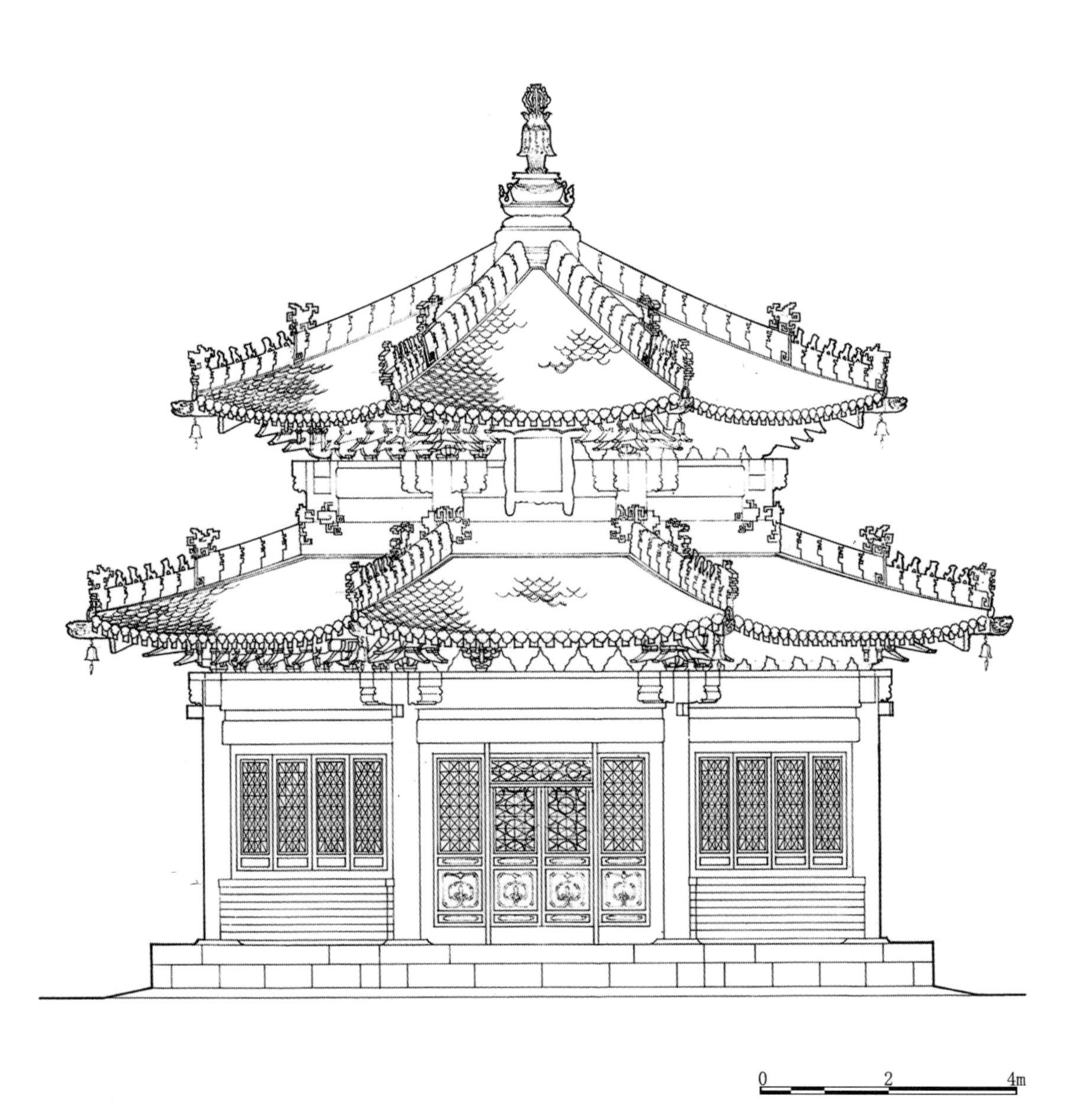

普陀宗乘之庙权衡三界南立面图
South elevation of Quanheng sanjie of Putuo zongcheng Temple

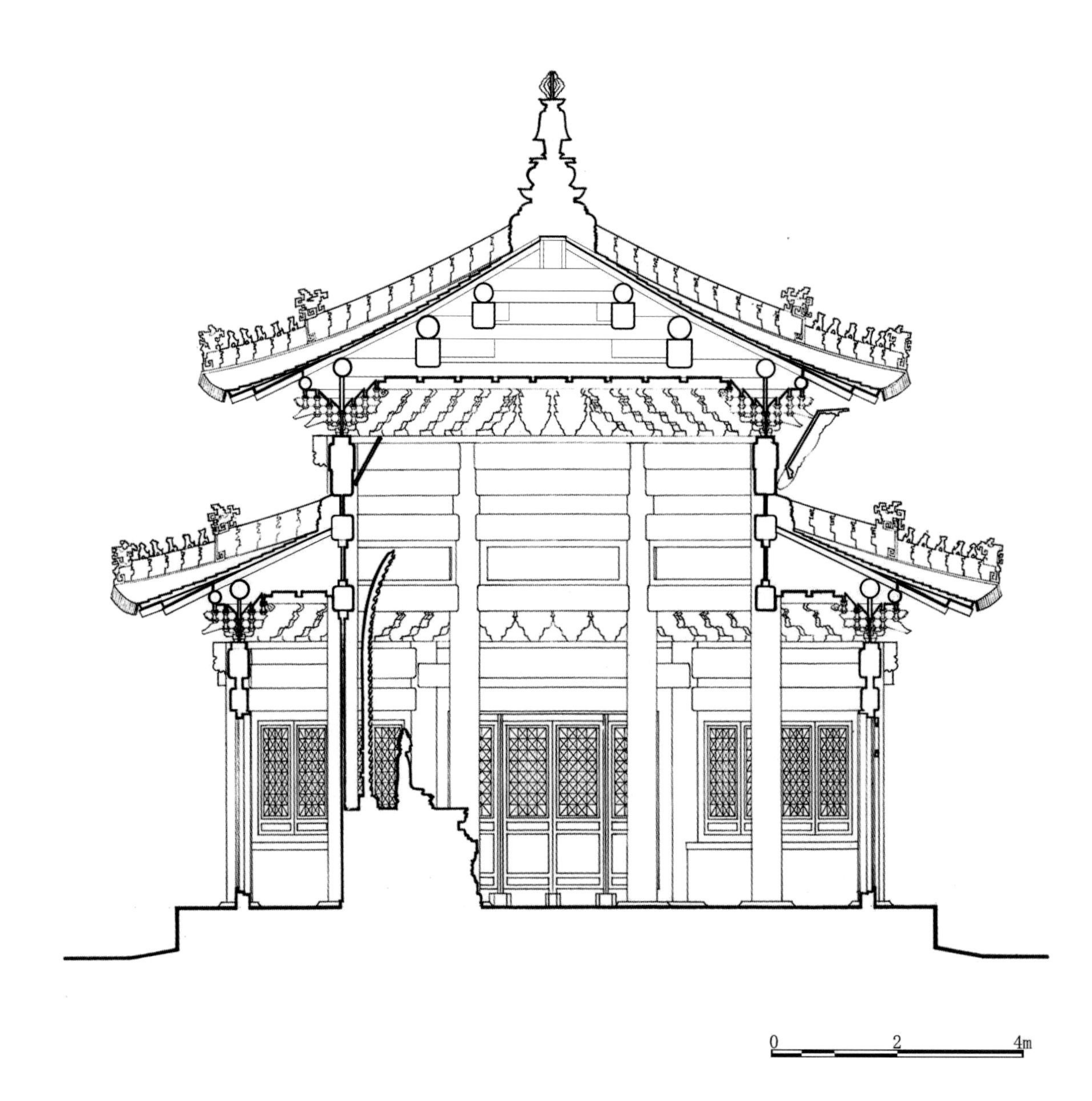

普陀宗乘之庙权衡三界纵剖面图
Longitudinal section of Quanheng sanjie of Putuo zongcheng Temple

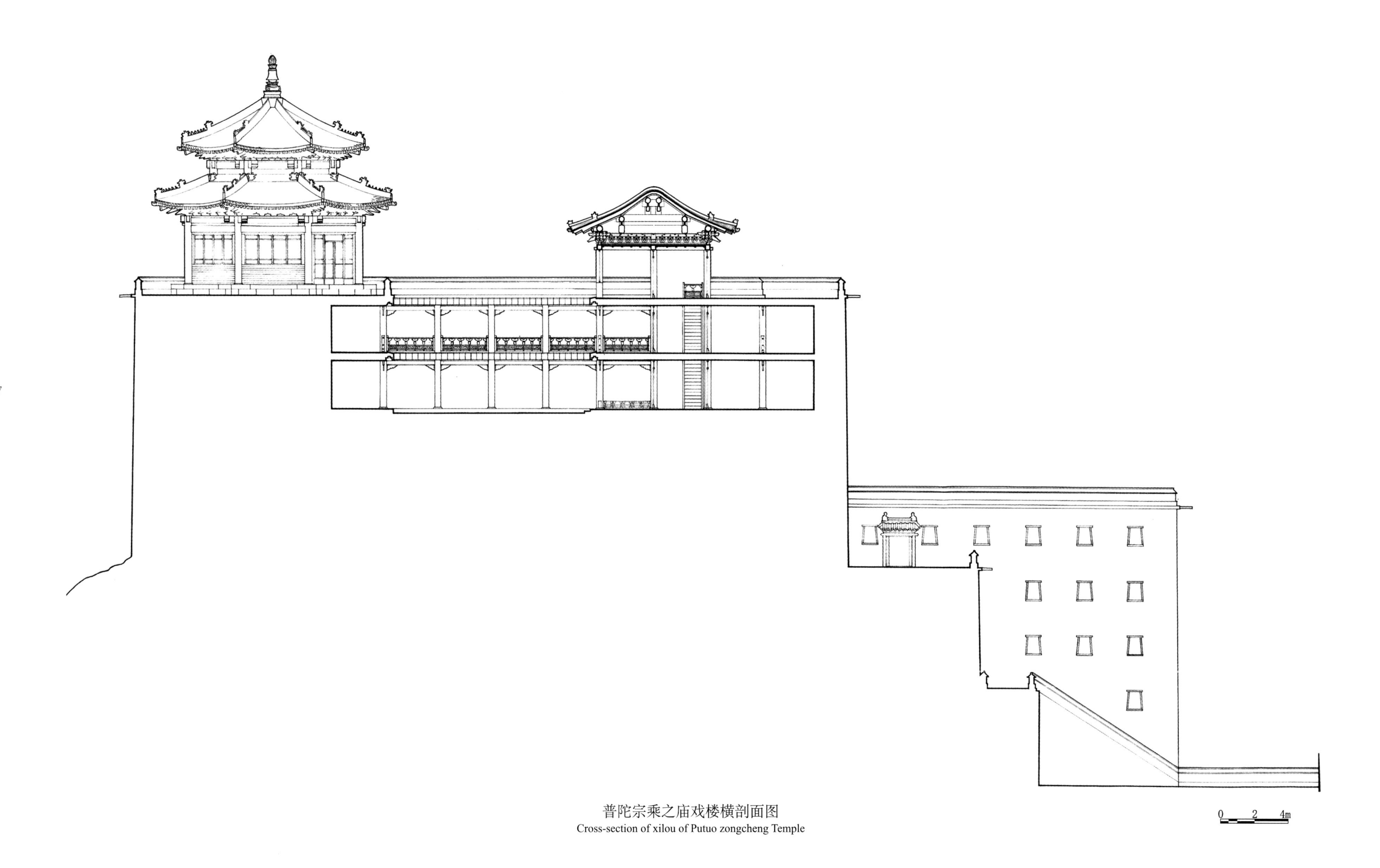

普陀宗乘之庙戏楼横剖面图

Cross-section of xilou of Putuo zongcheng Temple

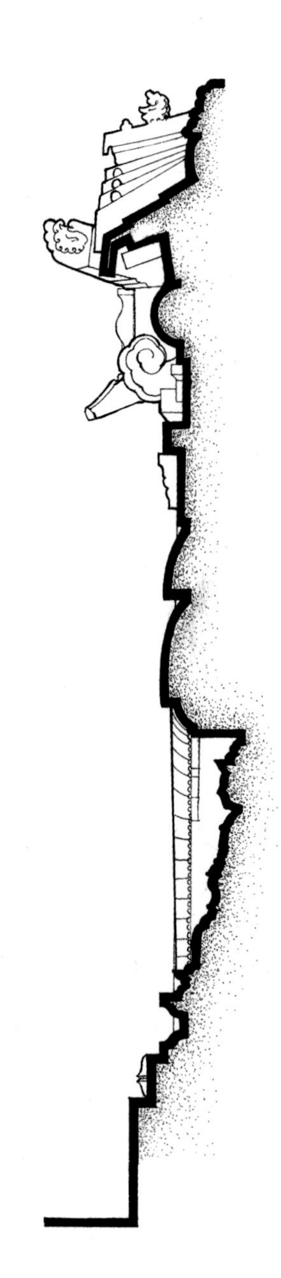

0 0.2 0.4m

普陀宗乘之庙琉璃窗大样图

Color-glazed window of Putuo zongcheng Temple

须弥福寿之庙
Xumi fushou Temple

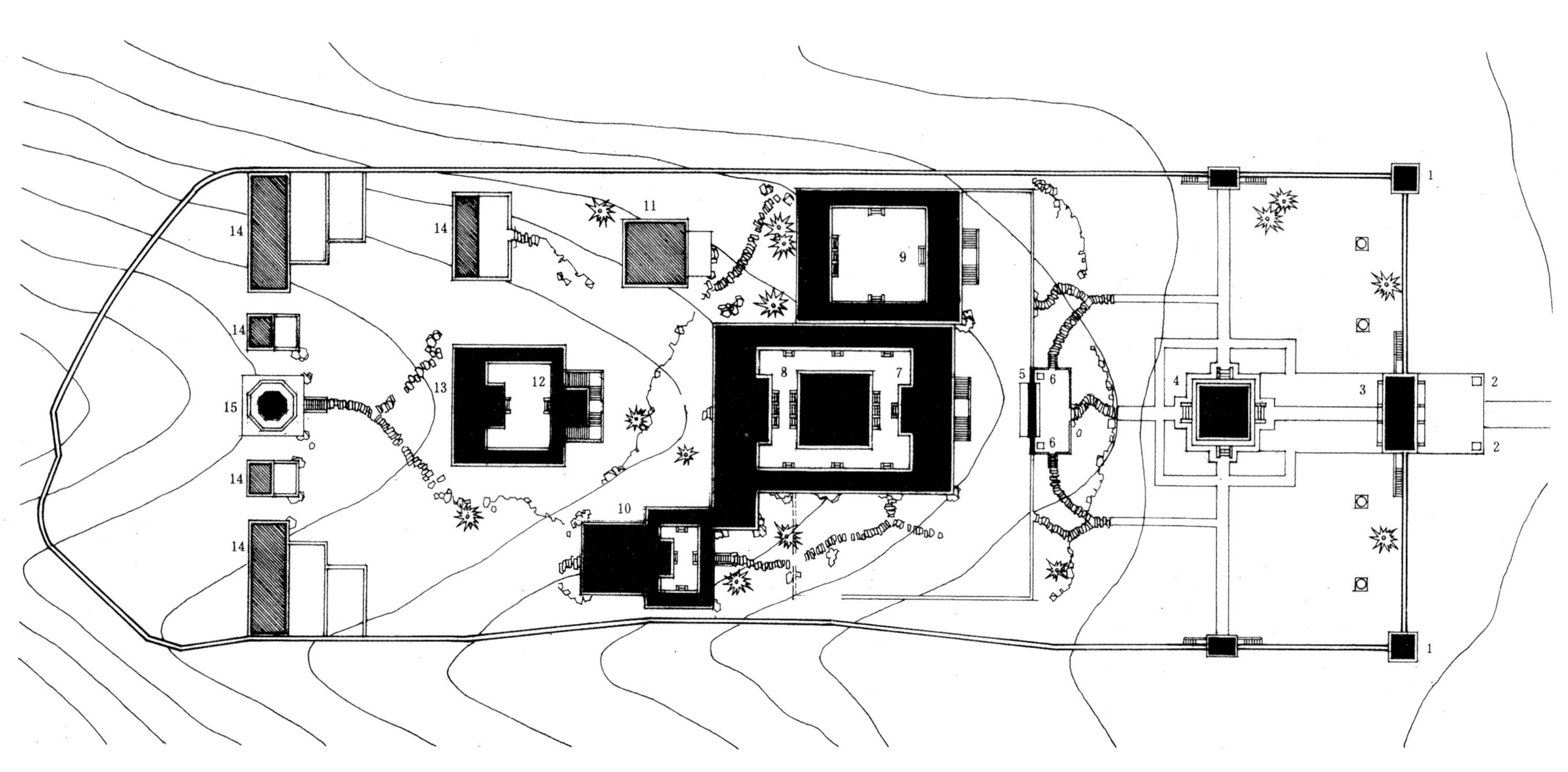

1 角楼 Corner Towers
2 石狮 Stone Lions
3 山门 Gate Hall
4 碑亭 Stone-tablet Pavilion
5 琉璃牌楼 Liuli pailou (Glaze Archway)
6 石象 Stone Elephants
7 大红台 Great Red Platform
8 妙高庄严殿 Miaogao zhuangyandian
9 东红台 East Red Platform
10 吉祥法喜殿 Jixiang faxidian
11 生欢喜心殿（遗址） Sheng huanxixindian (Site)
12 金贺堂 Jinhetang
13 万法宗源殿 Wanfa zongyuandian
14 白台（遗址） White Platforms (Site)
15 琉璃宝塔 Liuli baota (Glaze Pagoda)

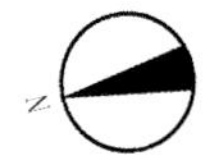

须弥福寿之庙总平面图
Site plan of Xumi fushou Temple

0 20 40m

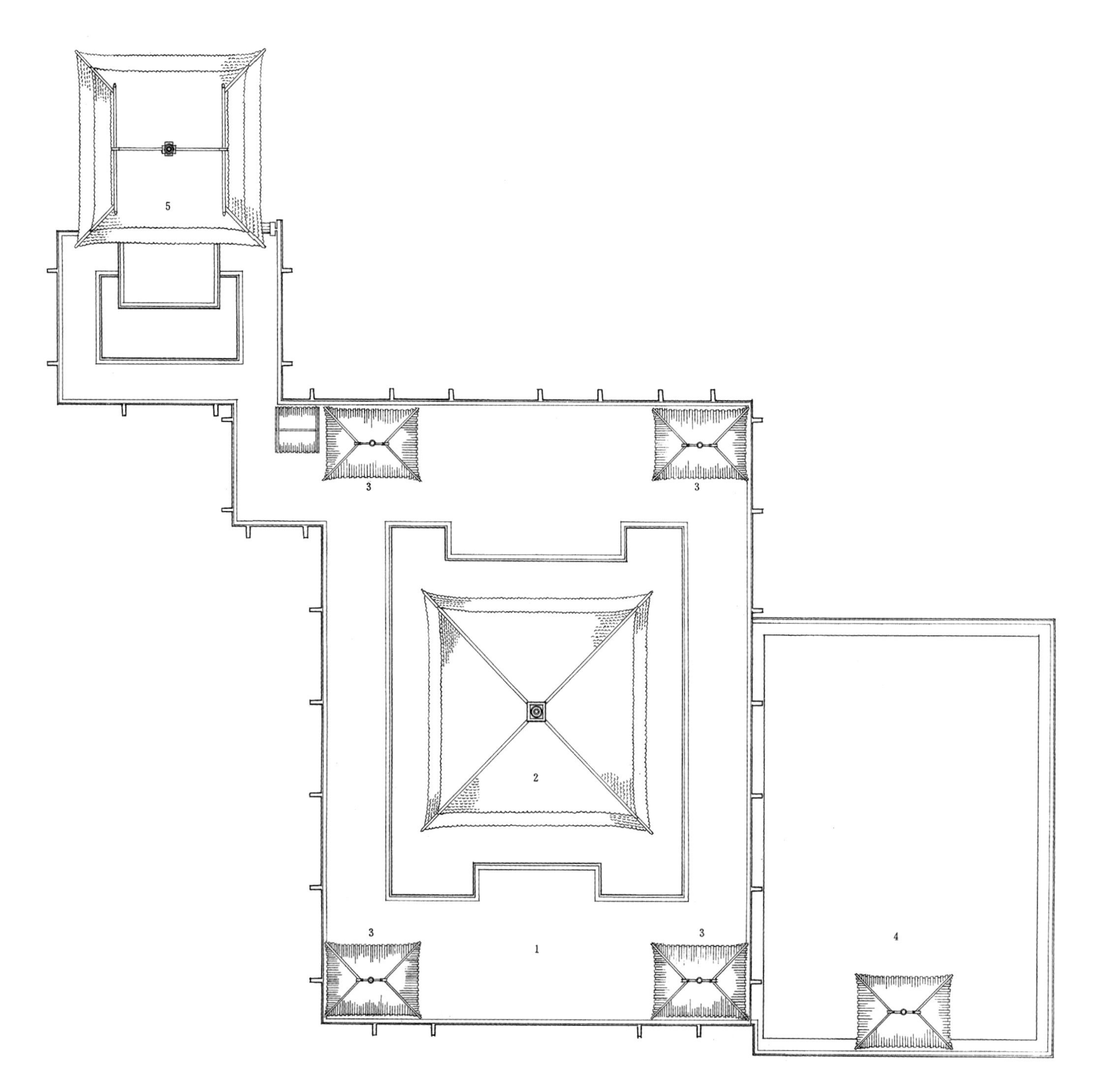

1 大红台　Great Red Platform
2 妙高庄严殿　Miaogao zhuangyandian
3 角楼　Corner towers
4 东红台　East Red Platform
5 吉祥法喜殿　Jixiang faxidian

0 4 8m

须弥福寿之庙大红台、东红台、吉祥法喜殿组群屋顶平面图

Roof plan of building cluster around Great Red Platform, East Red Platform, and Jixiang faxidian of Xumi fushou Temple

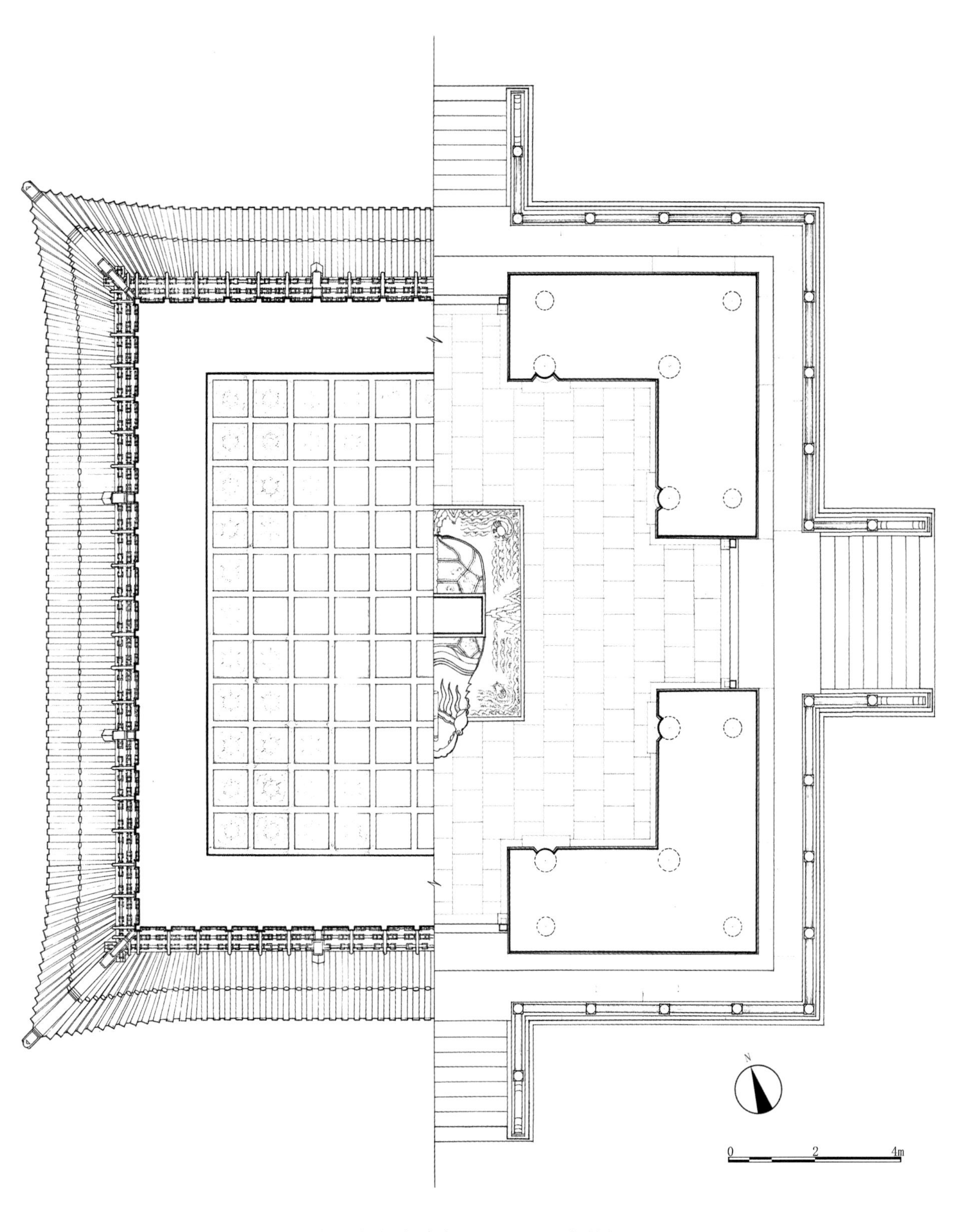

须弥福寿之庙碑亭平面图、天花仰视图

Plan and reflected ceiling plan of beiting of Xumi fushou Temple

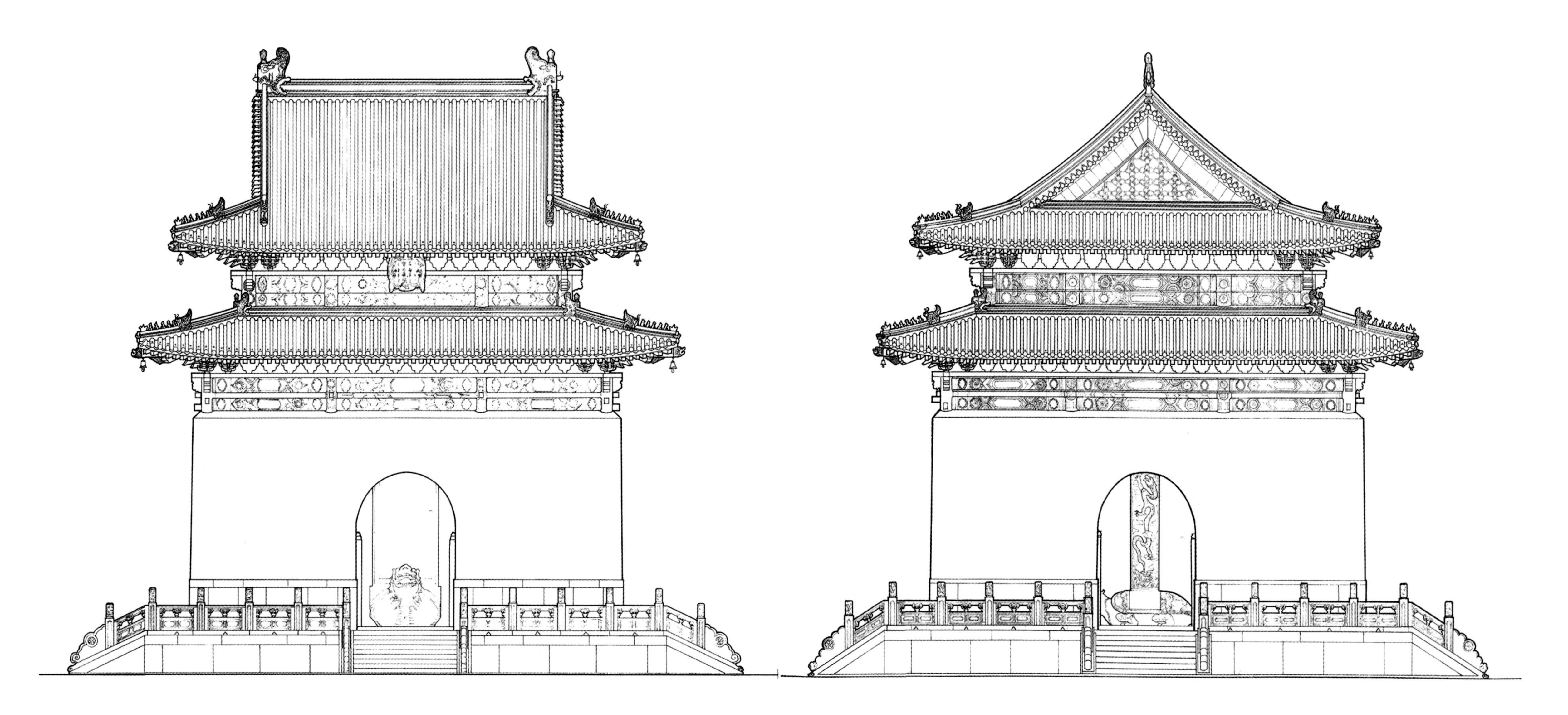

须弥福寿之庙碑亭正立面图
Front elevation of beiting of Xumi fushou Temple

须弥福寿之庙碑亭侧立面图
Side elevation of beiting of Xumi fushou Temple

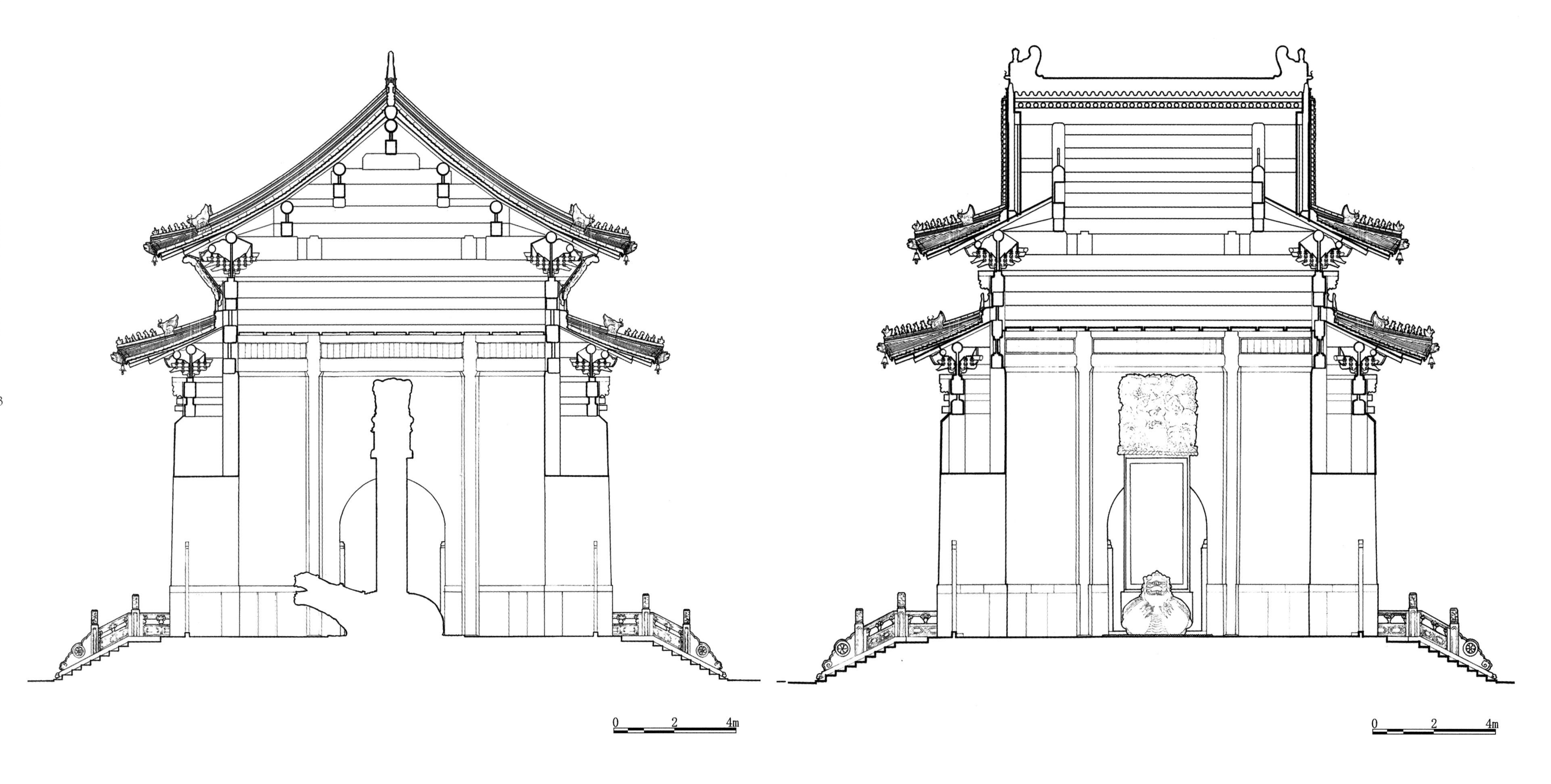

须弥福寿之庙碑亭横剖面图
Cross-section of beiting of Xumi fushou Temple

须弥福寿之庙碑亭纵剖面图
Longitudinal section of beiting of Xumi fushou Temple

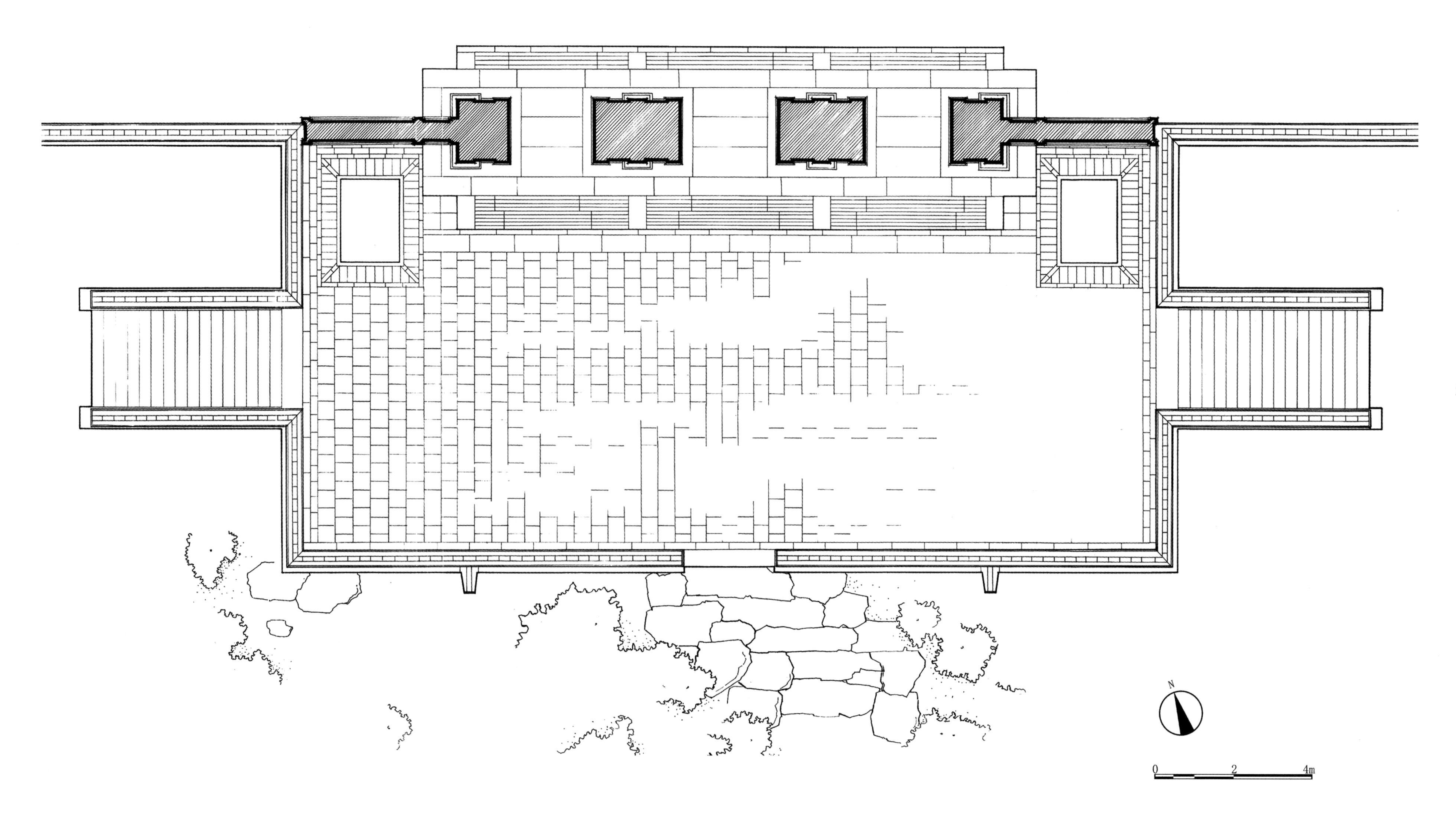

须弥福寿之庙琉璃牌楼平面图
Plan of Liuli pailou of Xumi fushou Temple

须弥福寿之庙琉璃牌楼正立面图

Front elevation of Liuli pailou of Xumi fushou Temple

须弥福寿庙琉璃牌楼侧立面图
Side elevation of Liuli pailou of Xumi fushou Temple

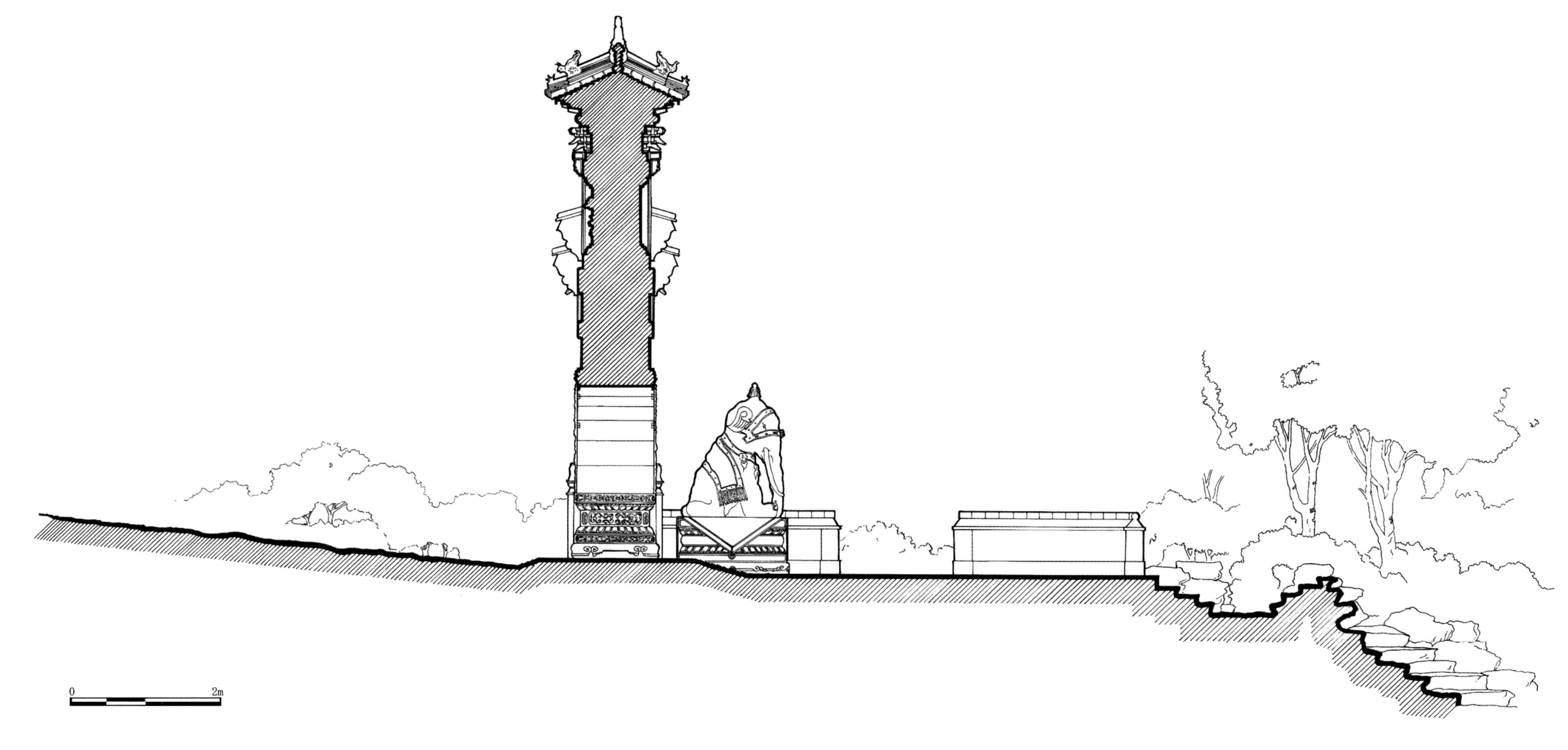

须弥福寿庙琉璃牌楼剖面图

Section of Liuli pailou of Xumi fushou Temple

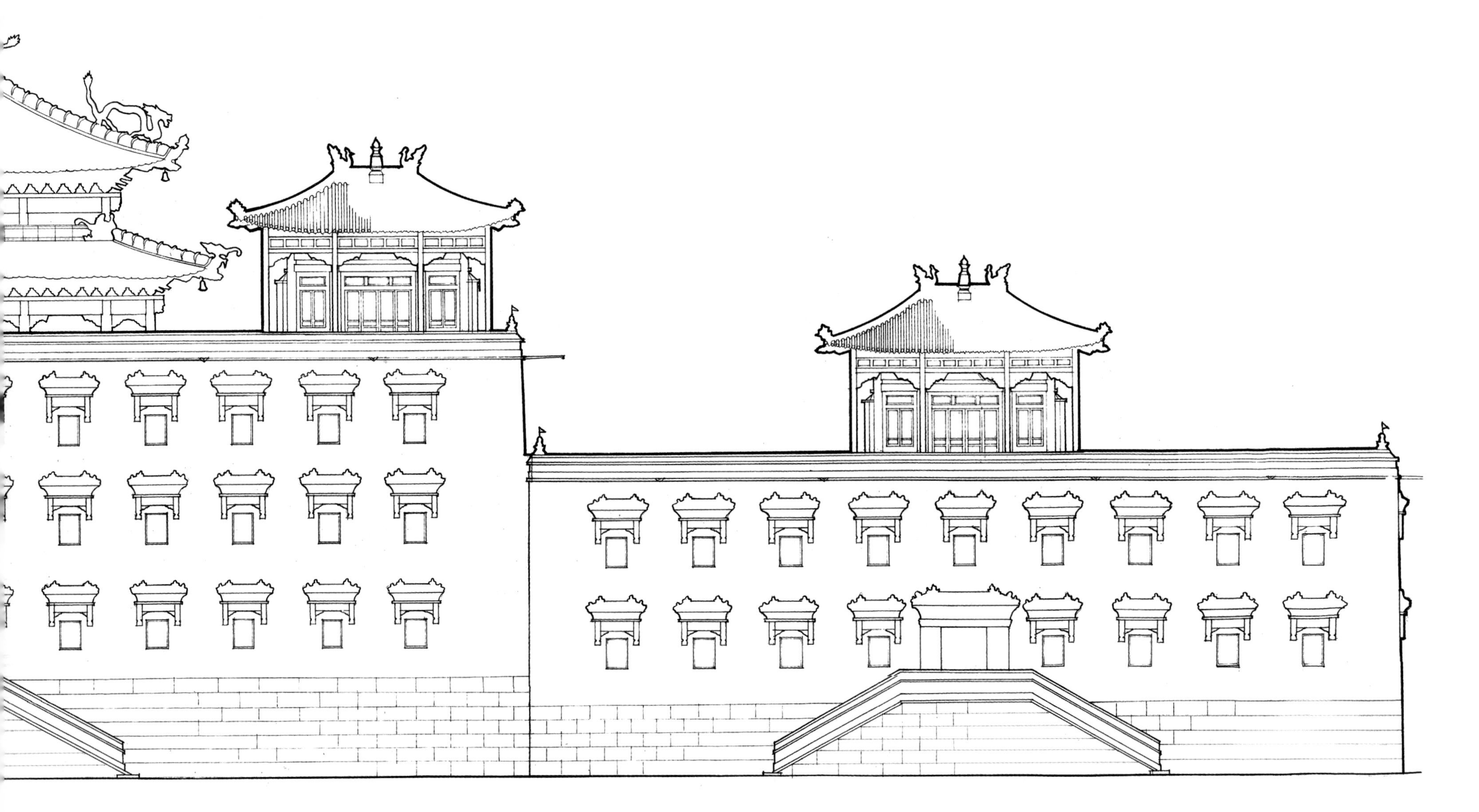

须弥福寿之庙大红台、东红台、吉祥法喜殿组群南立面图

South elevation of building cluster around Great Red Platform, East Red Platform, and Jixiang faxidian of Xumi fushou Temple

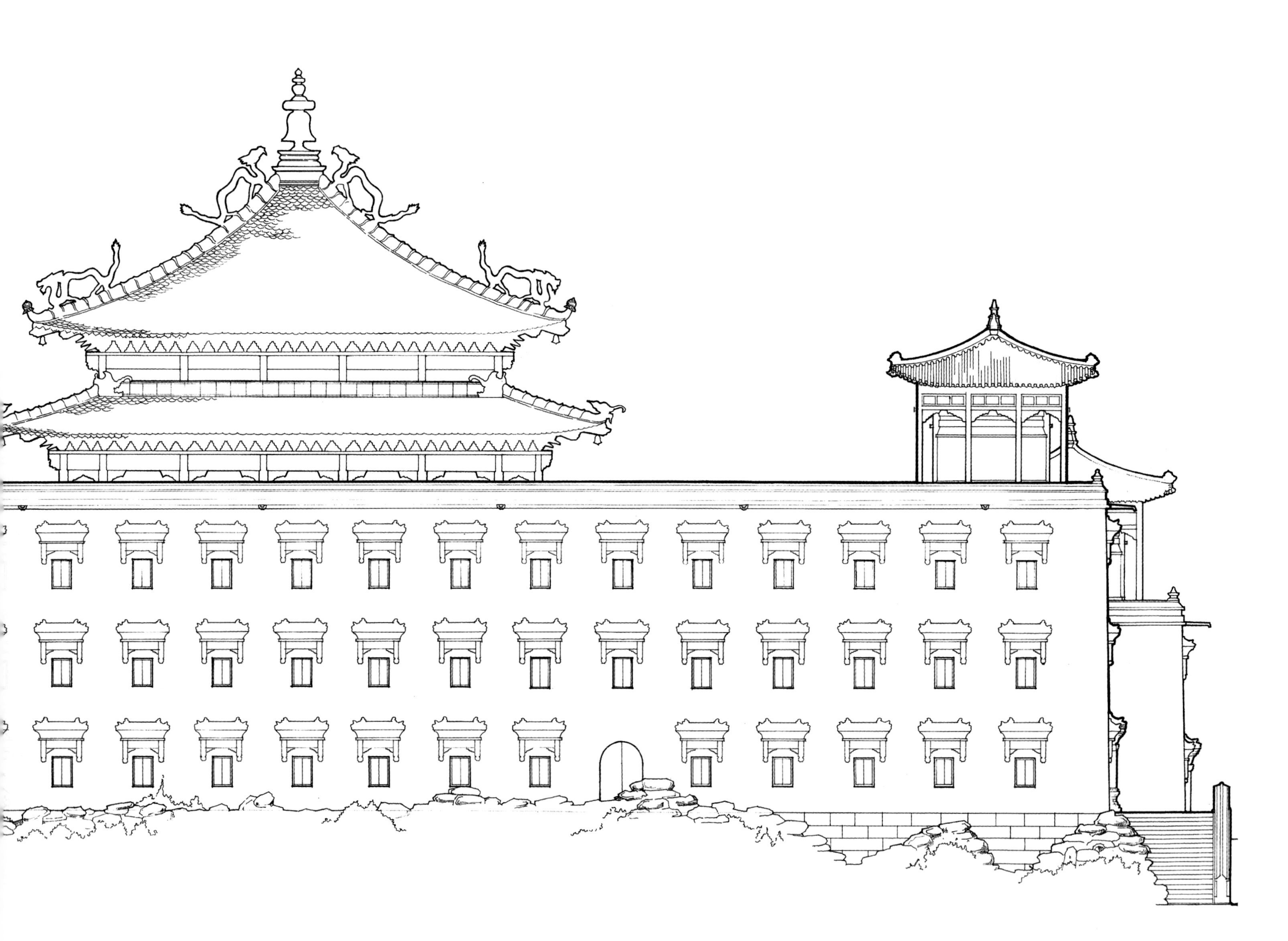

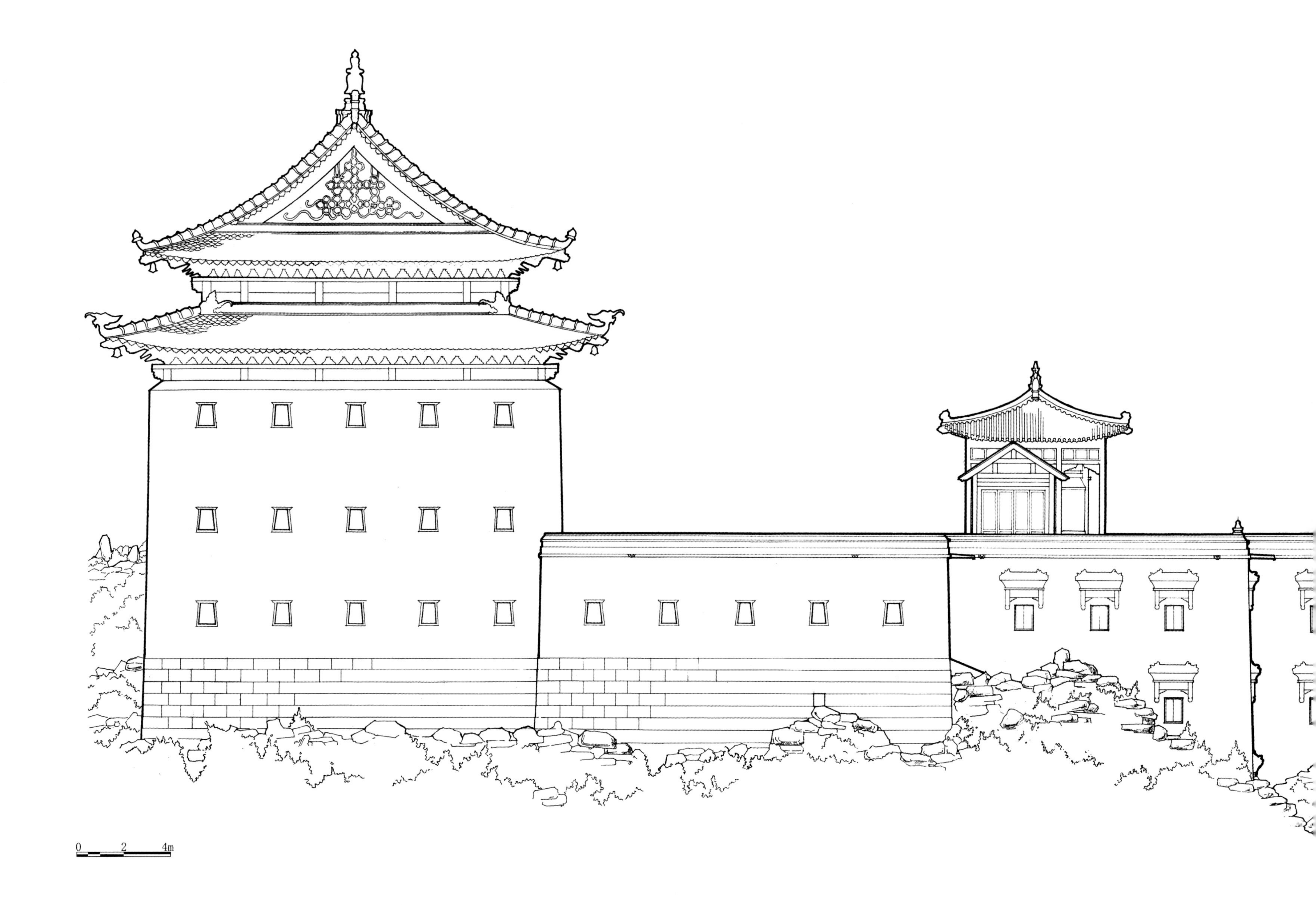

须弥福寿之庙大红台、东红台、吉祥法喜殿组群西立面图

West elevation of building cluster around Great Red Platform, East Red Platform, and Jixiang faxidian of Xumi fushou Temple

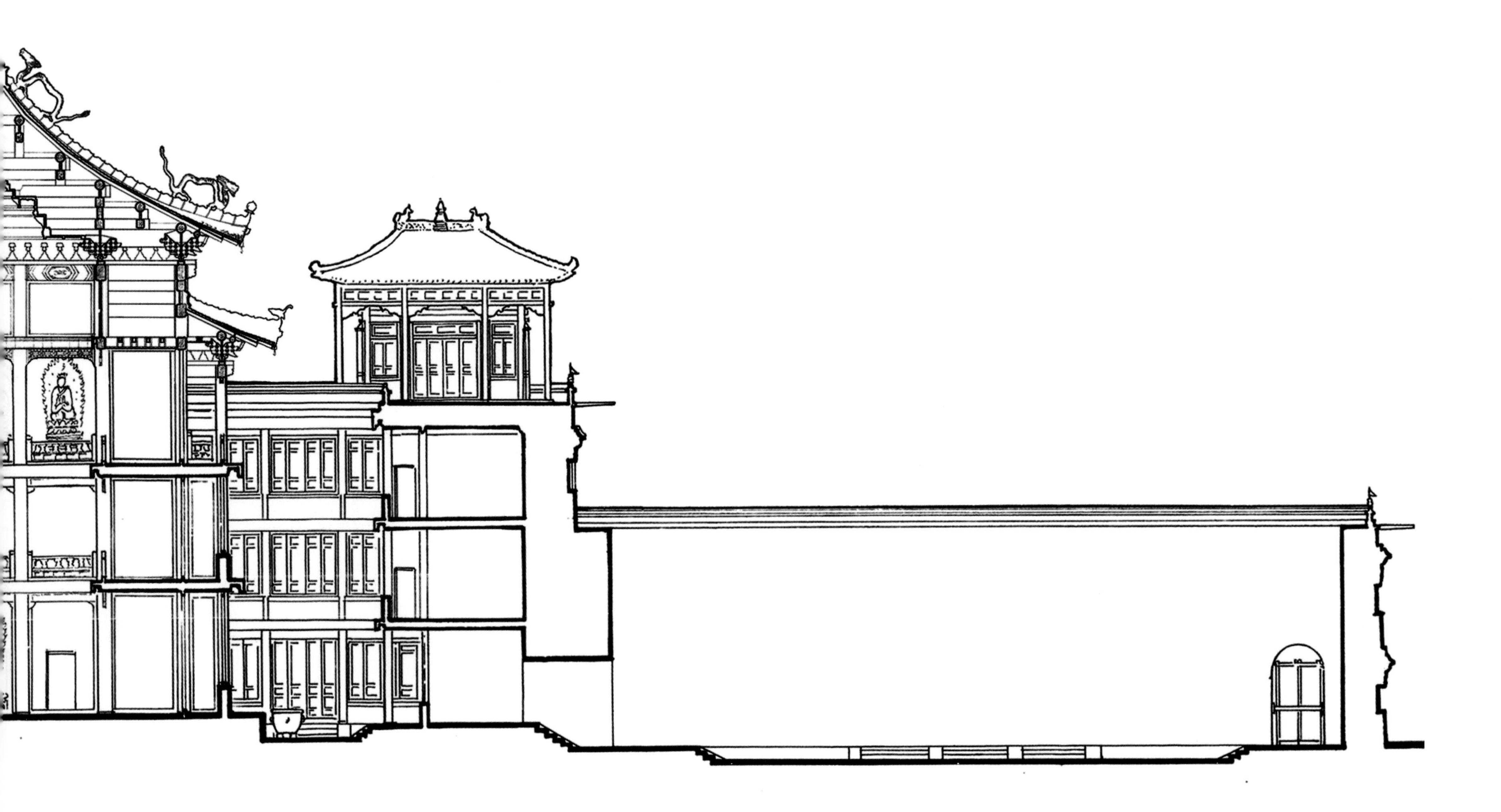

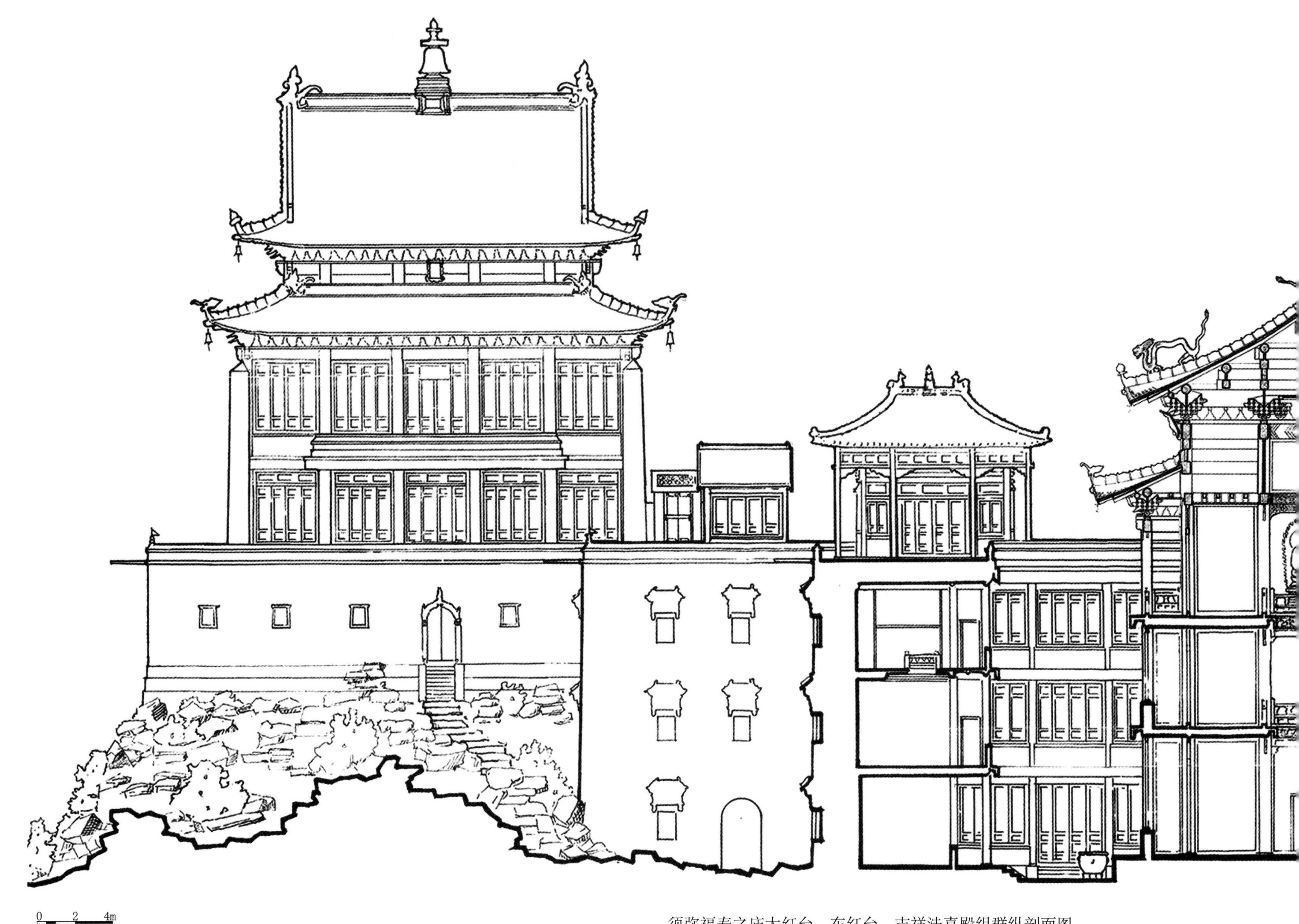

须弥福寿之庙大红台、东红台、吉祥法喜殿组群纵剖面图

Longitudinal section of building cluster around Great Red Platform, East Red Platform, and Jixiang faxidian of Xumi fushou Temple

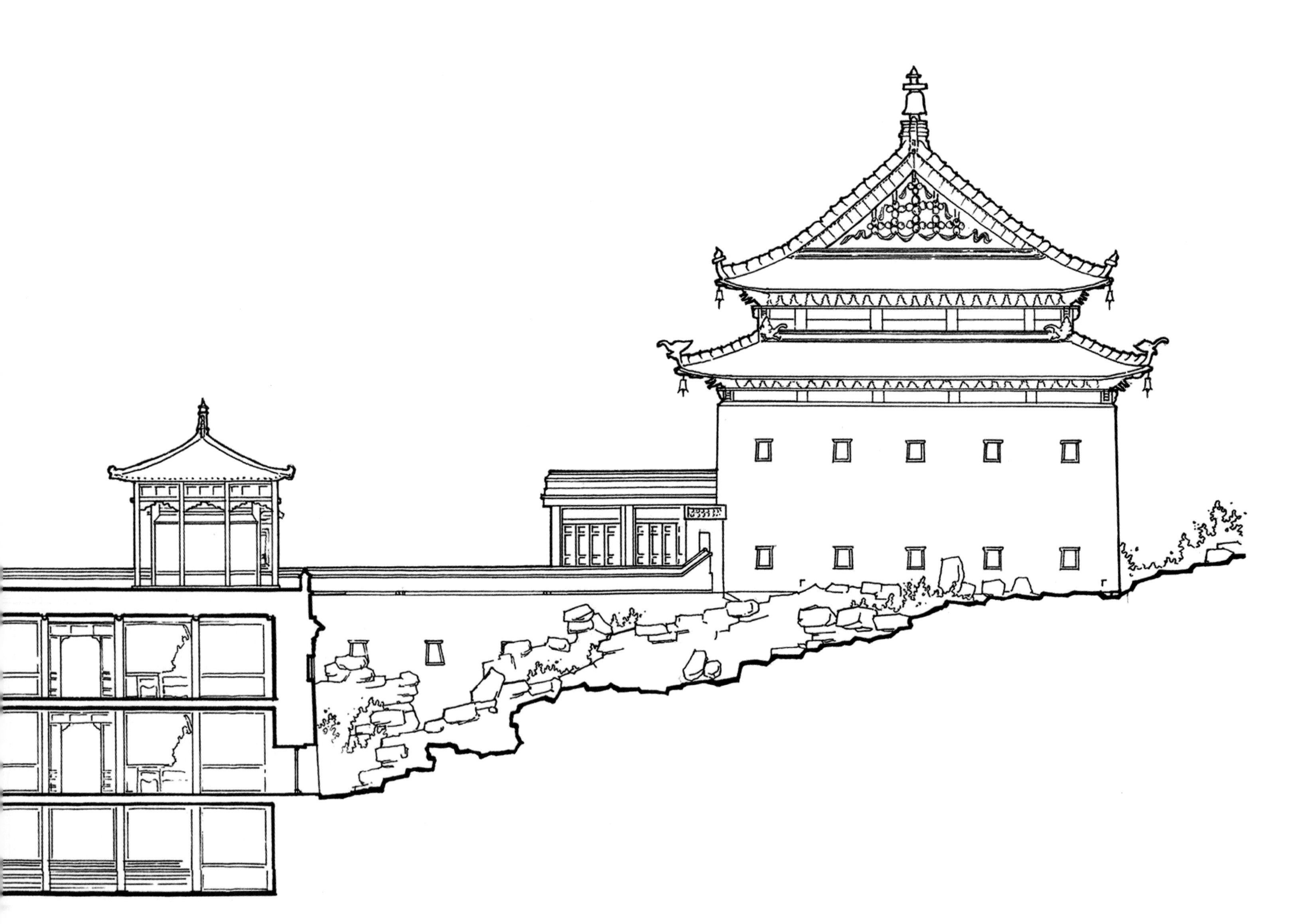

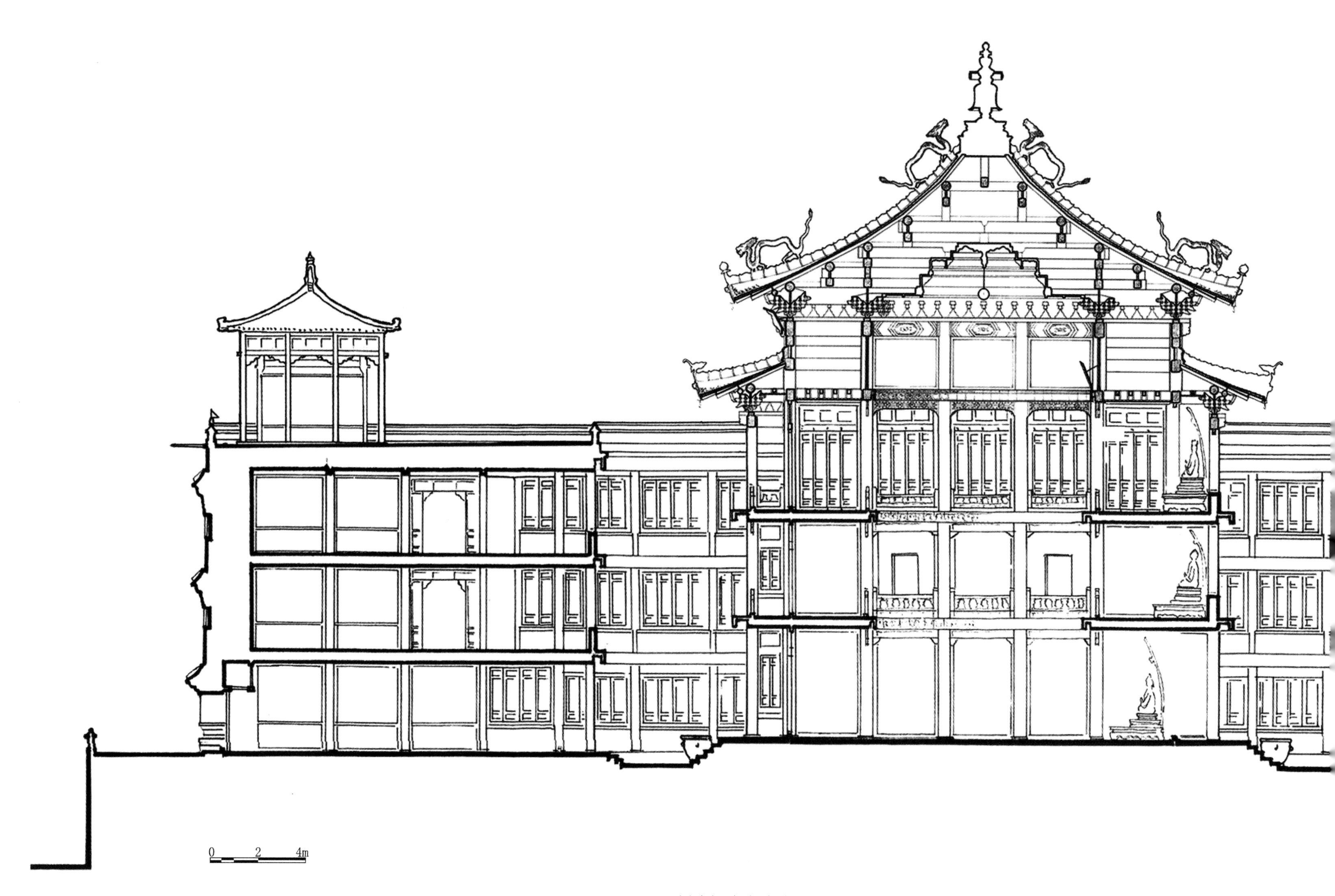

须弥福寿之庙大红台、东红台、吉祥法喜殿组群横剖面图

Cross-section of building cluster around Great Red Platform, East Red Platform, and Jixiang faxidian of Xumi fushou Temple

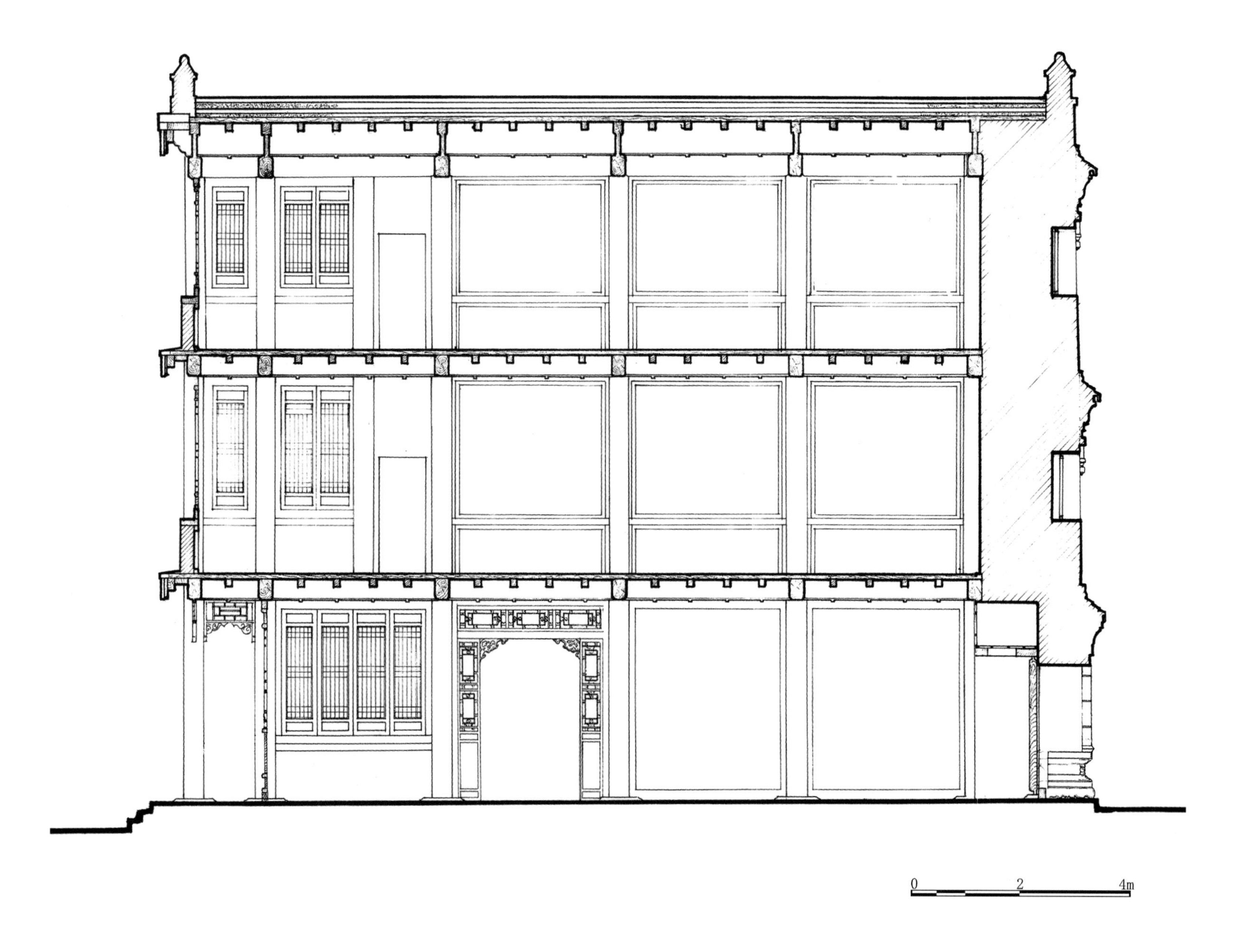

须弥福寿之庙大红台群楼横剖面图一
Cross-section of qunlou of Great Red Platform of Xumi fushou Temple (Ⅰ)

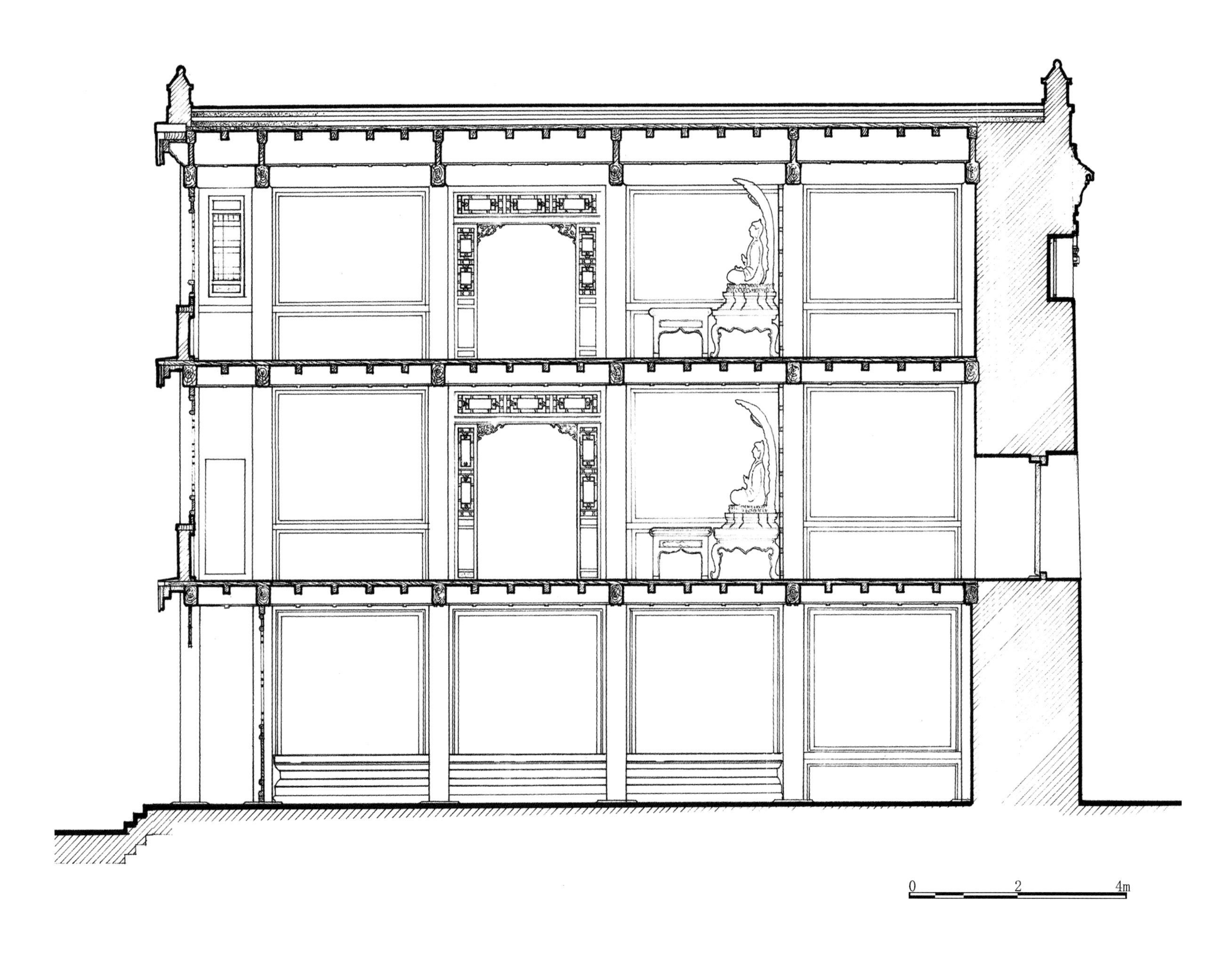

须弥福寿之庙大红台群楼横剖面图二

Cross-section of qunlou of Great Red Platform of Xumi fushou Temple (Ⅱ)

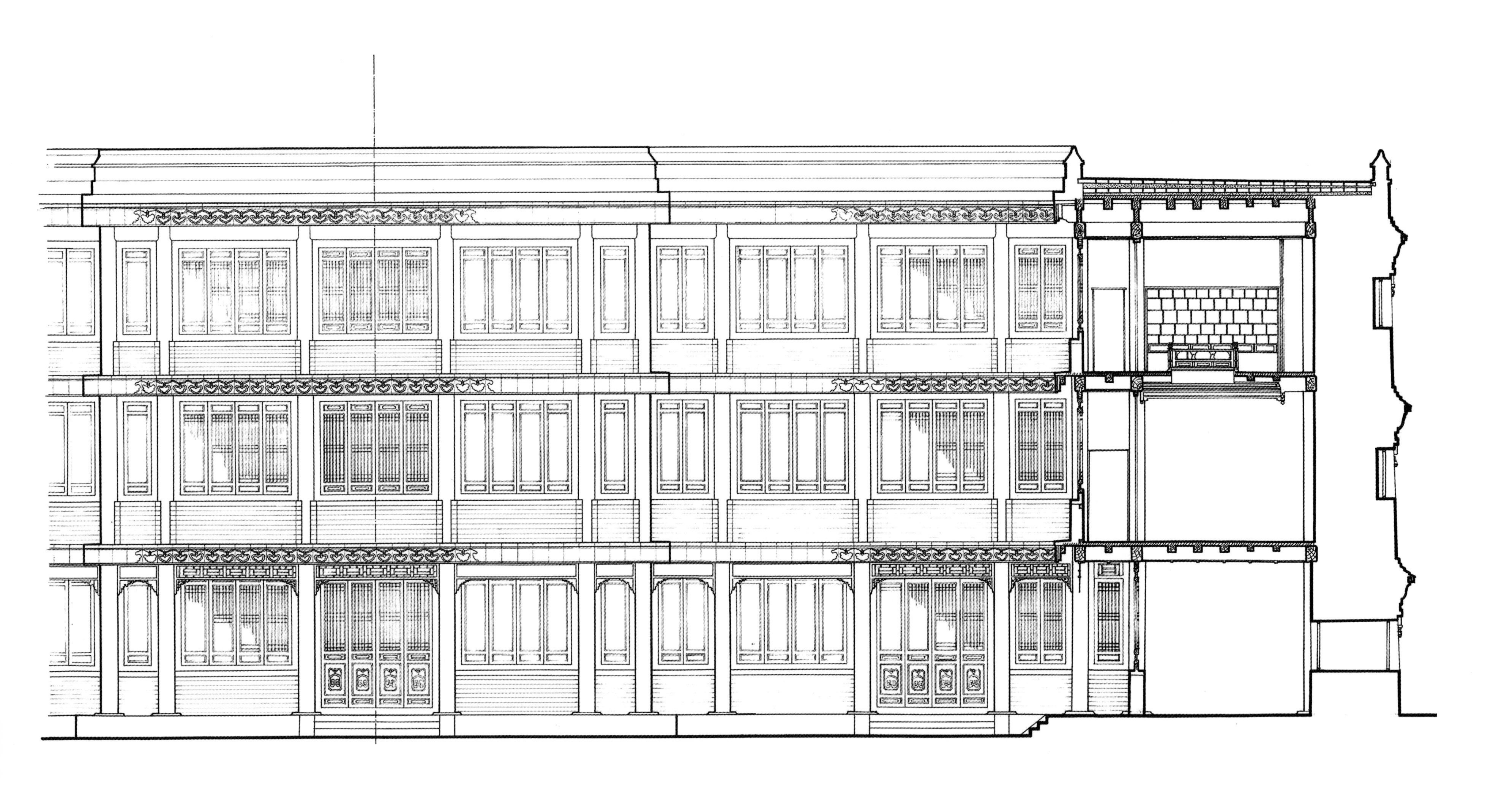

须弥福寿之庙大红台群楼立面、剖面图
Elevation and section of qunlou of Great Red Platform of Xumi fushou Temple

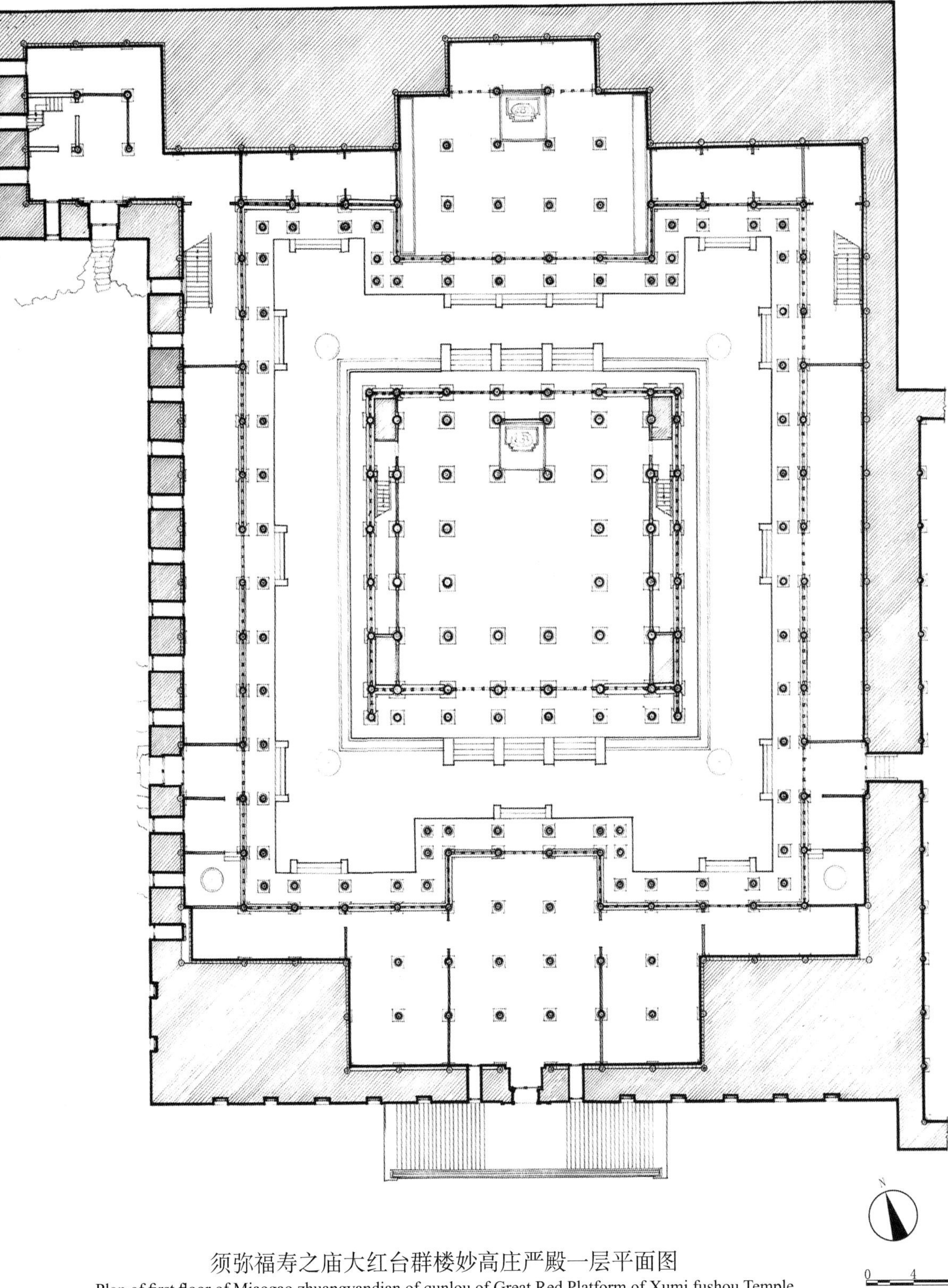

须弥福寿之庙大红台群楼妙高庄严殿一层平面图
Plan of first floor of Miaogao zhuangyandian of qunlou of Great Red Platform of Xumi fushou Temple

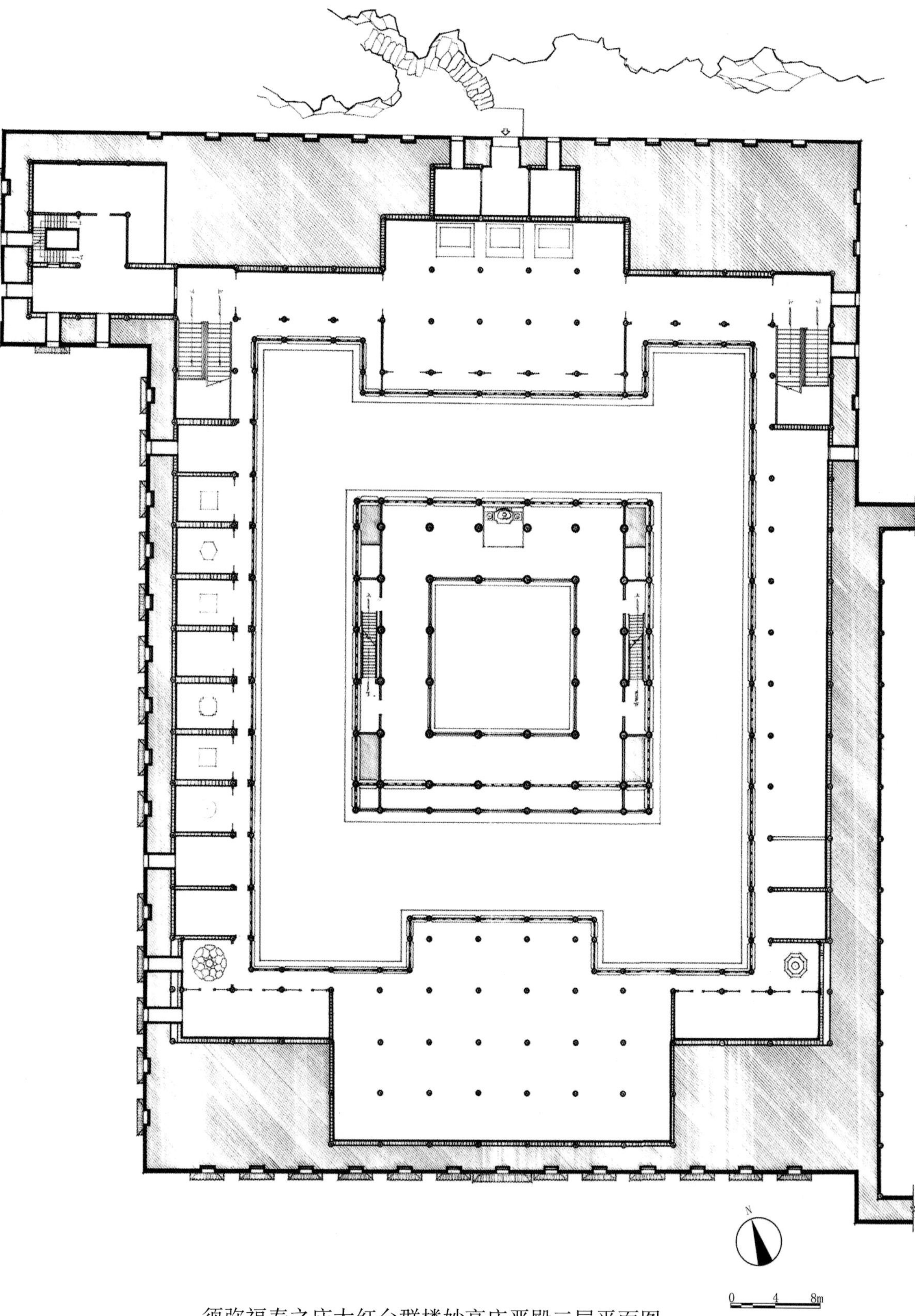

须弥福寿之庙大红台群楼妙高庄严殿二层平面图

Plan of second floor of Miaogao zhuangyandian of qunlou of Great Red Platform of Xumi fushou Temple

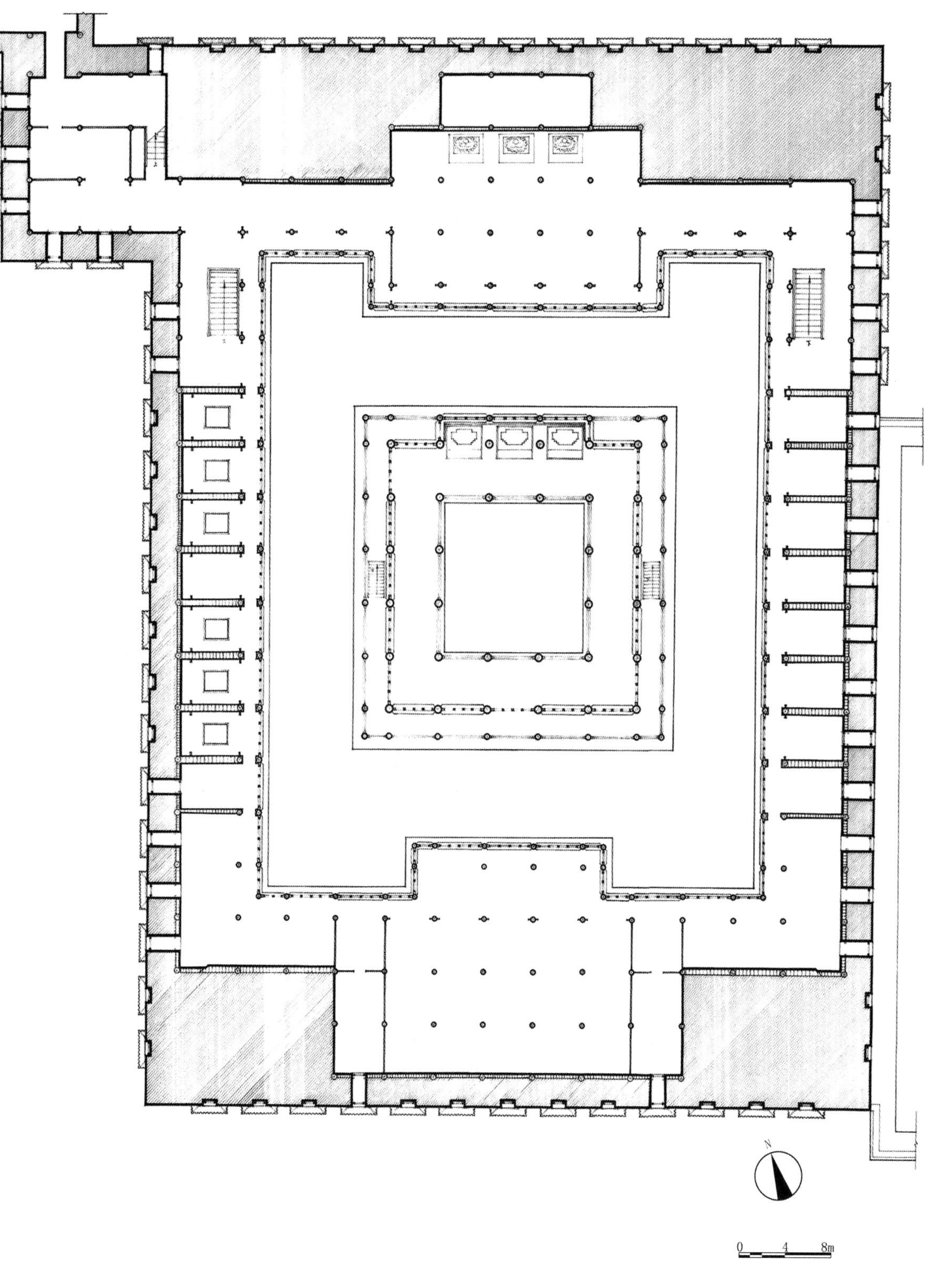

须弥福寿之庙大红台群楼妙高庄严殿三层平面图

Plan of third floor of Miaogao zhuangyandian of qunlou of Great Red Platform of Xumi fushou Temple

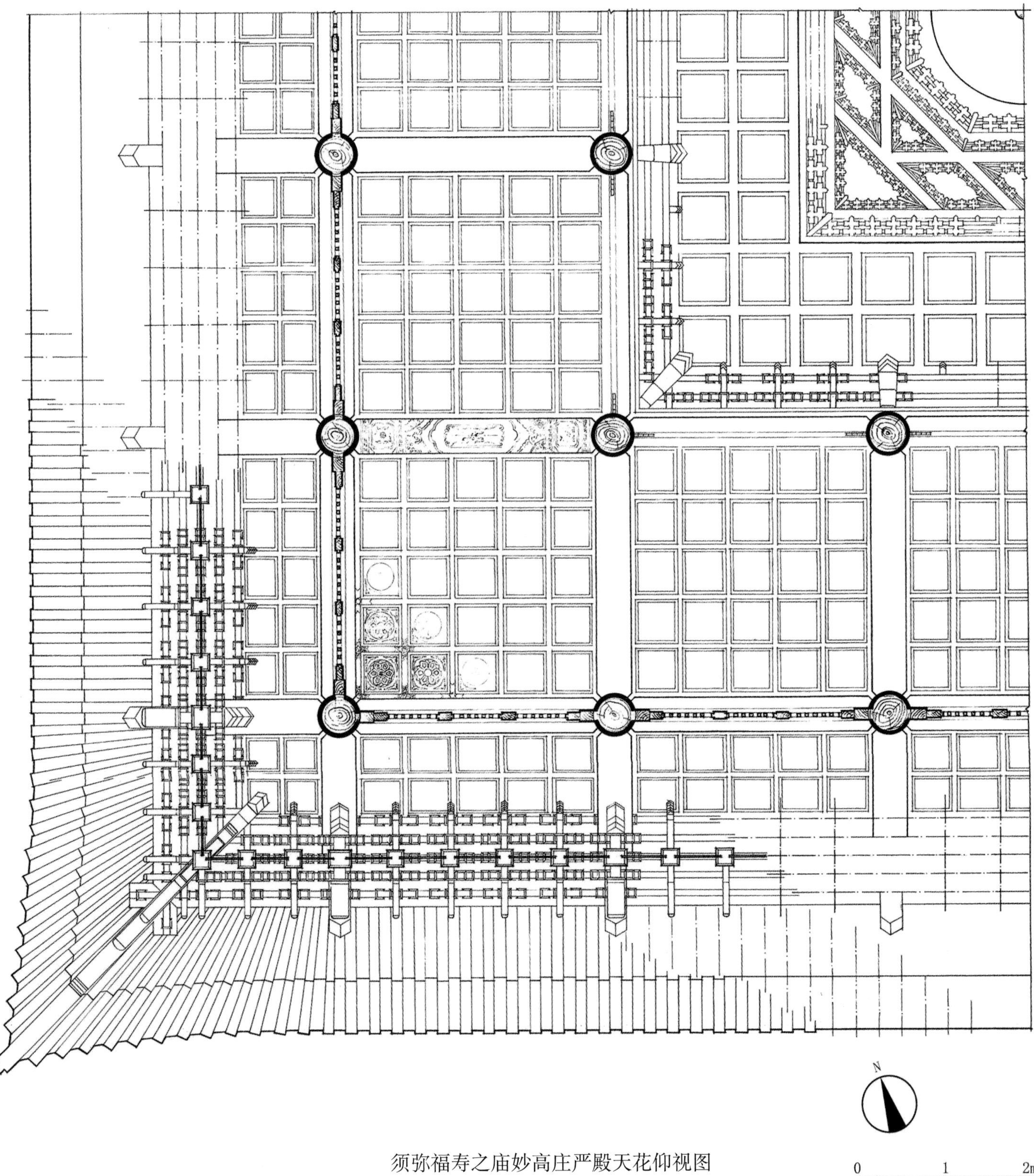

须弥福寿之庙妙高庄严殿天花仰视图

Reflected ceiling plan of Miaogao zhuangyandian of Xumi fushou Temple

须弥福寿之庙妙高庄严殿南立面图
South elevation of Miaogao zhuangyandian of Xumi fushou Temple

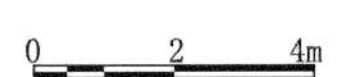

须弥福寿之庙妙高庄严殿剖面图
Section of Miaogao zhuangyandian of Xumi fushou Temple

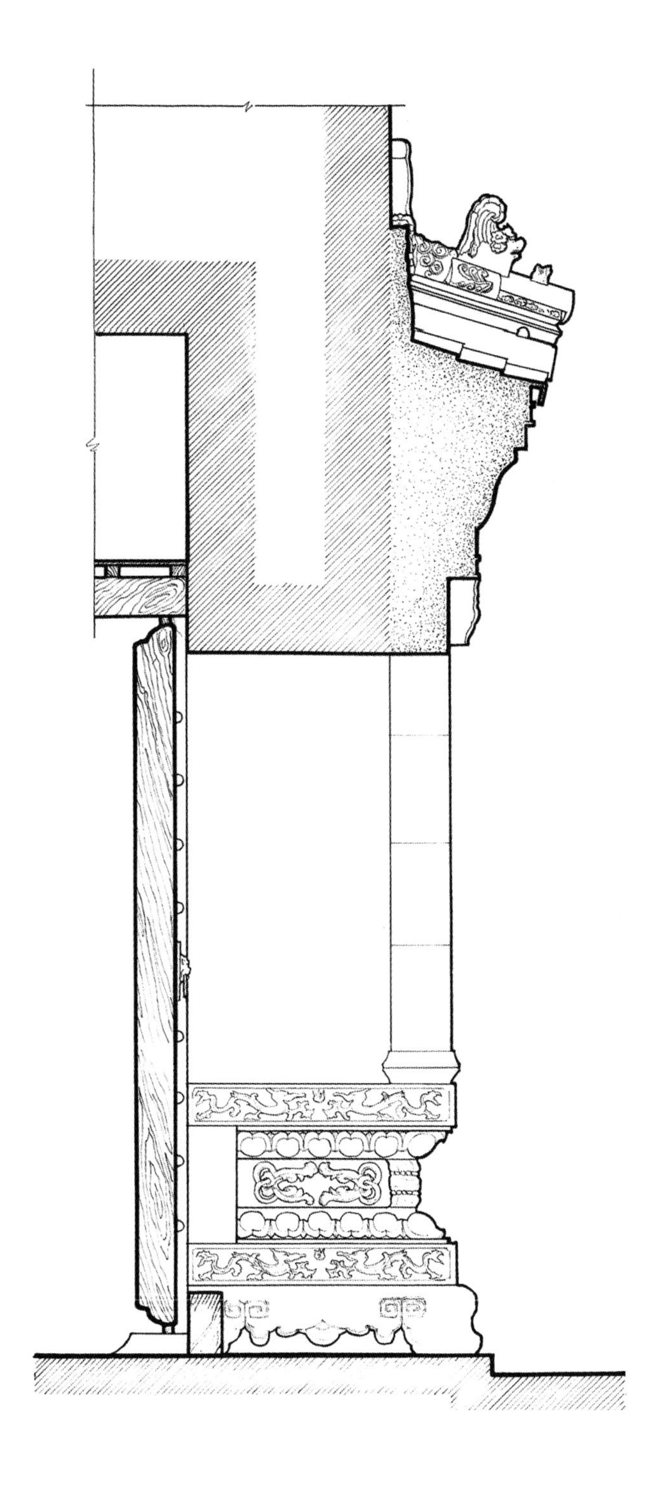

须弥福寿之庙大红台琉璃门大样图
Color-glazed door of Great Red Platform of Xumi fushou Temple

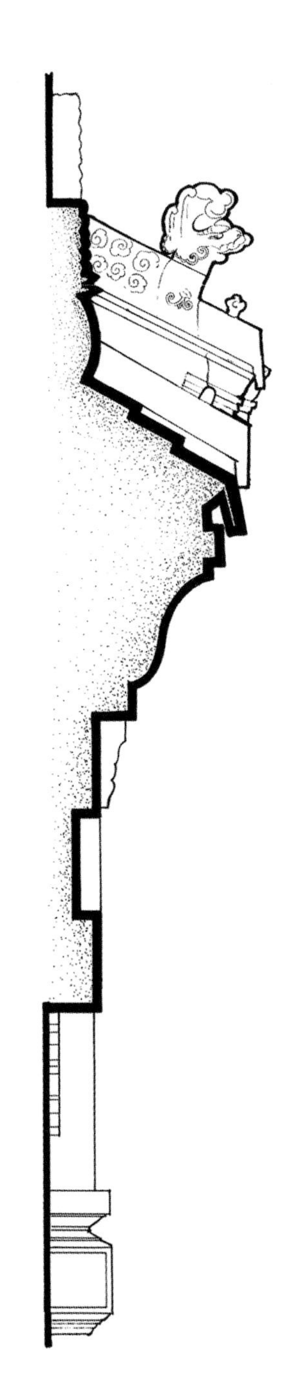

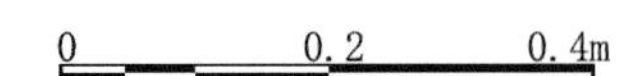

须弥福寿之庙大红台琉璃窗大样图

Color-glazed door and window of Great Red Platform of Xumi fushou Temple

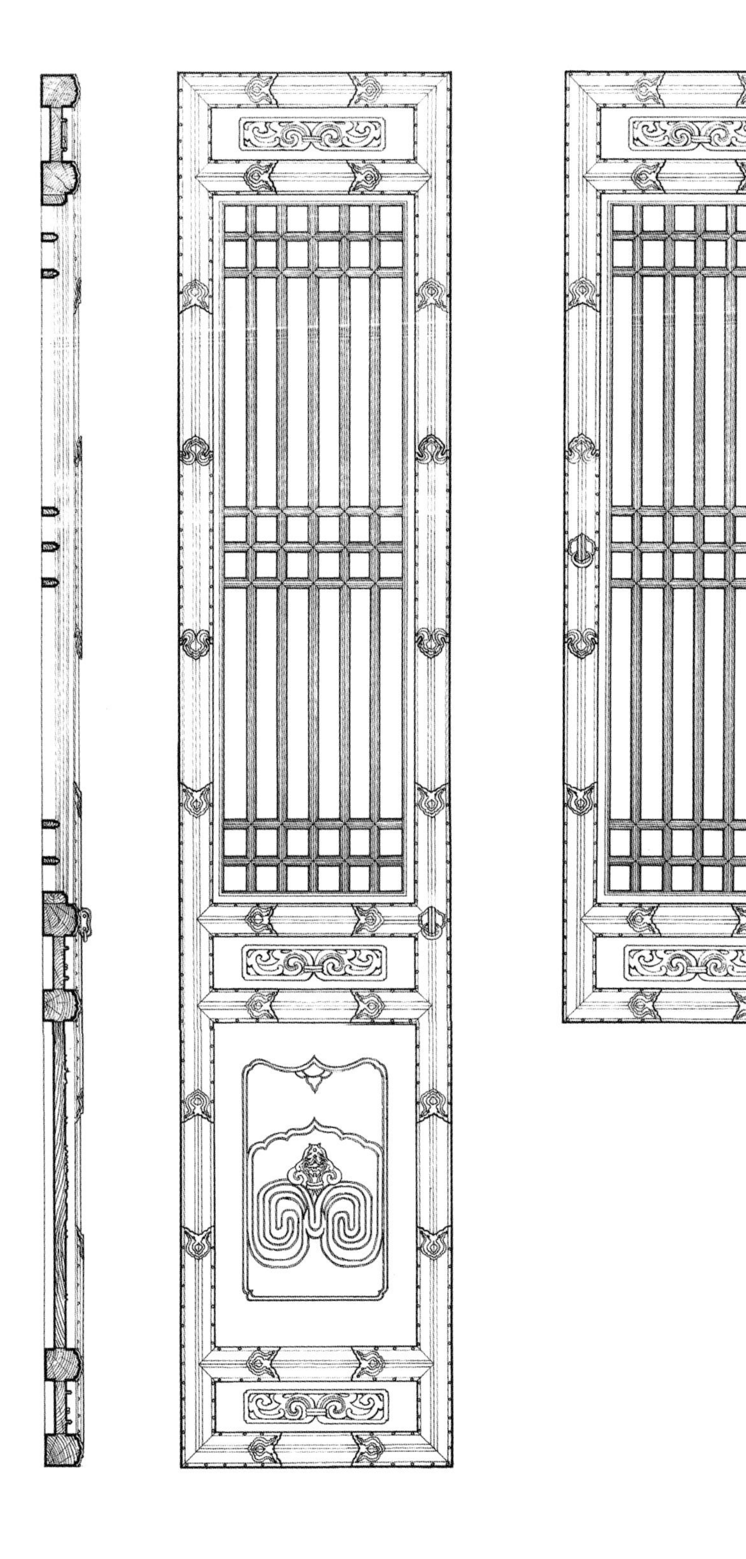

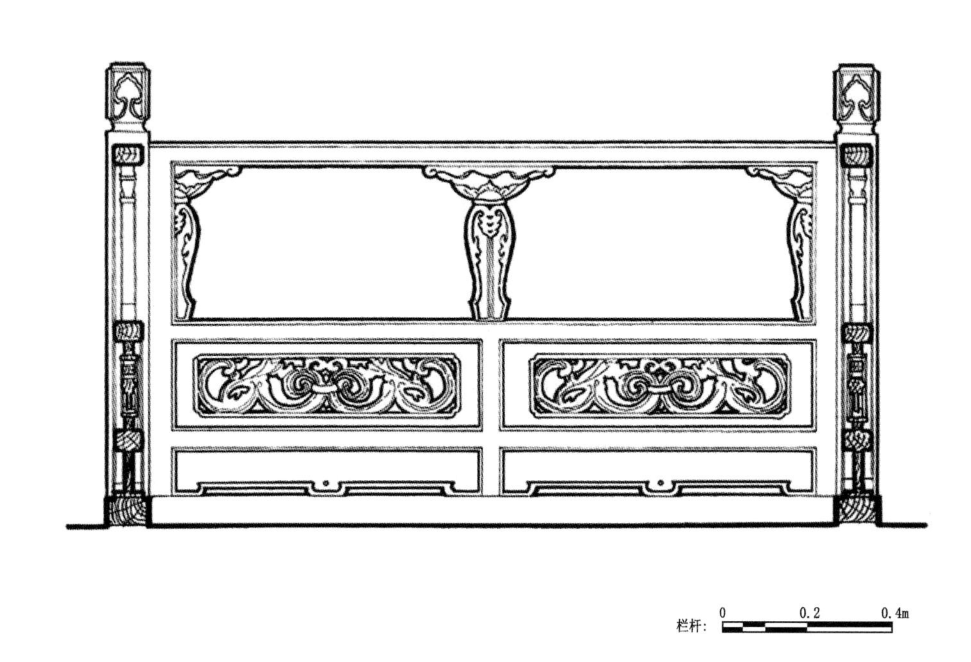

须弥福寿之庙大红台群楼隔扇栏杆大样图
Railing of qunlou of Great Red Platform of Xumi fushou Temple

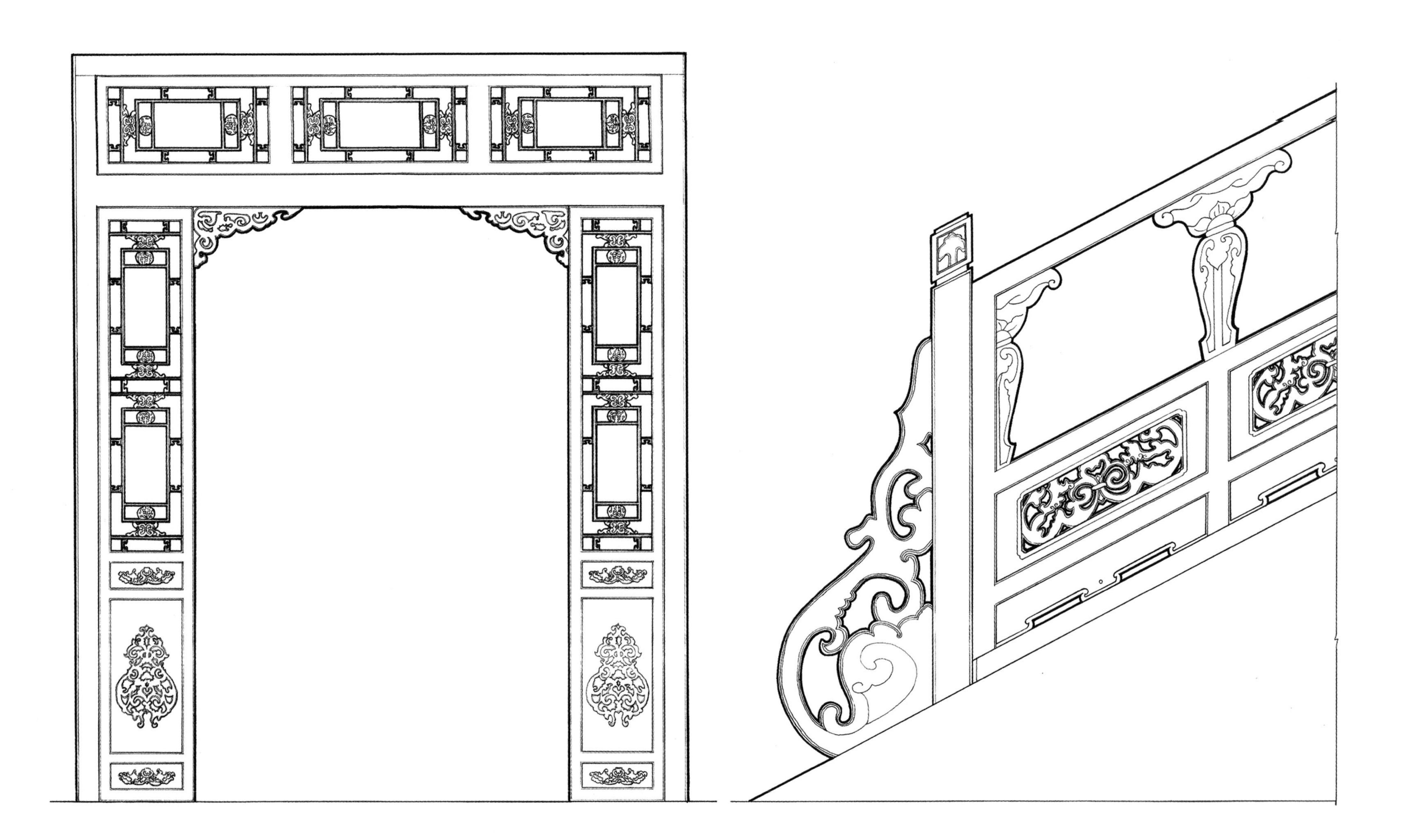

须弥福寿之庙大红台群楼楠木罩、楼梯栏杆大样图
Nanmuzhao and staircase railing of qunlou of Great Red Platform of Xumi fushou Temple

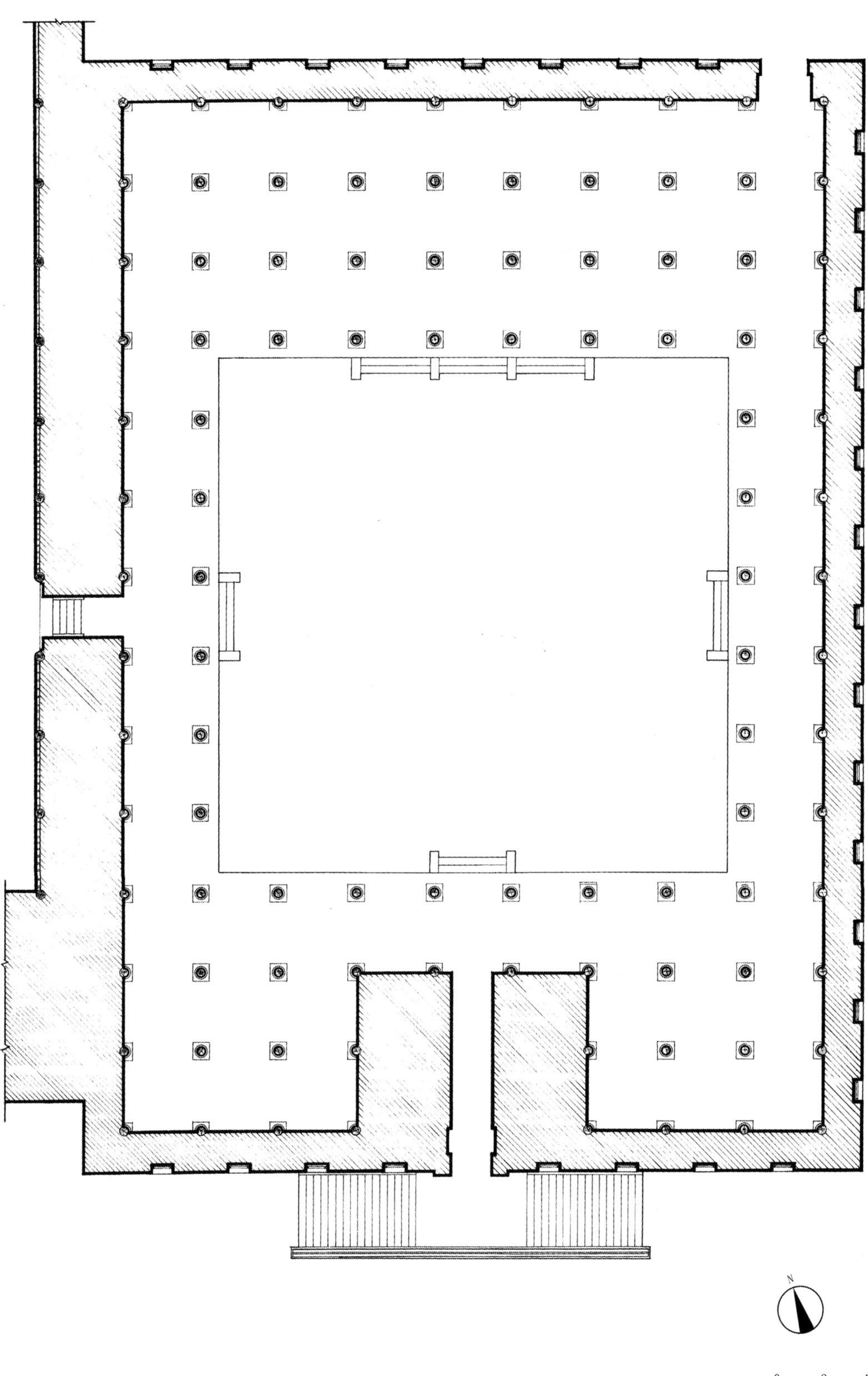

须弥福寿之庙东红台平面图
Plan of East Red Platform of Xumi fushou Temple

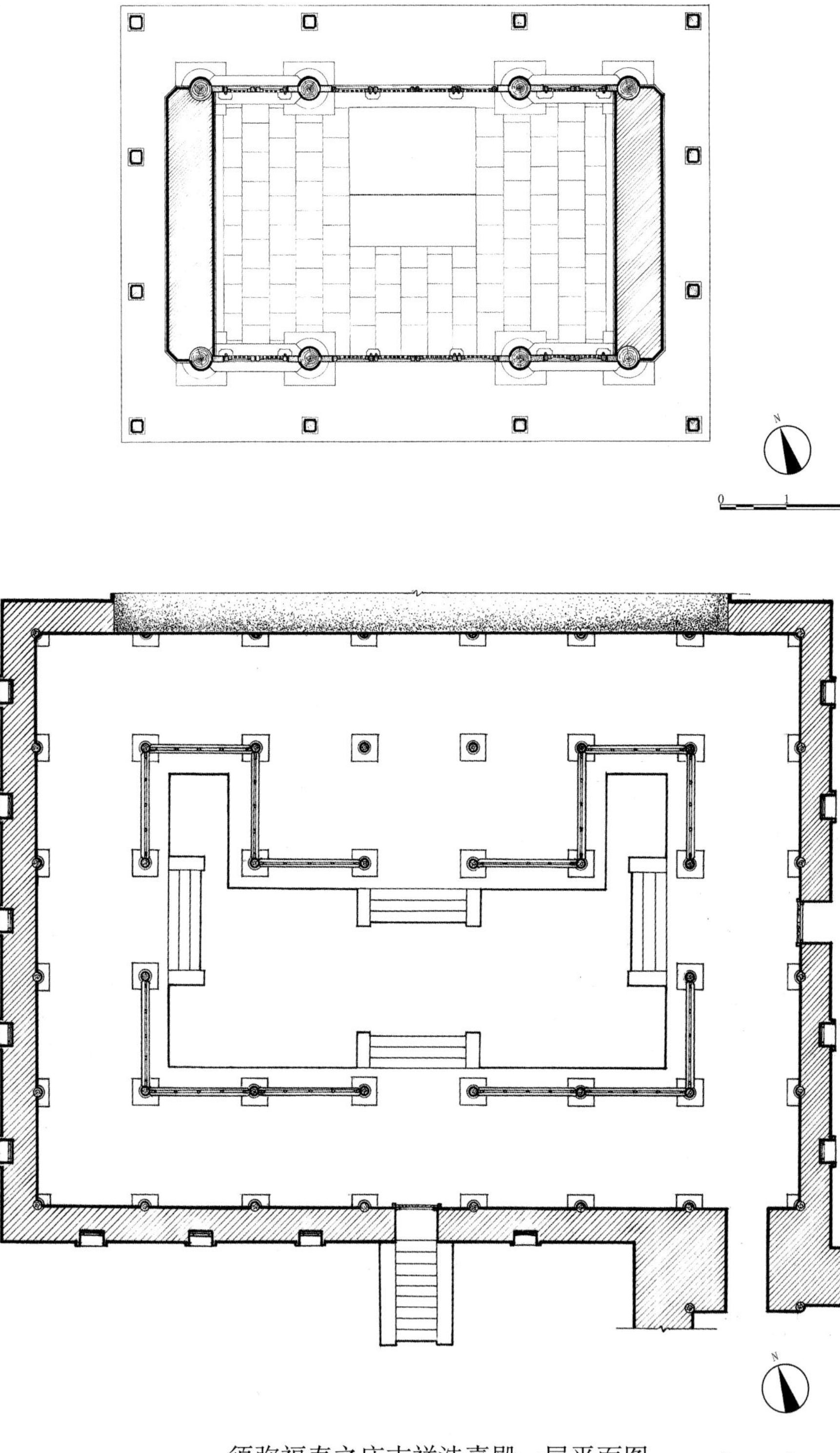

须弥福寿之庙吉祥法喜殿一层平面图
Plan of first floor of Jixiang faxidian of Xumi fushou Temple

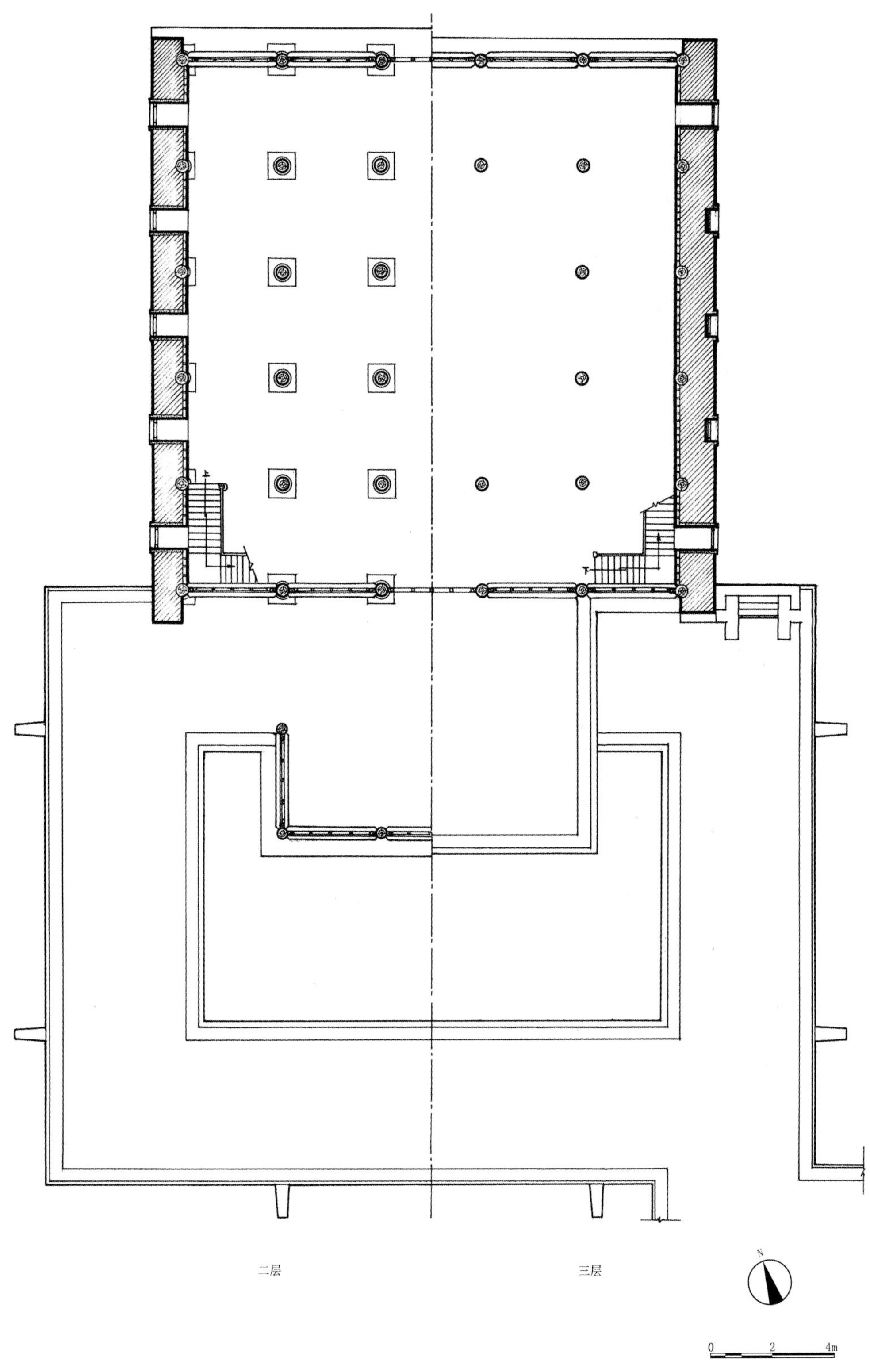

须弥福寿之庙吉祥法喜殿二层、三层平面图

Plan of second floor and third floor of Jixiang faxidian of Xumi fushou Temple

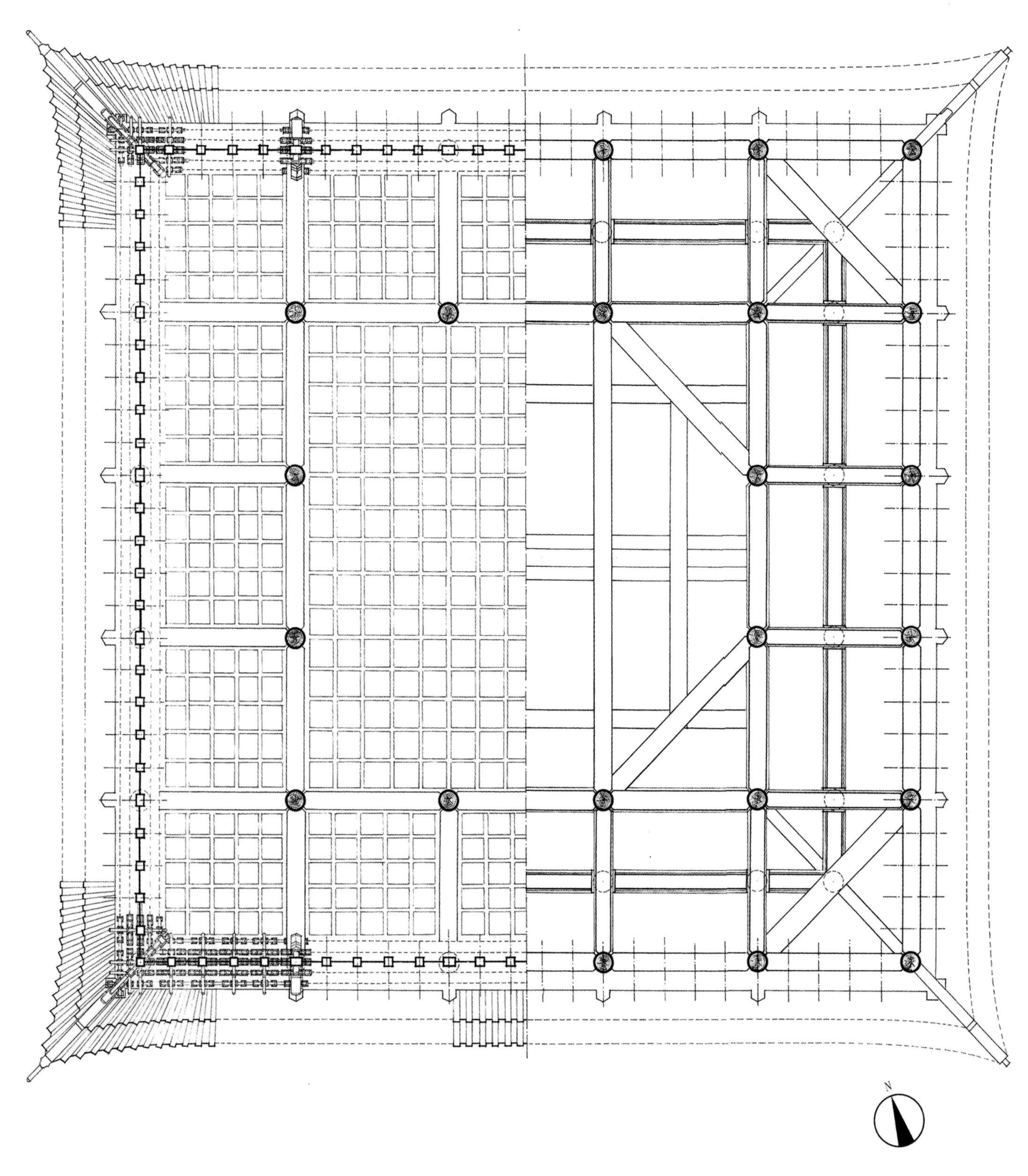

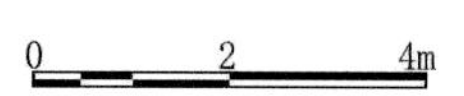

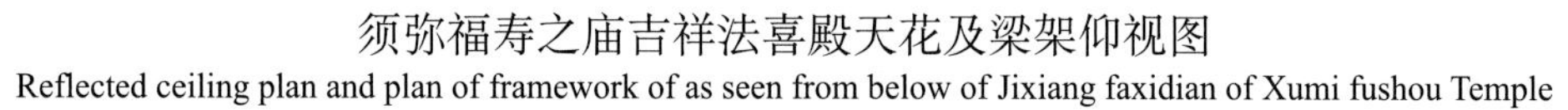

须弥福寿之庙吉祥法喜殿天花及梁架仰视图

Reflected ceiling plan and plan of framework of as seen from below of Jixiang faxidian of Xumi fushou Temple

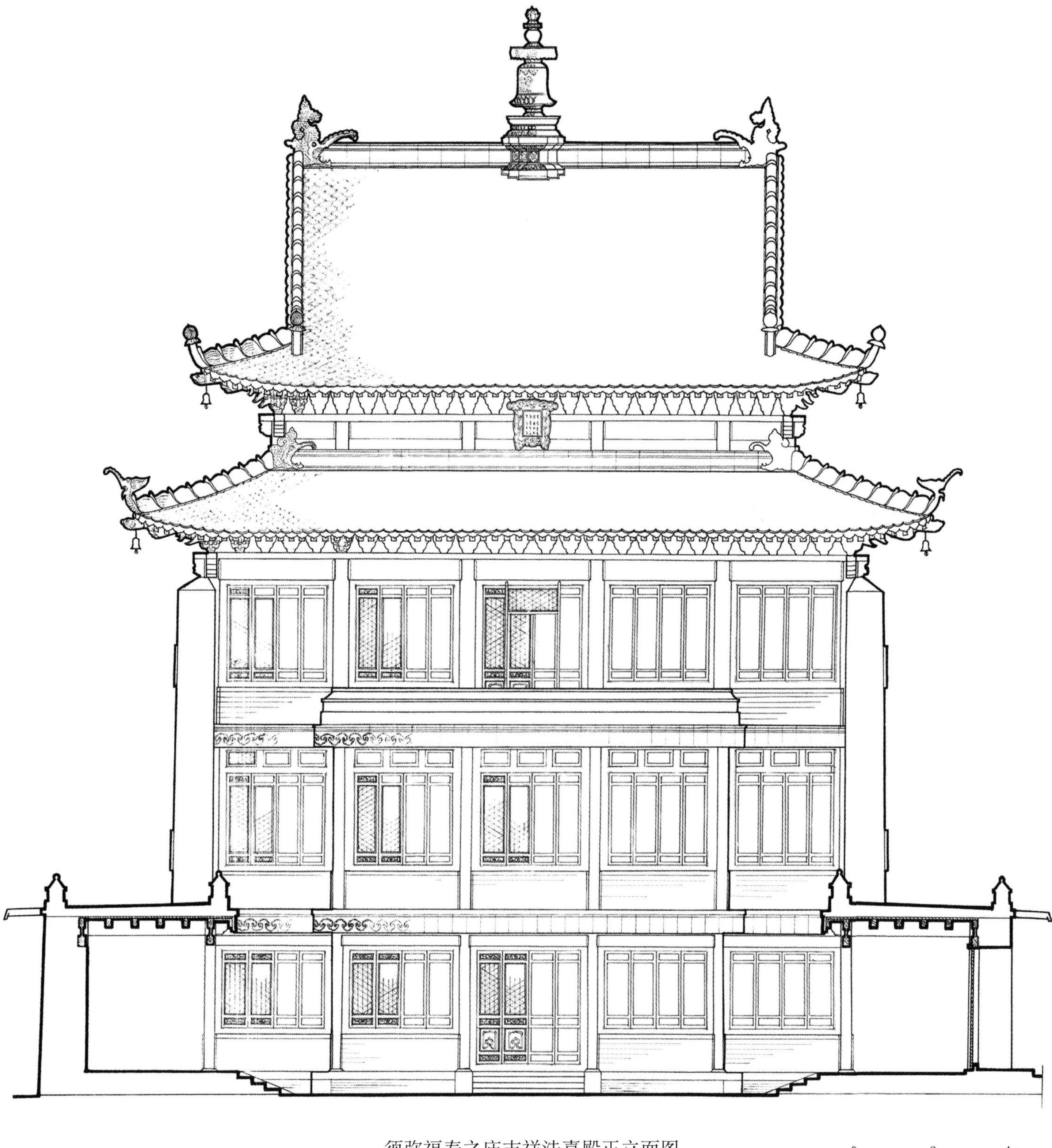

须弥福寿之庙吉祥法喜殿正立面图
Front elevation of Jixiang faxidian of Xumi fushou Temple

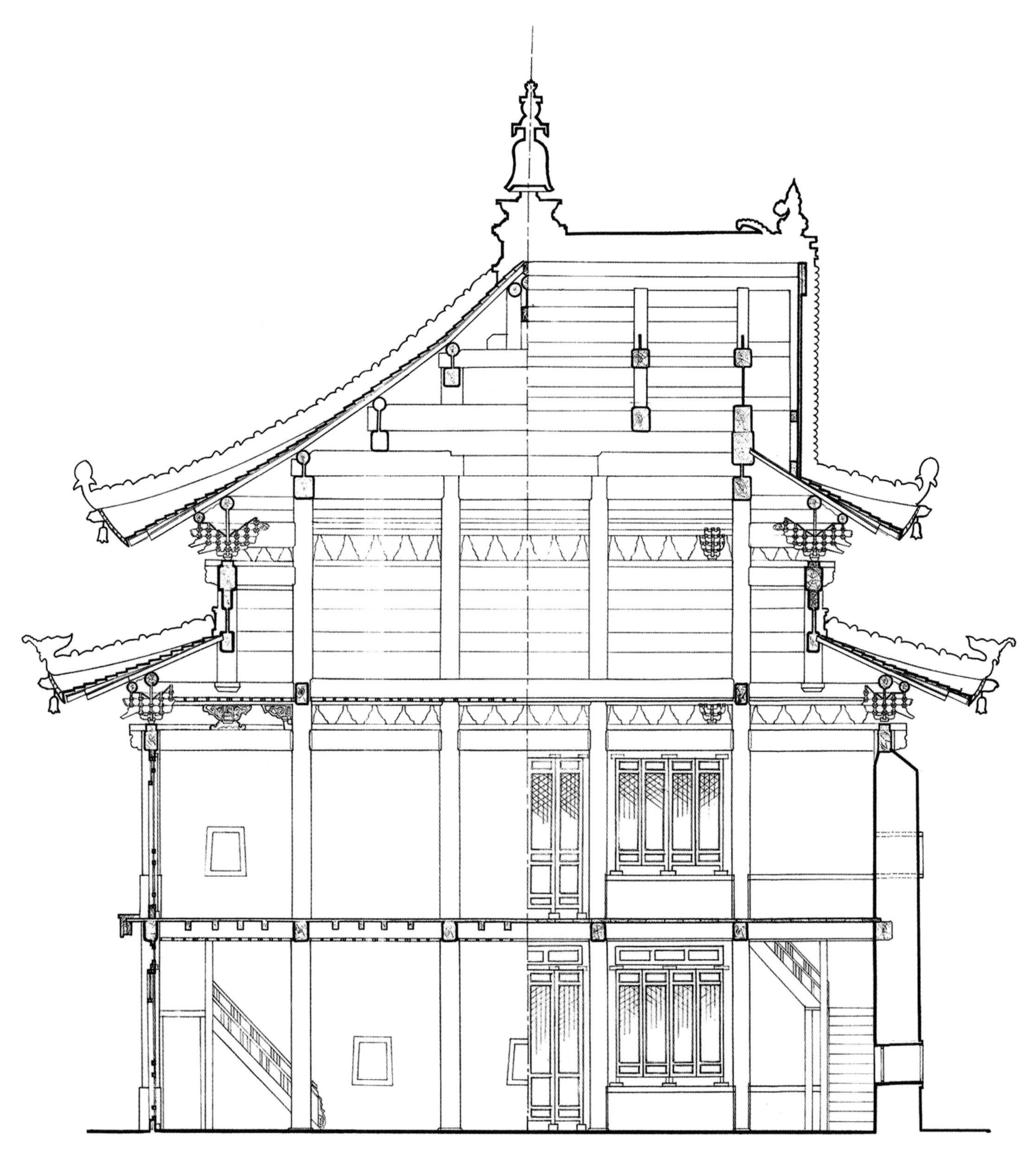

须弥福寿之庙吉祥法喜殿剖面图
Section of Jixiang faxidian of Xumi fushou Temple

0 2 4m

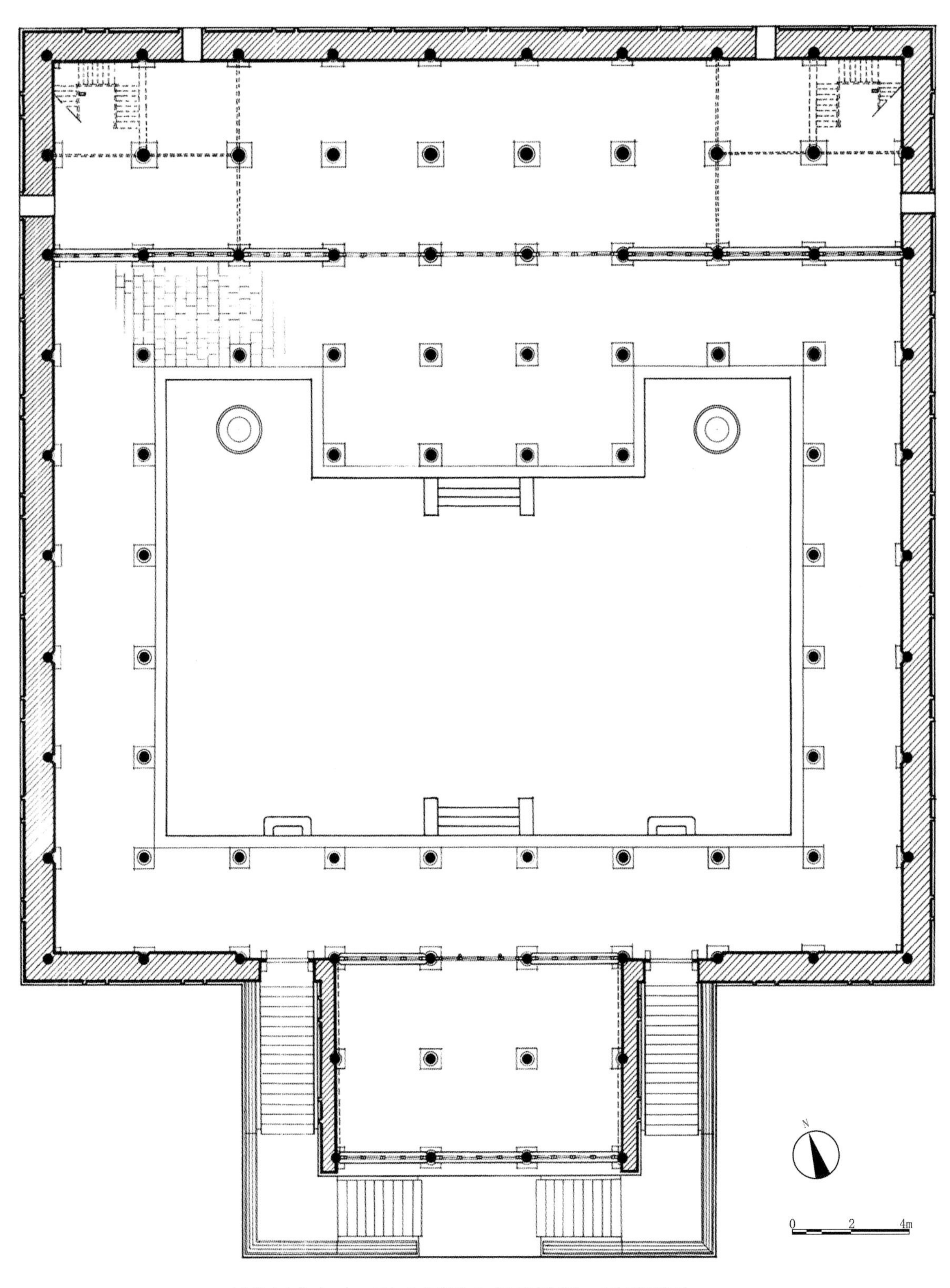

须弥福寿之庙万法宗源殿金贺堂组群一层平面图
Plan of first floor of building cluster around Jinhetang of Wanfa zongyuandian of Xumi fushou Temple

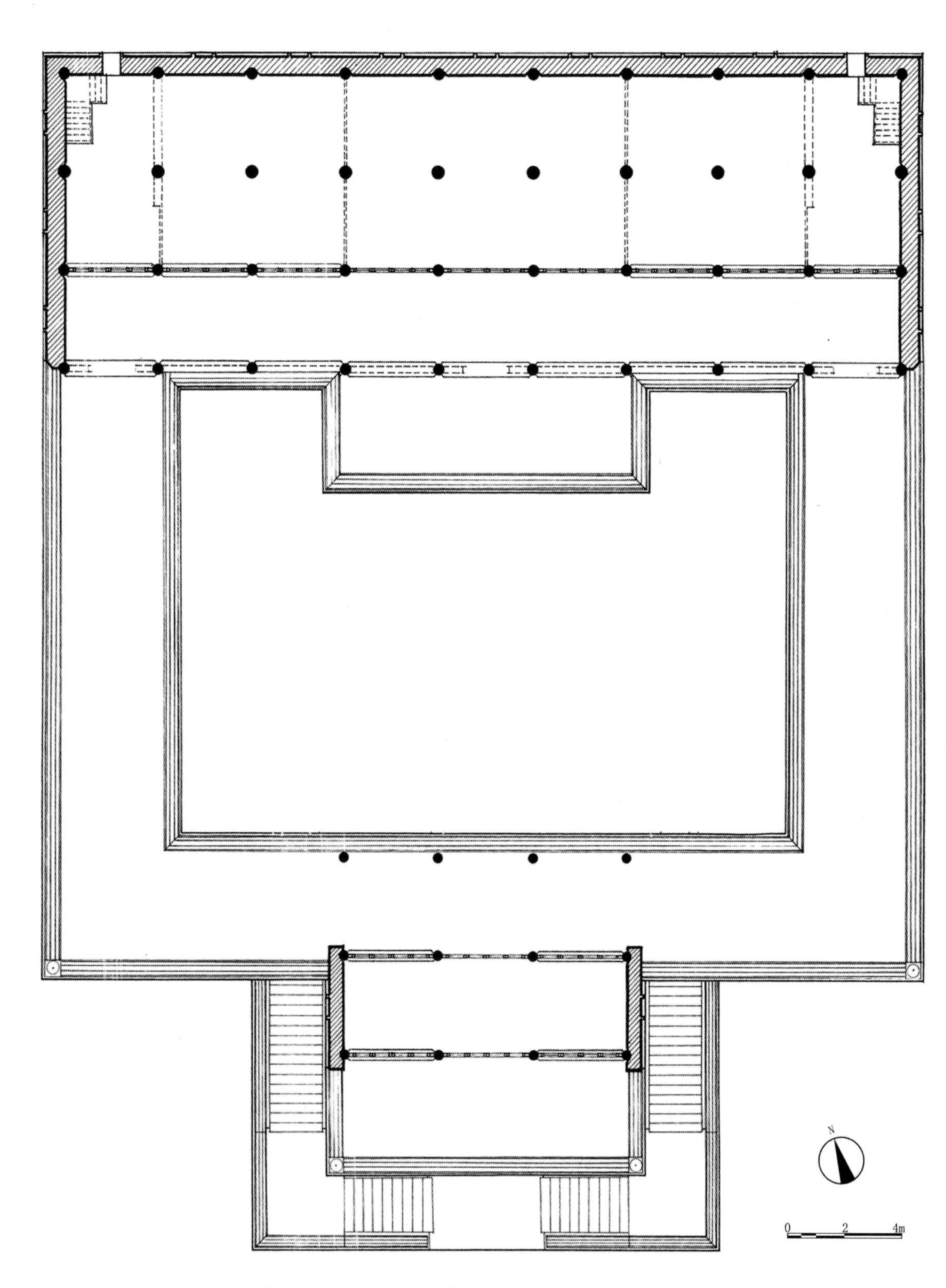

须弥福寿之庙万法宗源殿金贺堂组群二层平面图
Plan of second floor of building cluster around Jinhetang of Wanfa zongyuandian of Xumi fushou Temple

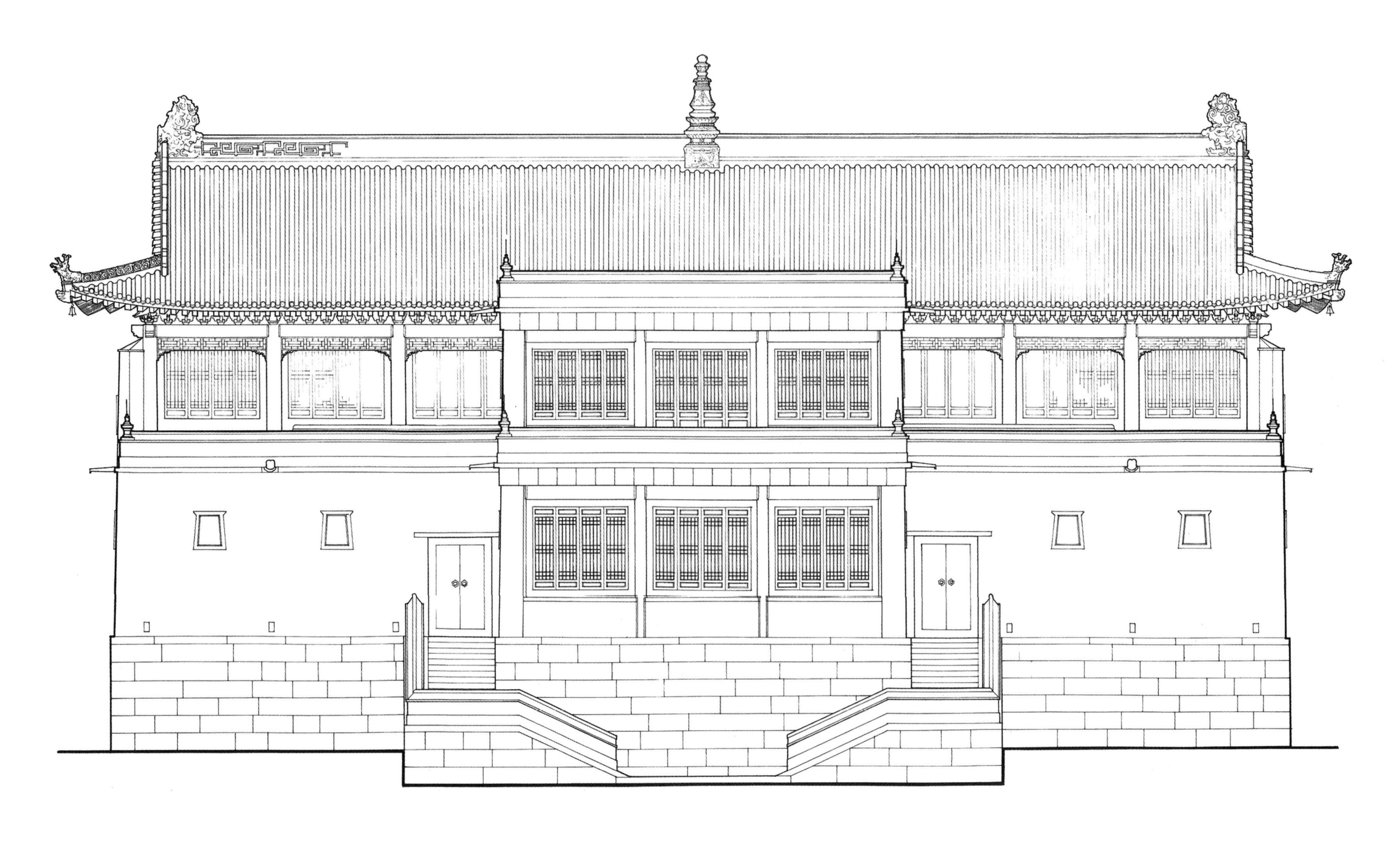

须弥福寿之庙万法宗源殿金贺堂组群正立面图

Front elevation of building cluster around Jinhetang of Wanfa zongyuandian of Xumi fushou Temple

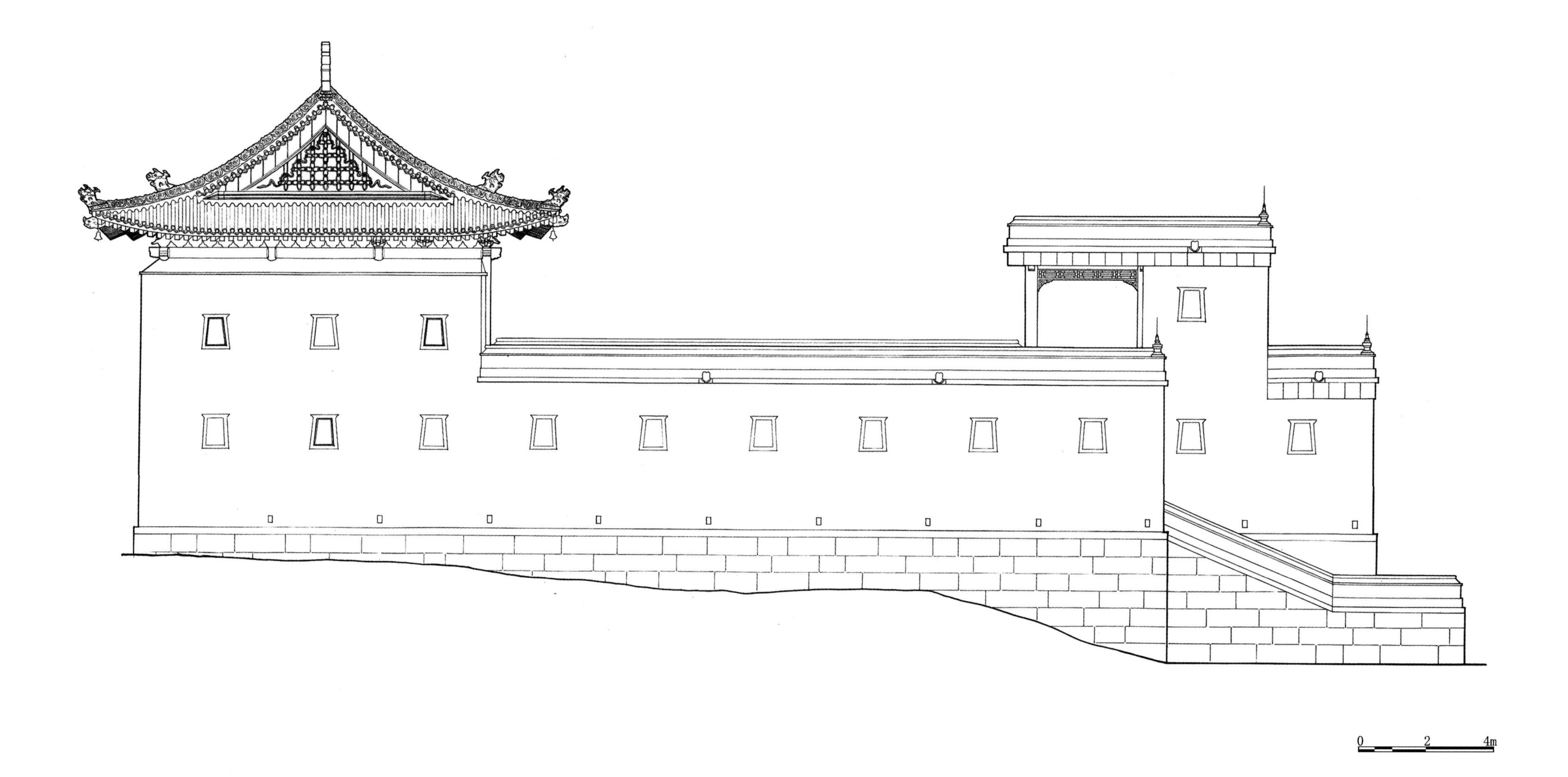

须弥福寿之庙万法宗源殿金贺堂组群侧立面图
Side elevation of building cluster around Jinhetang of Wanfa zongyuandian of Xumi fushou Temple

0 2 4m

须弥福寿之庙万法宗源殿金贺堂组群纵剖面北望图

Longitudinal section of building cluster around Jinhetang of Wanfa zongyuandian of Xumi fushou Temple looking north

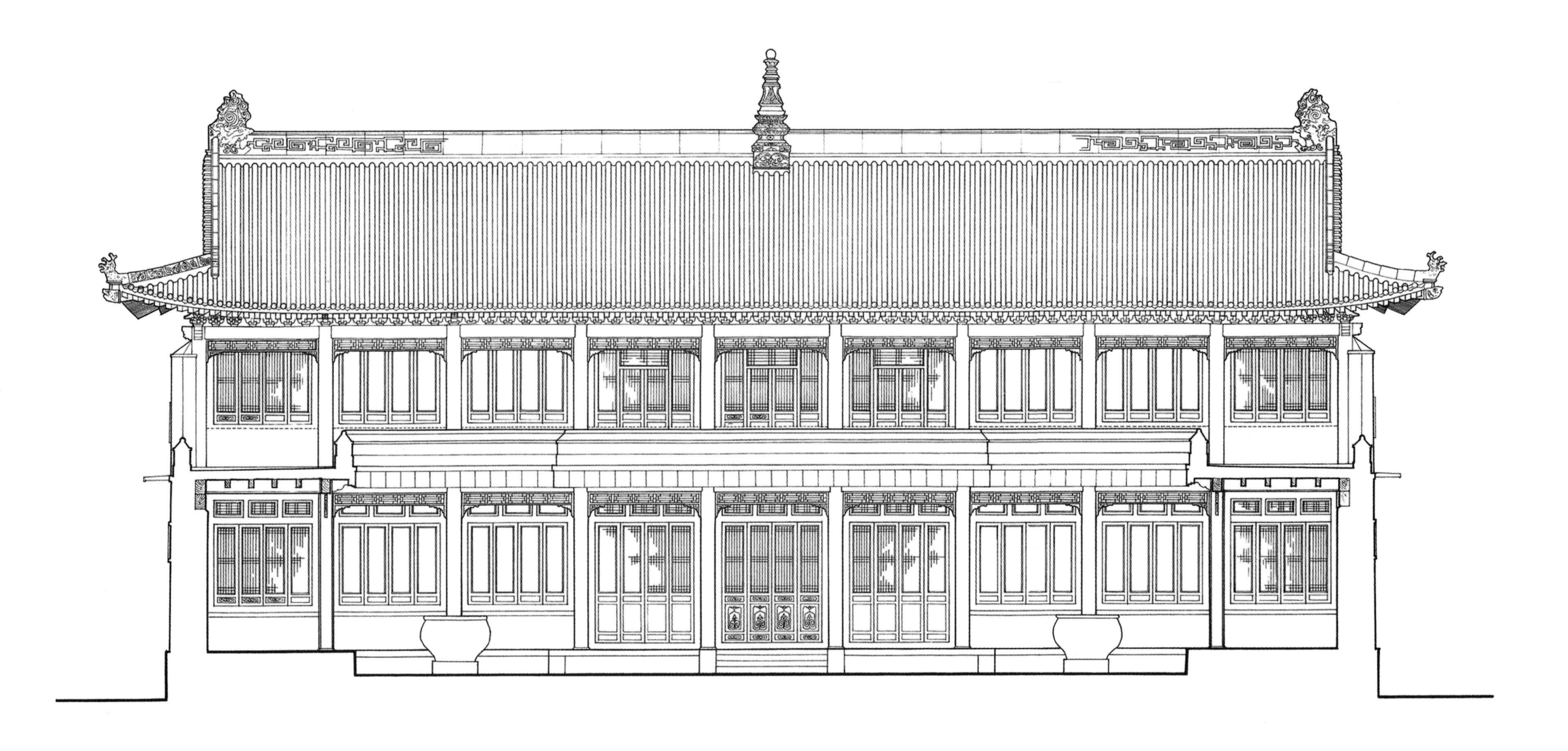

须弥福寿之庙万法宗源殿金贺堂组群纵剖面南望图
Longitudinal section of building cluster around Jinhetang of Wanfa zongyuandian of Xumi fushou Temple looking south

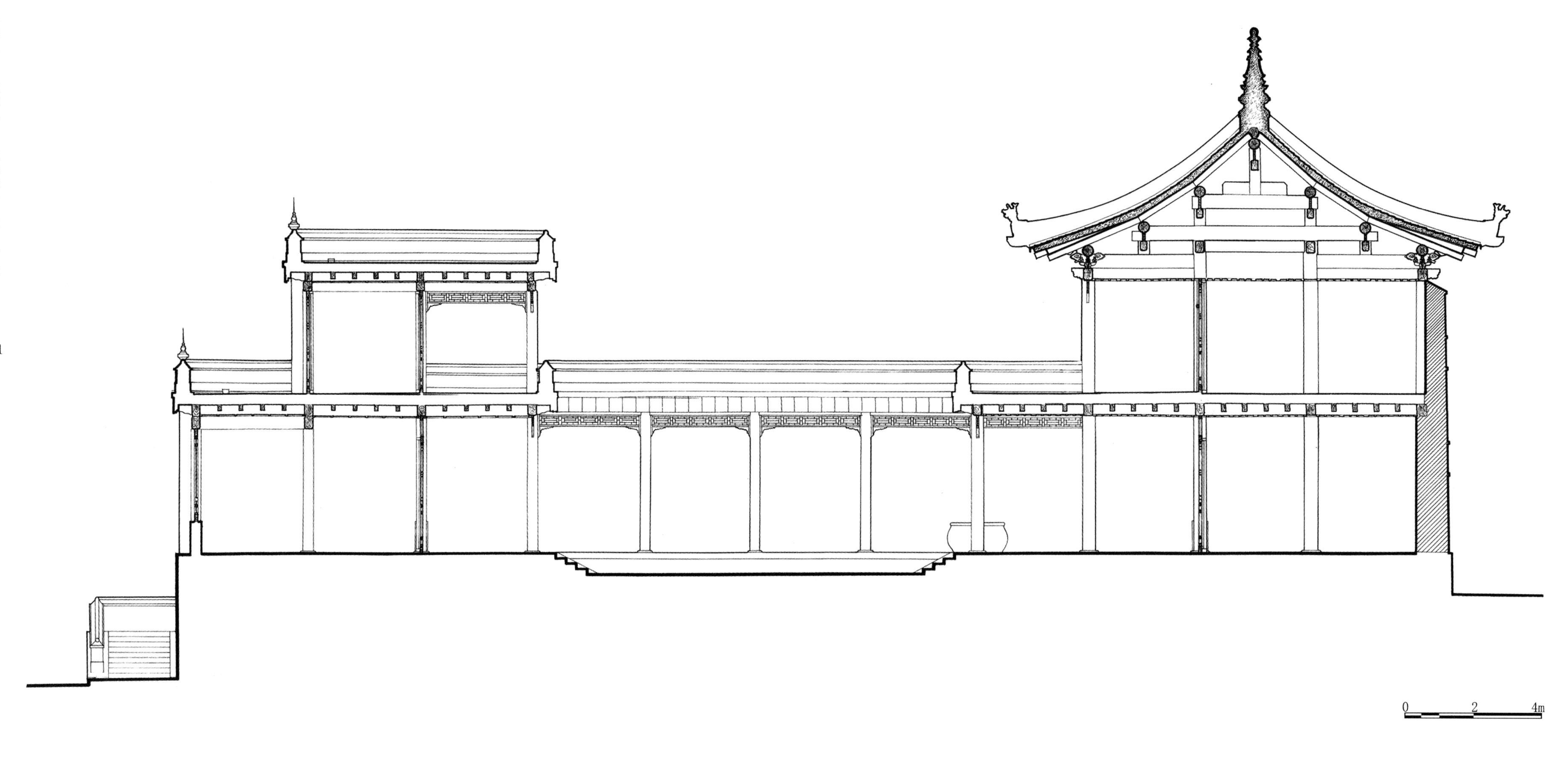

须弥福寿之庙万法宗源殿金贺堂组群明间剖面图
Section of central-bay of building cluster around Jinhetang of Wanfa zongyuandian of Xumi fushou Temple

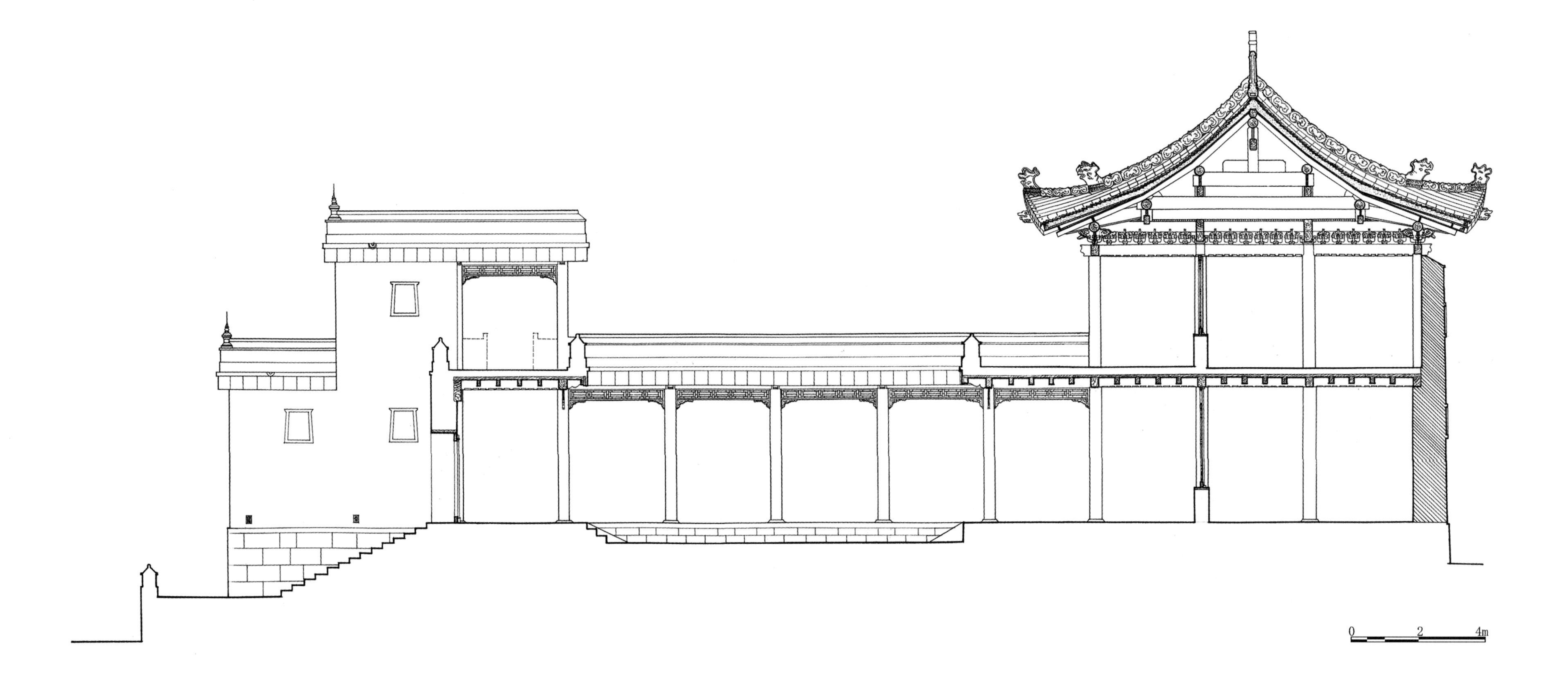

须弥福寿之庙万法宗源殿金贺堂组群梢间剖面图

Section of second-to-last-bay of building cluster around Jinhetang of Wanfa zongyuandian of Xumi fushou Temple

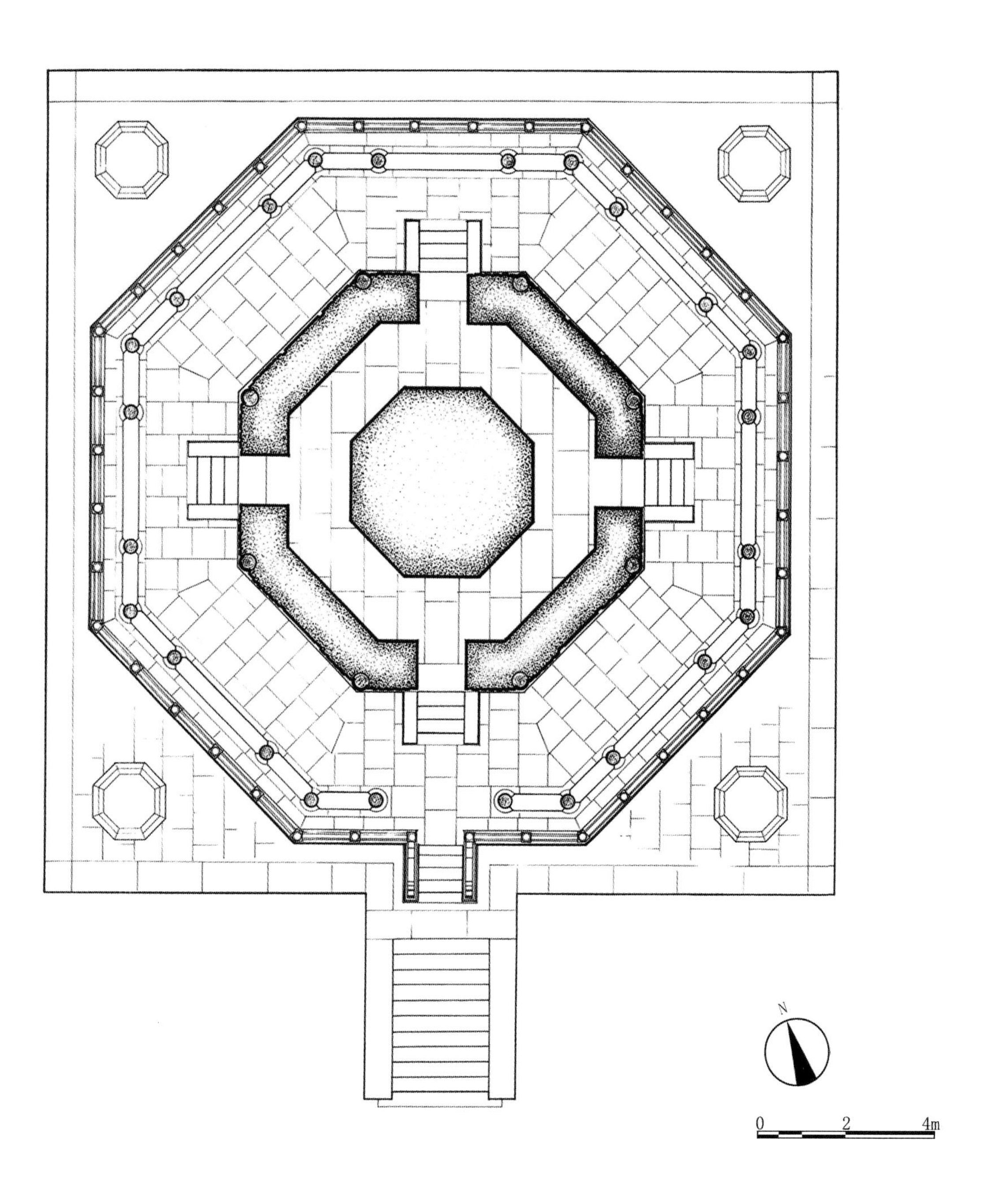

须弥福寿之庙琉璃宝塔底层平面图
Plan of basement of Liuli baota of Xumi fushou Temple

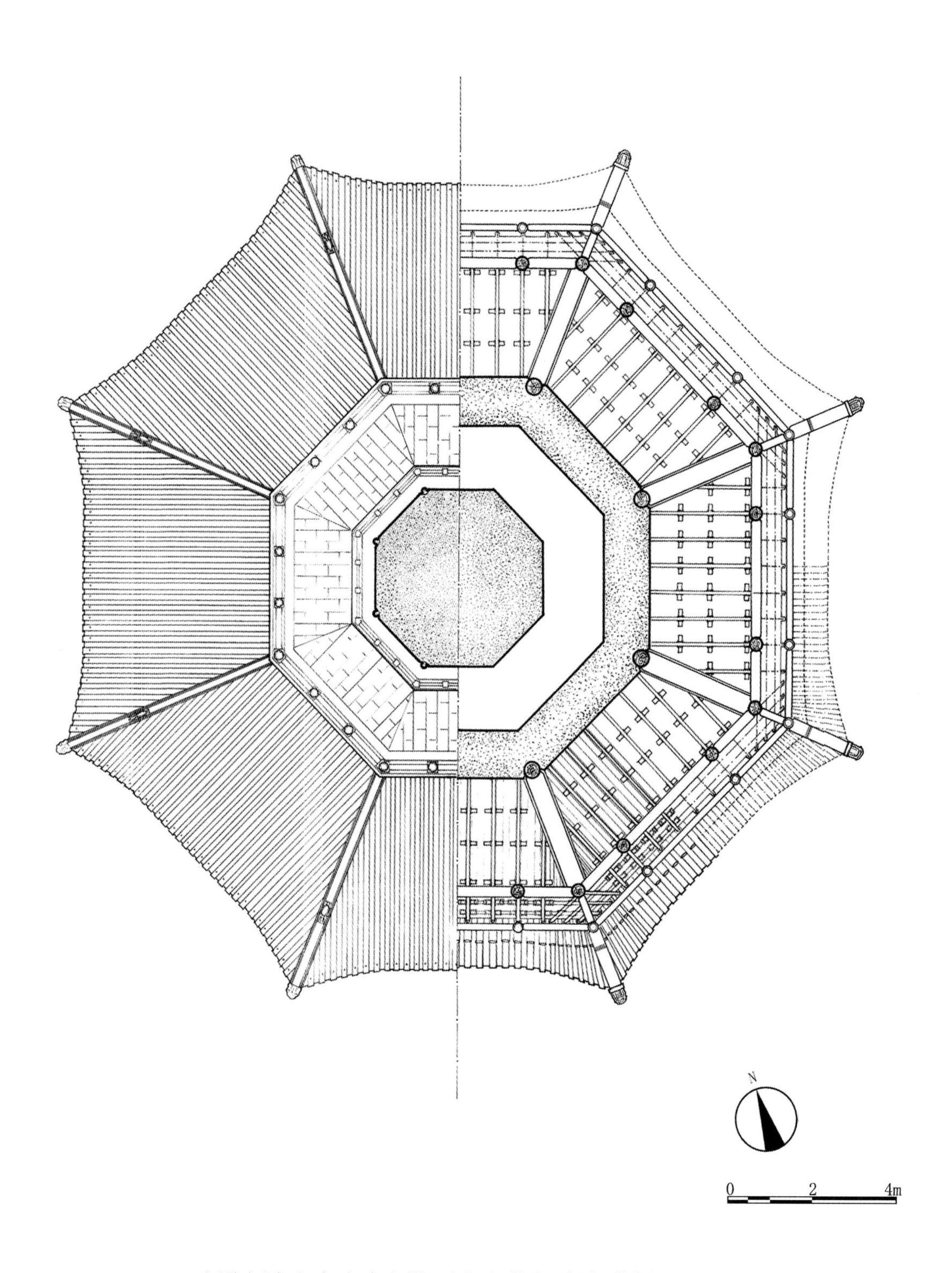

须弥福寿之庙琉璃宝塔二层平面图、梁架仰视图
Plan of second floor and plan of framework as seen from below of Liuli baota of Xumi fushou Temple

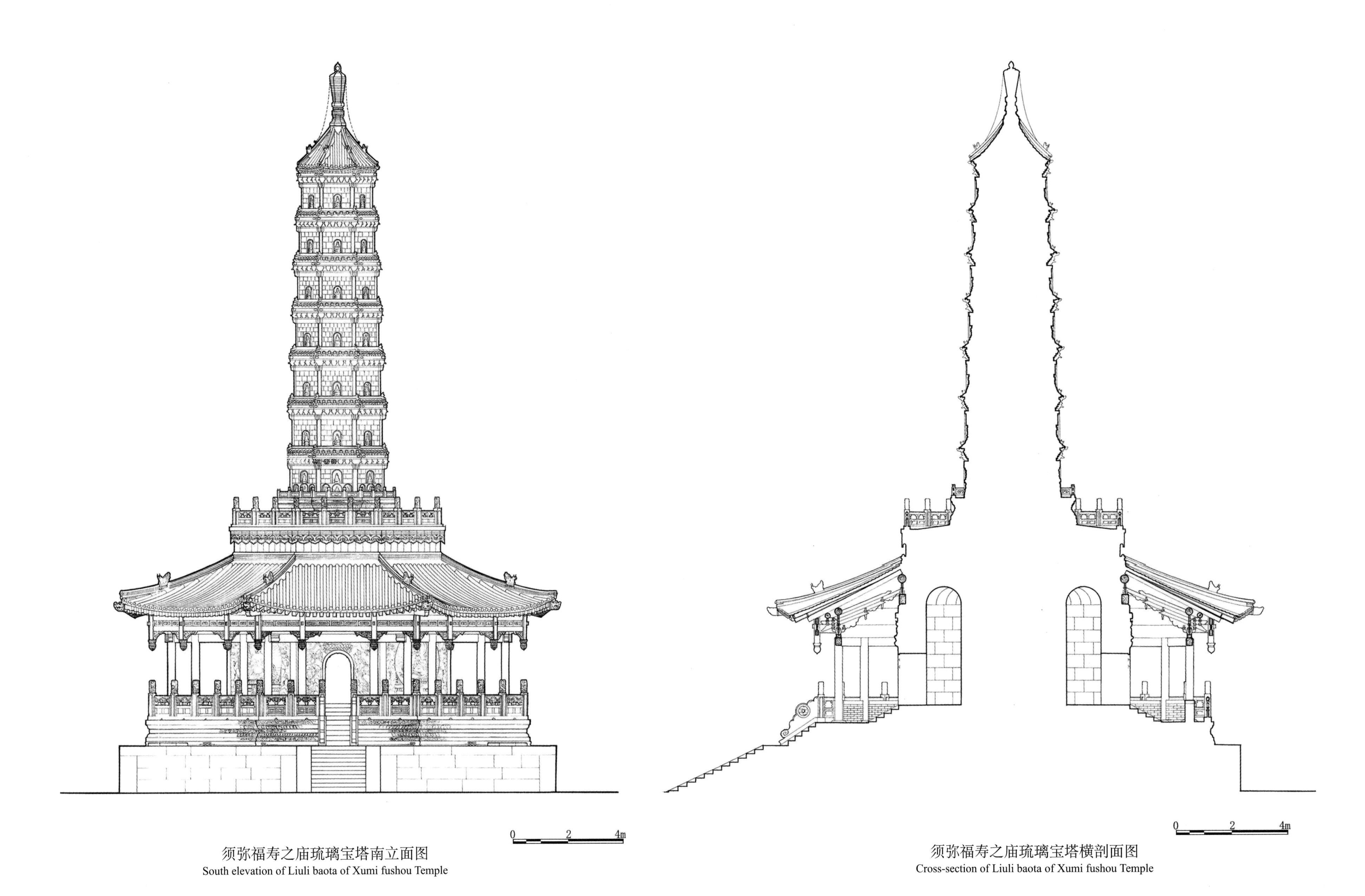

须弥福寿之庙琉璃宝塔南立面图
South elevation of Liuli baota of Xumi fushou Temple

须弥福寿之庙琉璃宝塔横剖面图
Cross-section of Liuli baota of Xumi fushou Temple

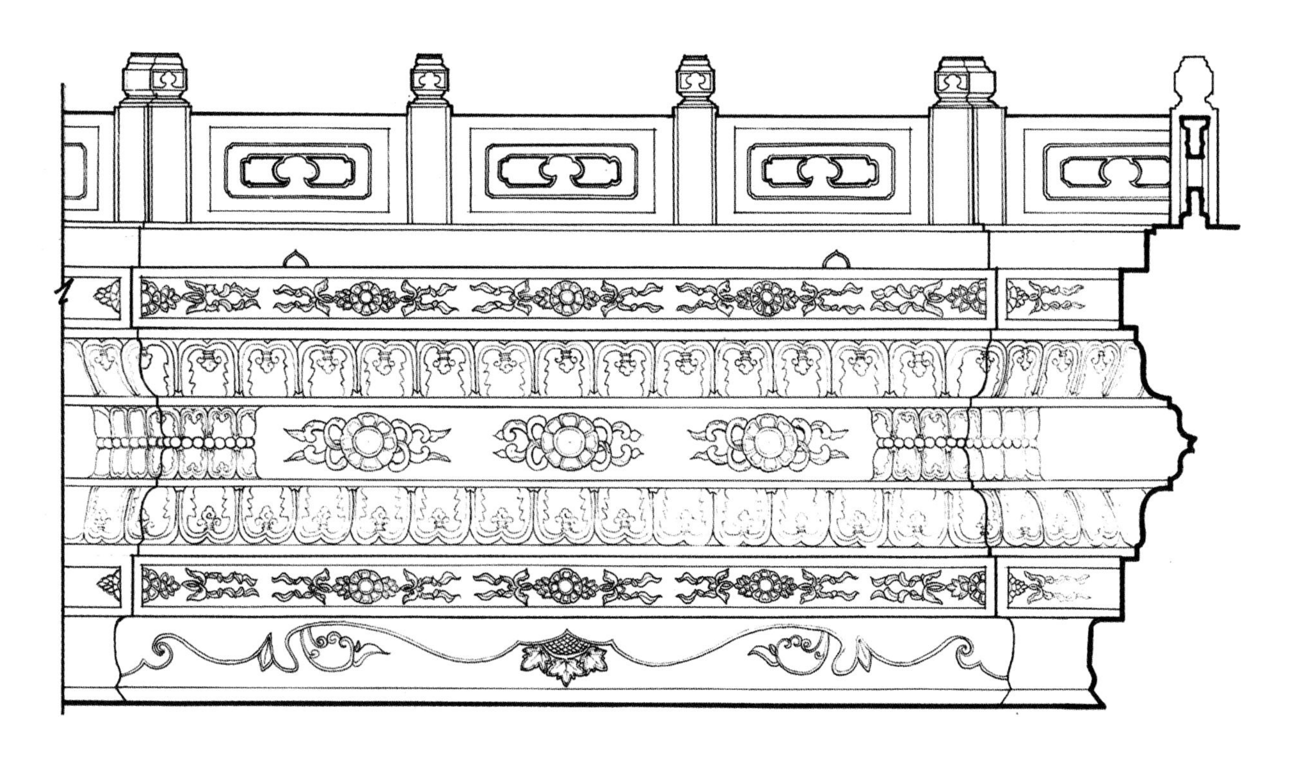

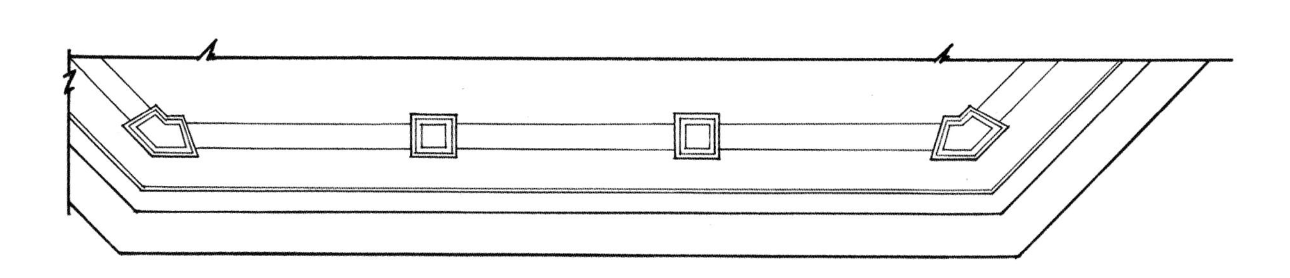

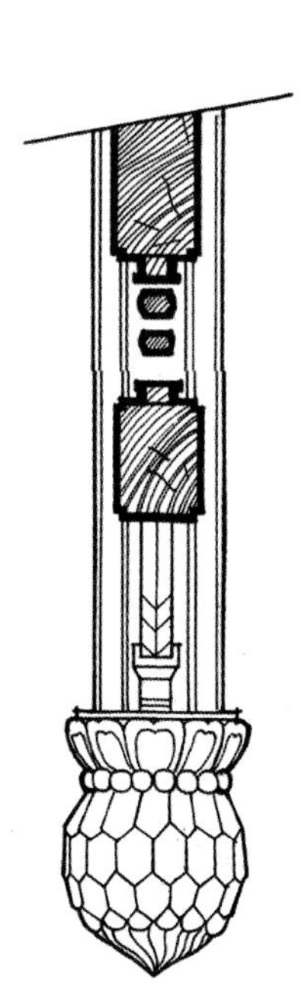

须弥福寿之庙琉璃宝塔大样图

Liuli baota of Xumi fushou Temple

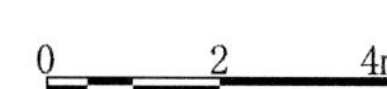

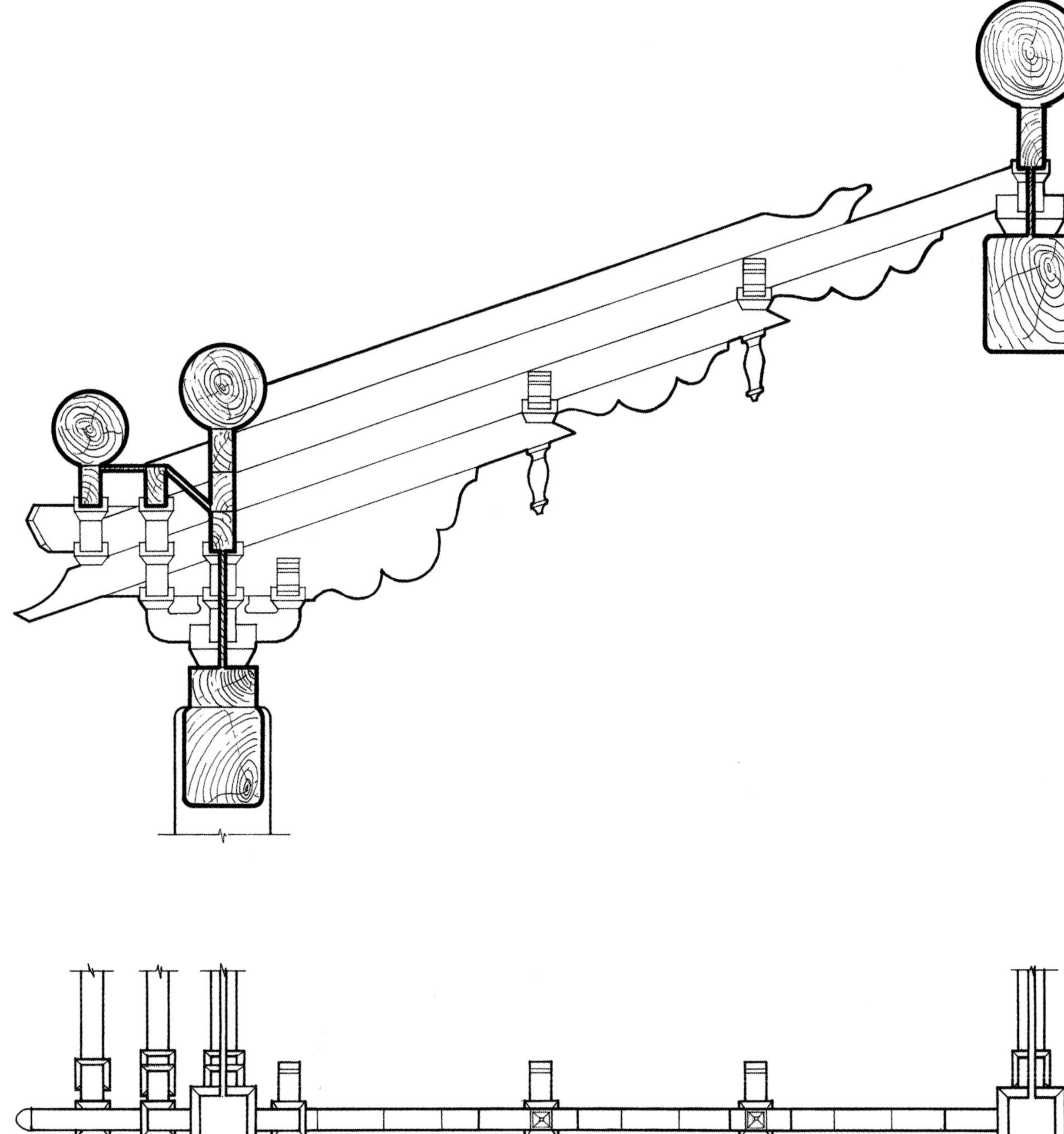

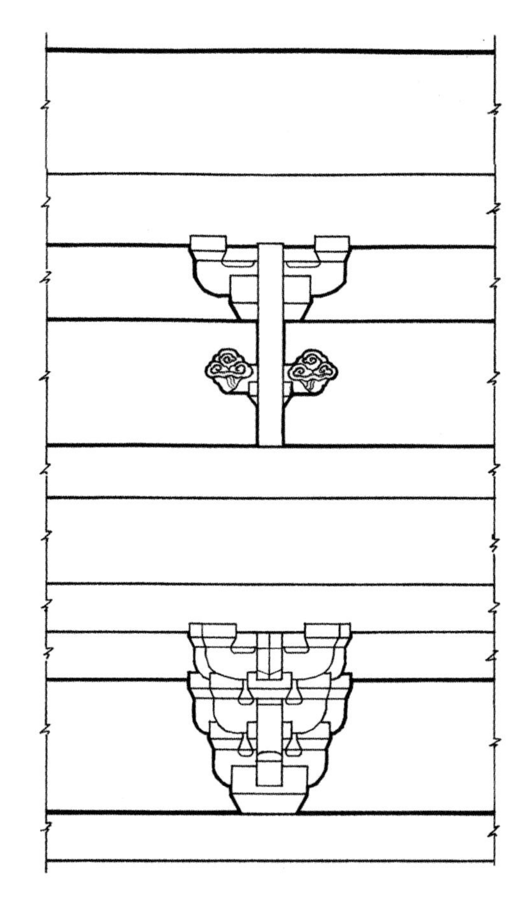

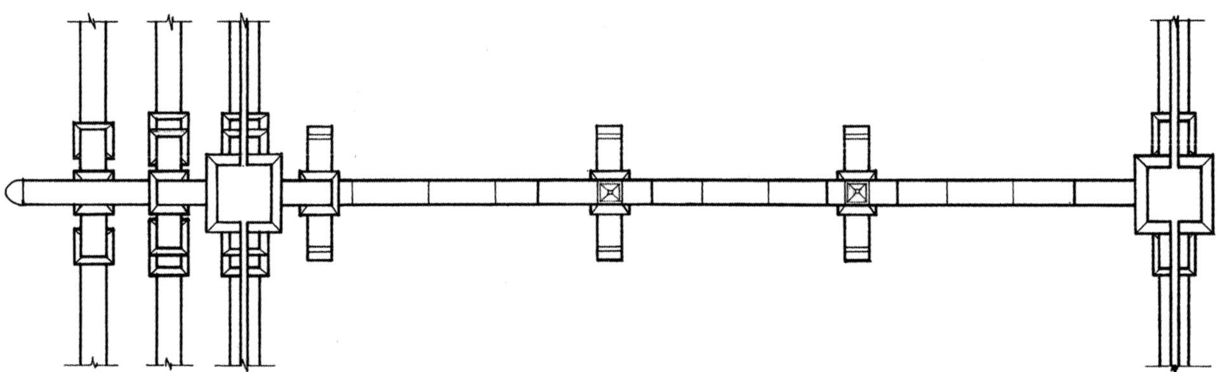

须弥福寿之庙琉璃宝塔溜金斗栱大样图

Liujin bracket set of Liuli baota of Xumi fushou Temple

殊像寺
Shuxiangsi

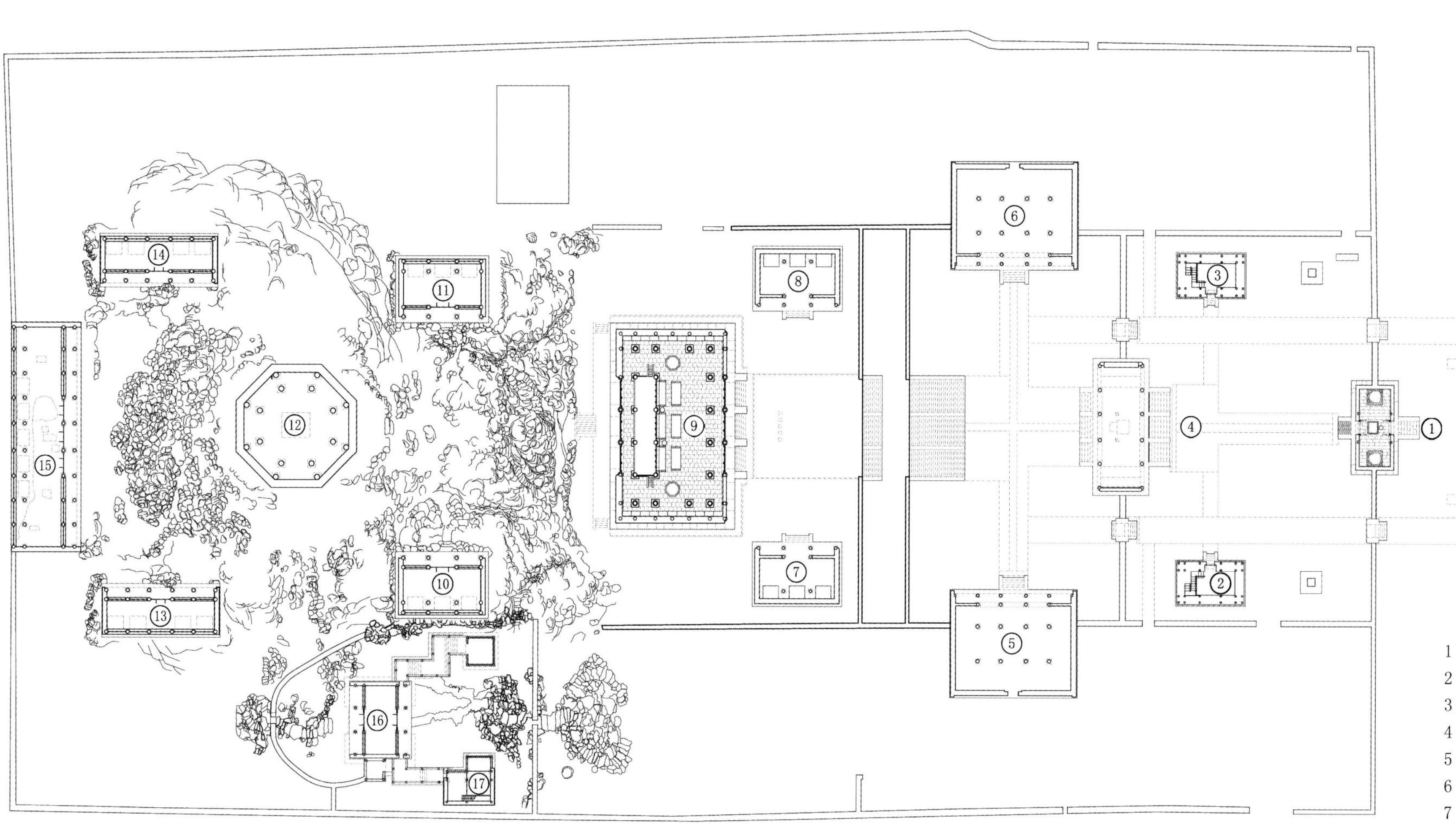

1 山门　Gate Hall
2 鼓楼　Gulou (Drum-tower)
3 钟楼　Zhonglou (Bell-tower)
4 天王殿（遗址）　Hall of Heavenly King (Site)
5 演梵堂（遗址）　Yanfantang Hall (Site)
6 馔香堂（遗址）　Zhuanxiangtang Hall (Site)
7 面月殿（遗址）　Mianyuedian Hall (Site)
8 指峰殿（遗址）　Zhifengdian Hall (Site)
9 会乘殿　Huichengdian Hall
10 雪静殿（遗址）　Xuejingdian Hall (Site)
11 云来殿（遗址）　Yunlaidian Hall (Site)
12 宝相阁　Baoxiangge Pavilion
13 慧喜殿（遗址）　Huixidian Hall (Site)
14 吉晖殿（遗址）　Jihuidian Hall (Site)
15 清凉楼（遗址）　Qinglianglou Pavilion
16 香林室（复原平面）　Xianglinshi Hall (Restored)
17 倚云楼（复原平面）　Yiyunlou Pavilion(Restored)

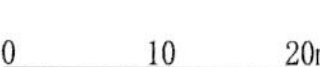

殊像寺总平面图（2015 年测绘）
Site plan of Shuxiangsi (2015 survey)

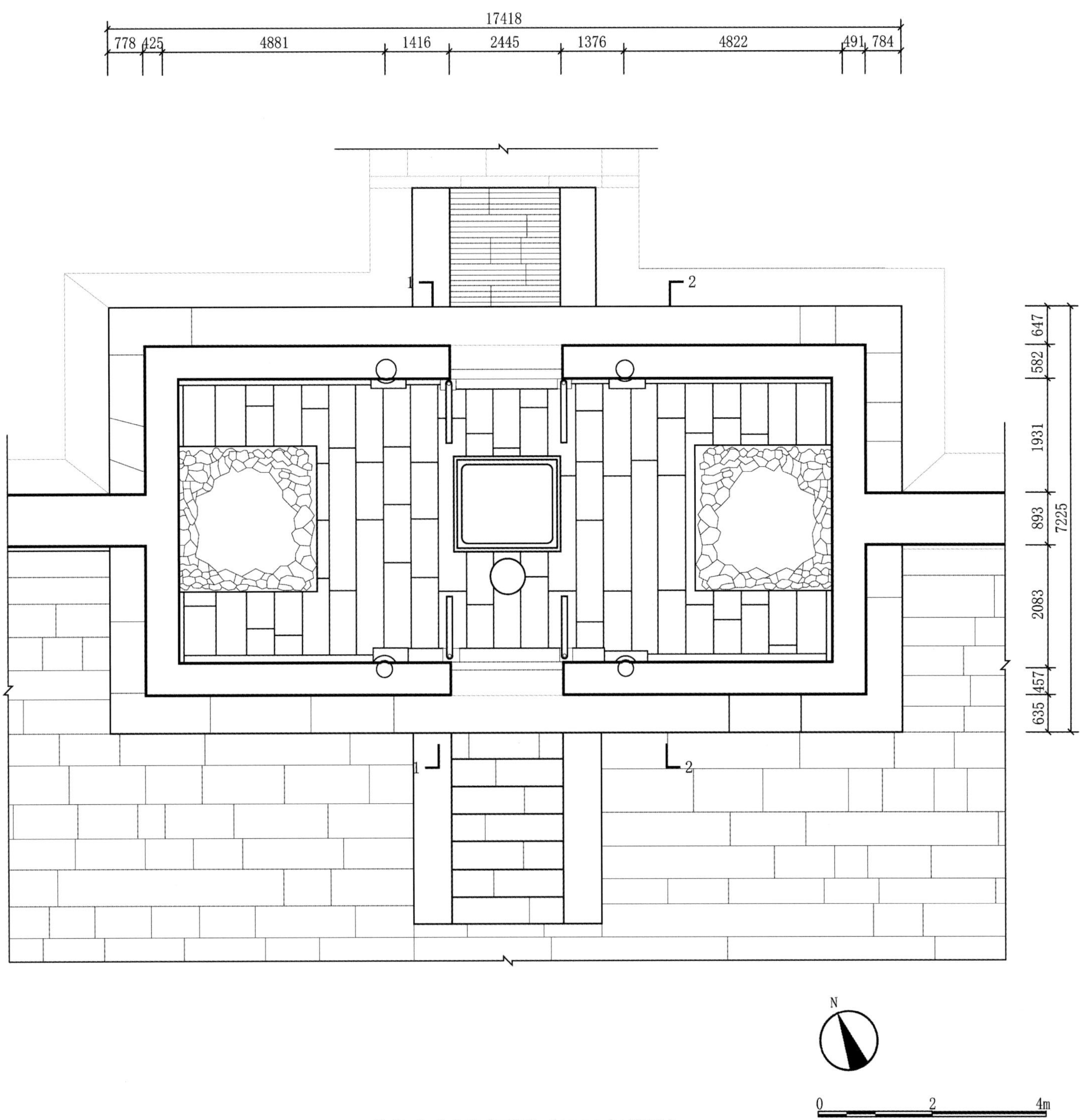

殊像寺山门平面图（2015 年测绘）
Plan of shanmen of Shuxiangsi (2015 survey)

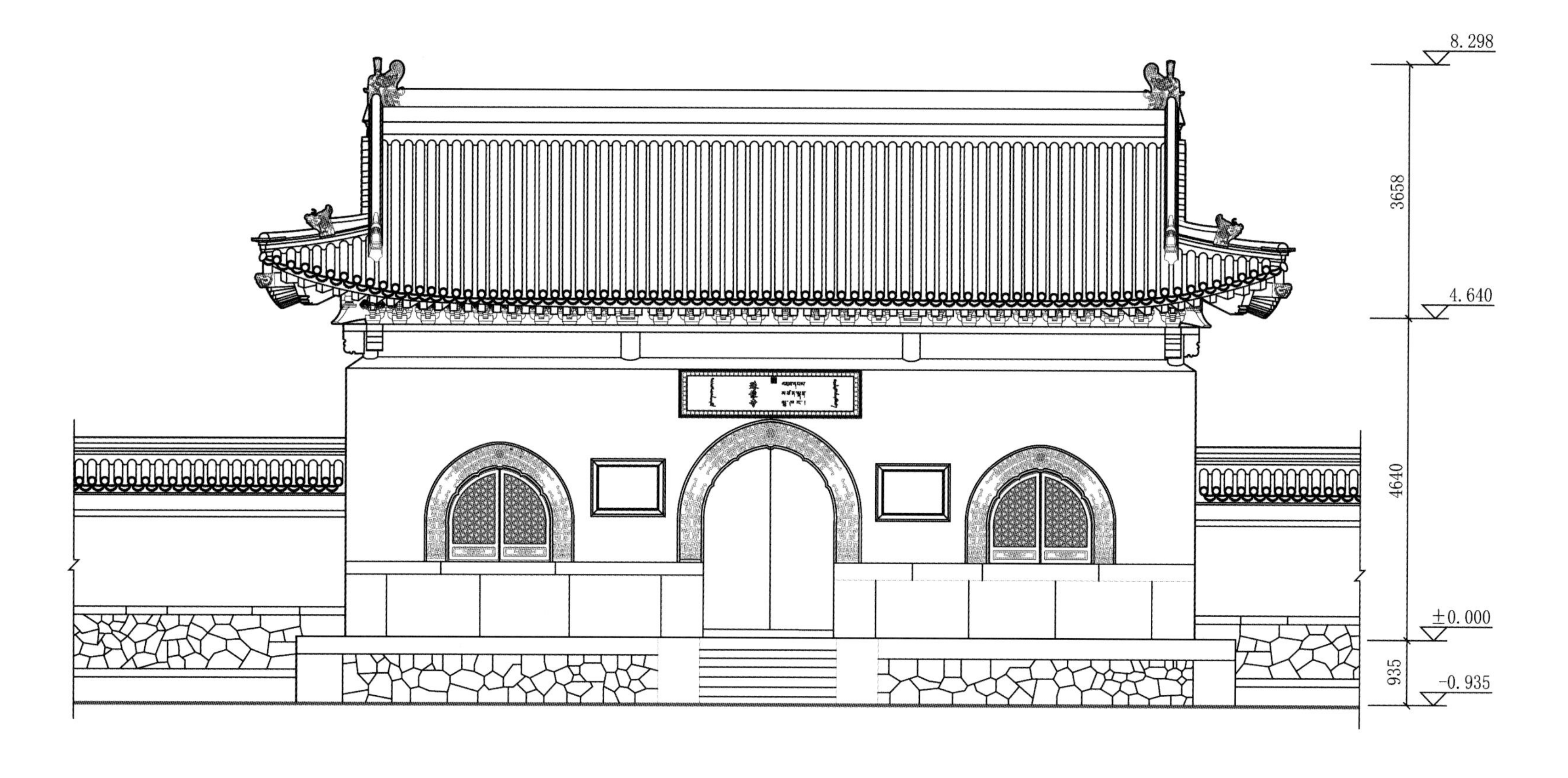

殊像寺山门正立面图（2015 年测绘）
Front elevation of shanmen of Shuxiangsi (2015 survey)

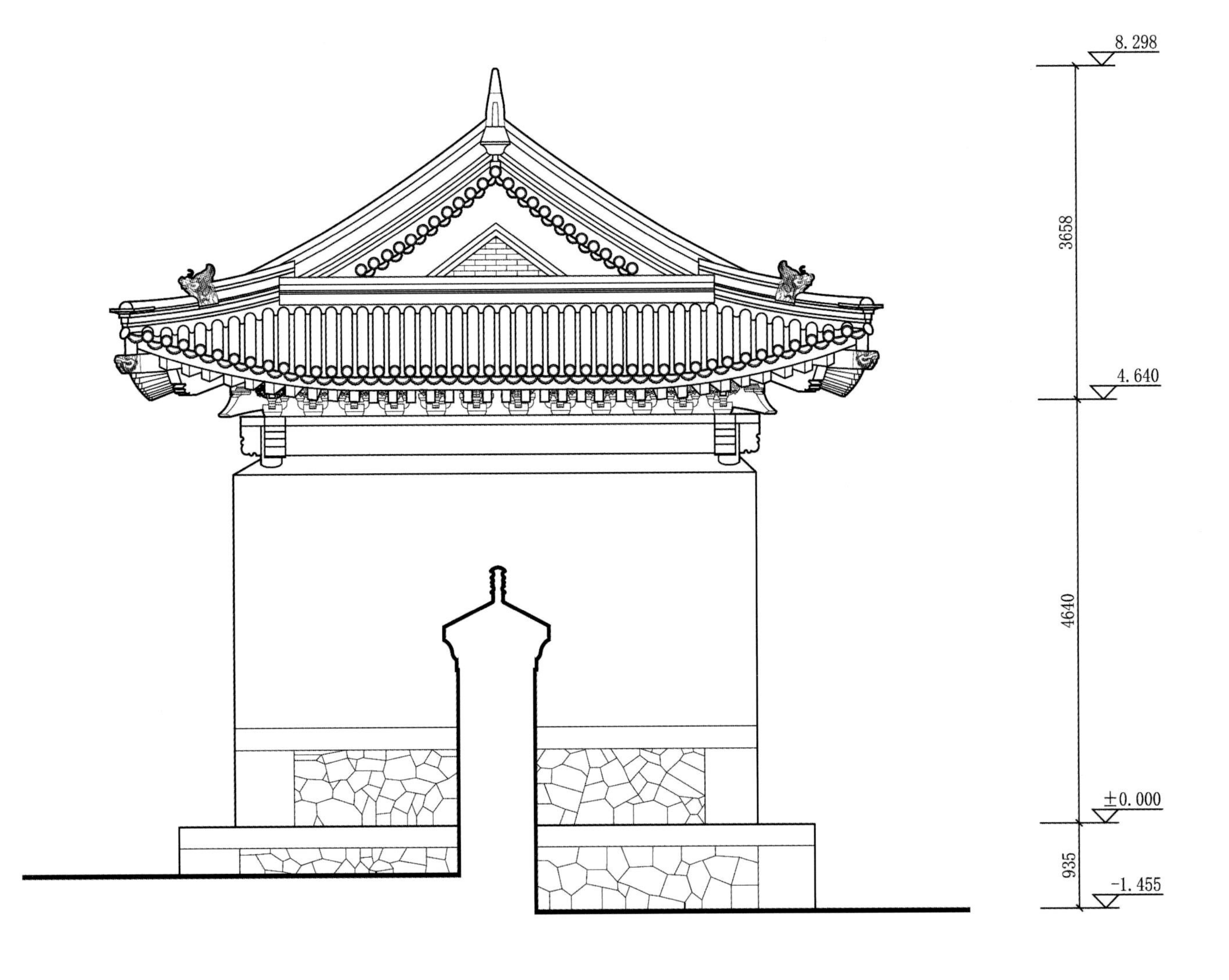

殊像寺山门侧立面图（2015 年测绘）
Side elevation of shanmen of Shuxiangsi (2015 survey)

殊像寺山门窗饰大样图（2015 年测绘）
Window decoration of shanmen of Shuxiangsi (2015 survey)

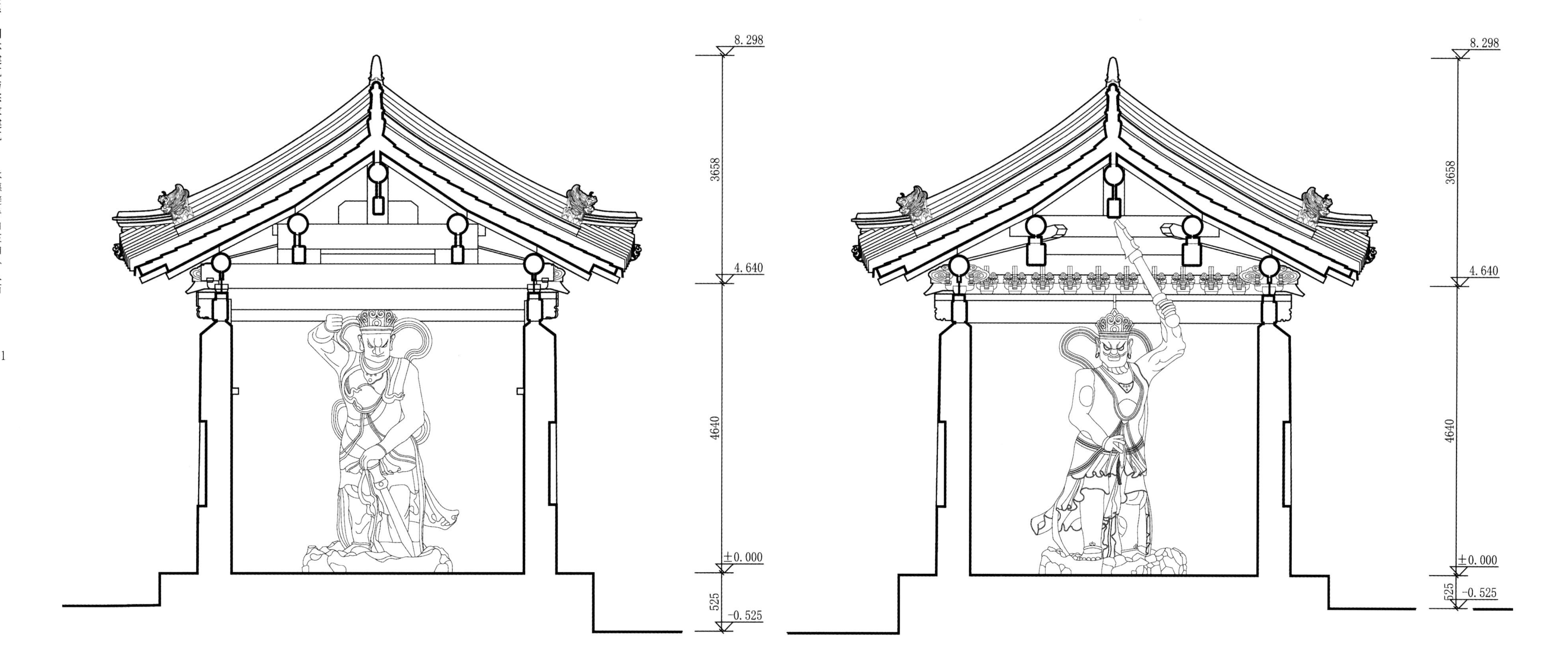

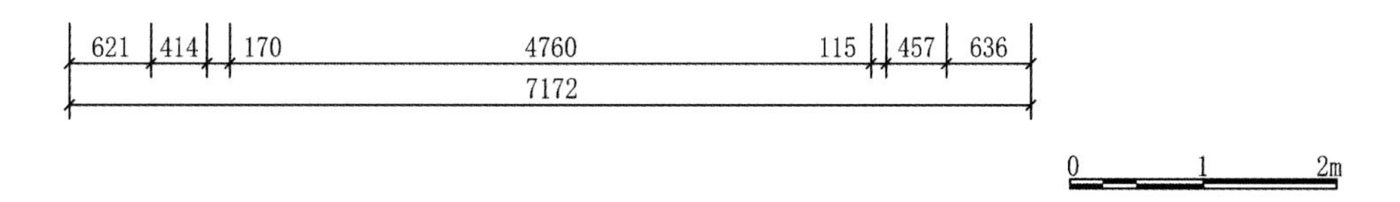

殊像寺山门明间横剖面图（2015 年测绘）
Cross-section of central-bay of shanmen of Shuxiangsi (2015 survey)

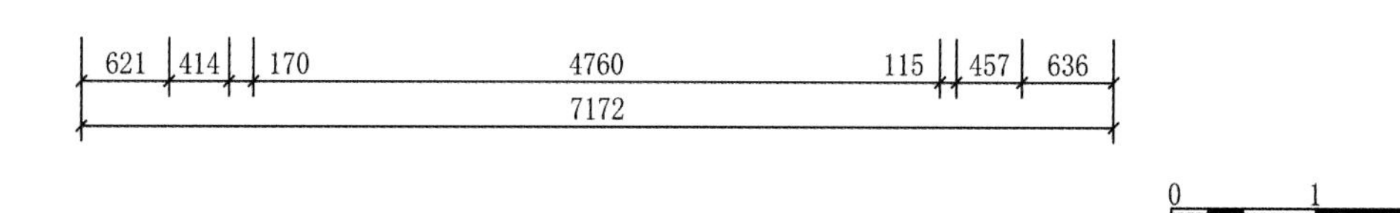

殊像寺山次间横剖面图（2015 年测绘）
Cross-section of side-bay of shanmen of Shuxiangsi (2015 survey)

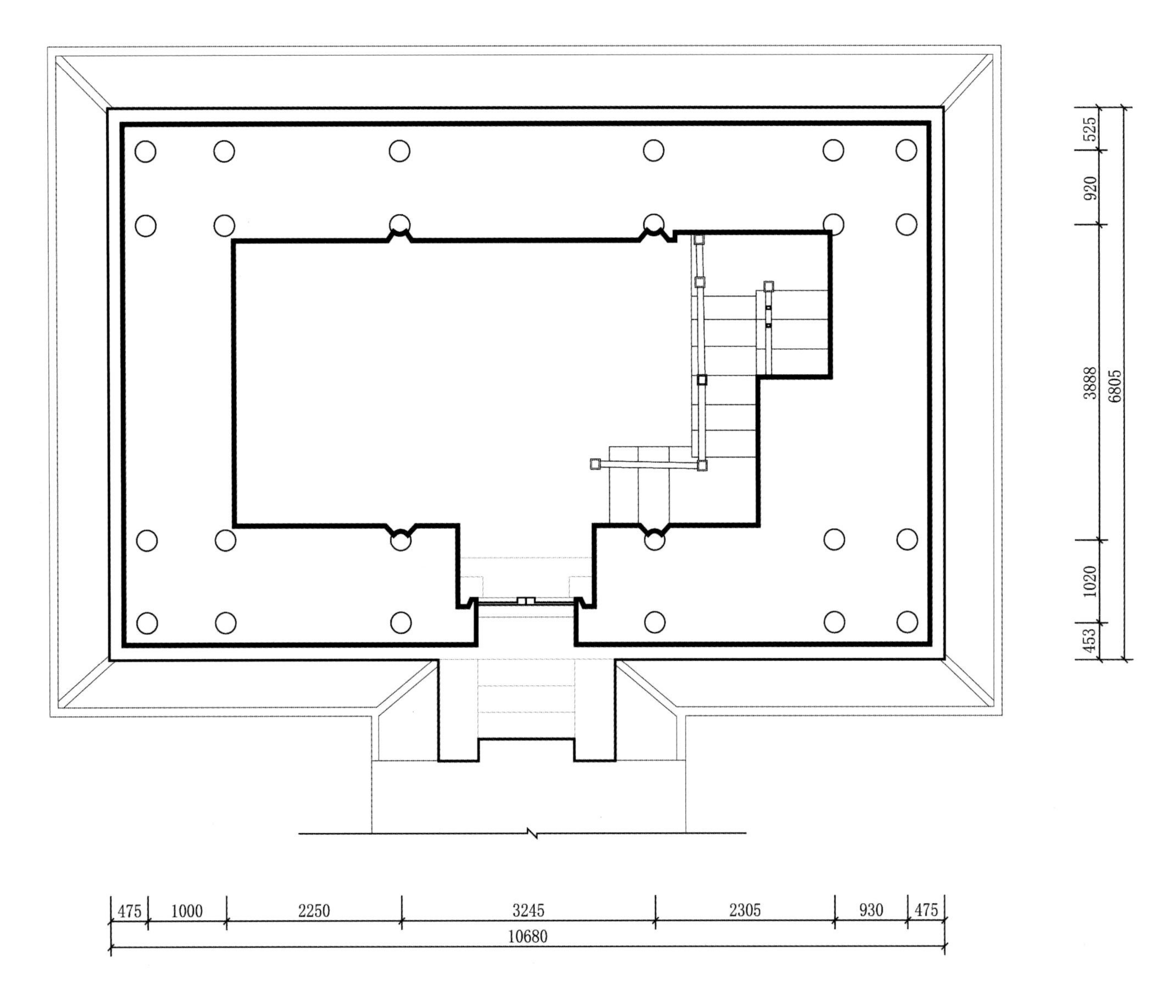

殊像寺钟鼓楼一层平面图（2015 年测绘）
Plan of first floor of zhonggulou of Shuxiangsi (2015 survey)

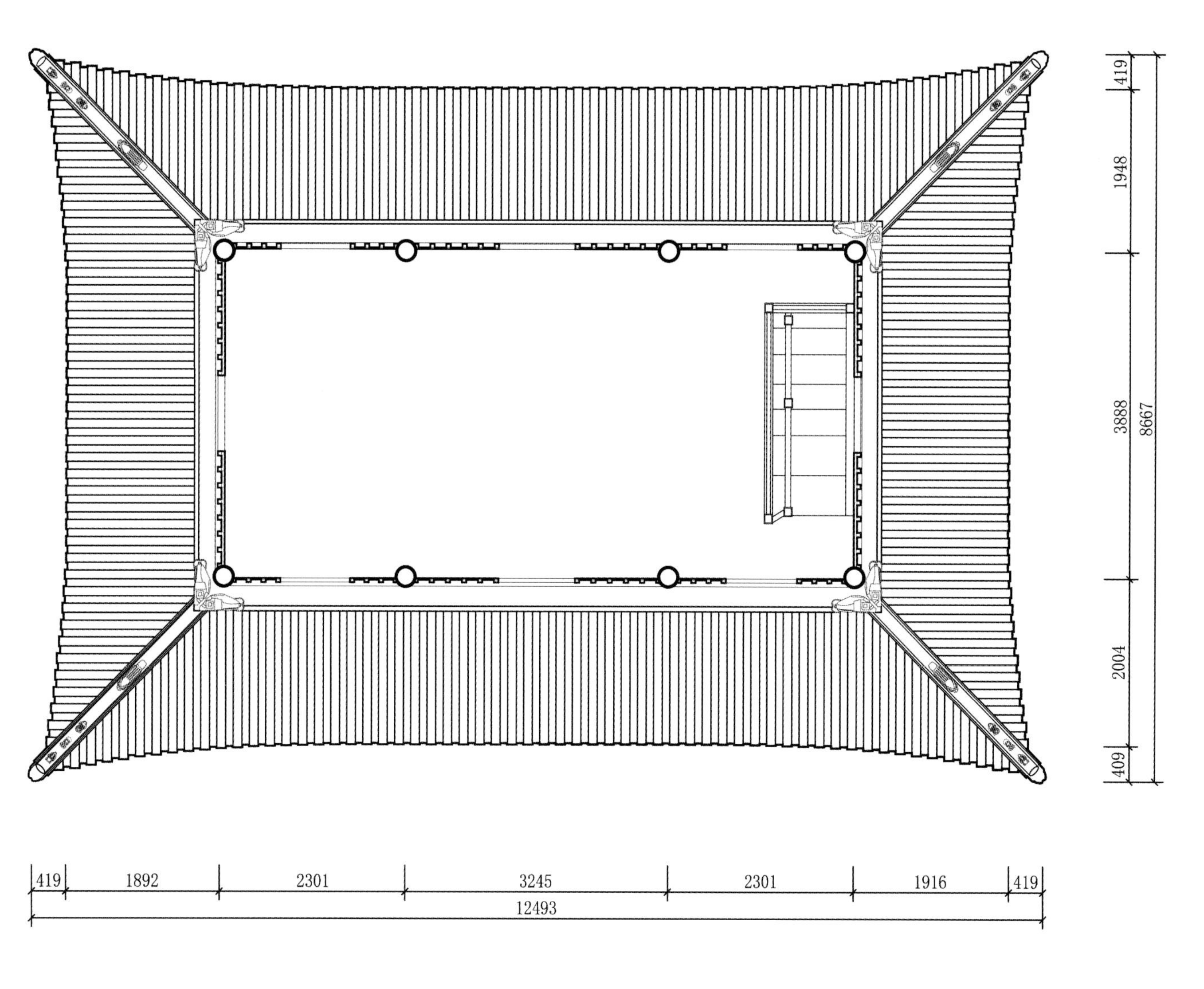

殊像寺钟鼓楼二层平面图（2015 年测绘）
Plan of second floor of zhonggulou of Shuxiangsi (2015 survey)

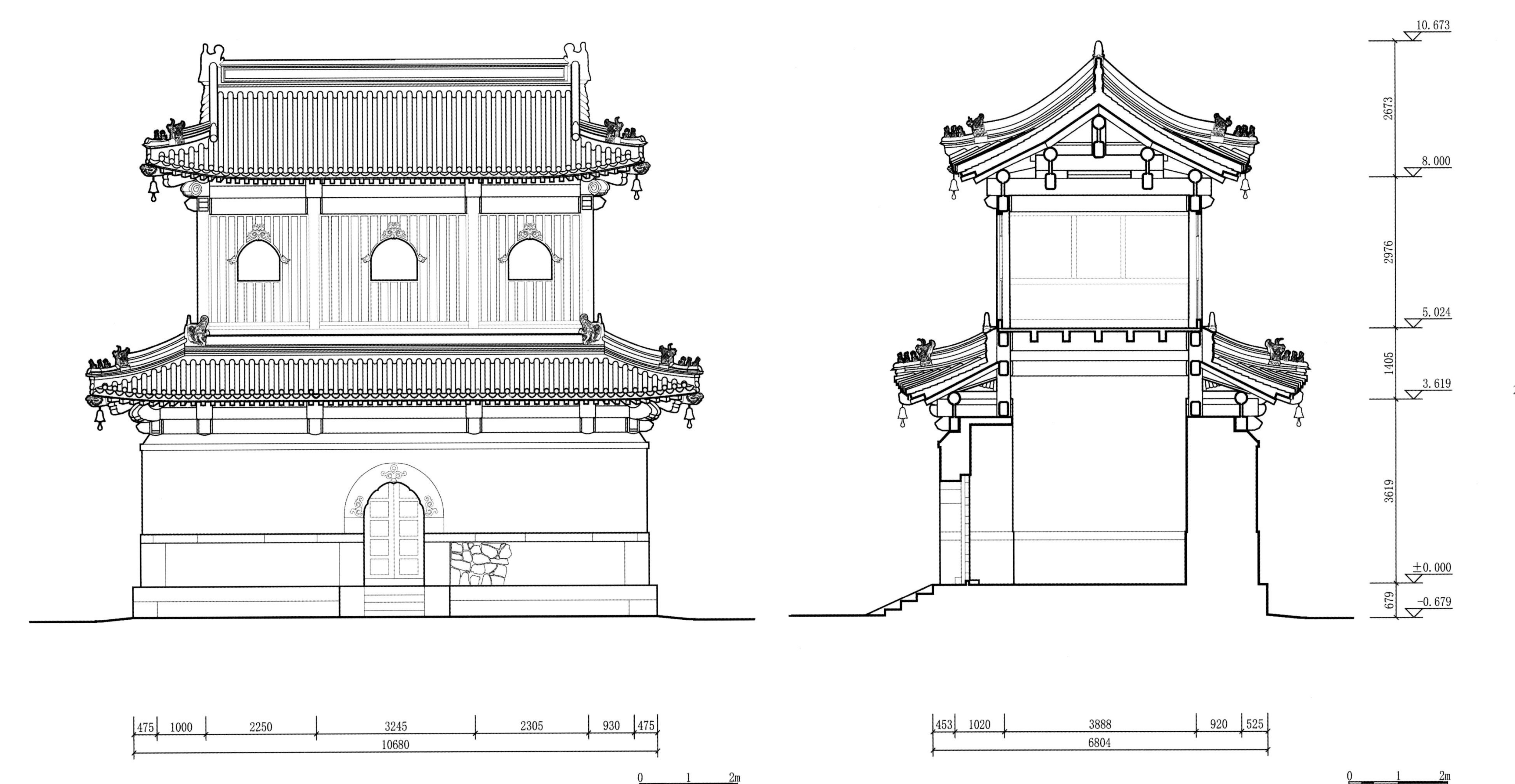

殊像寺钟鼓楼正立面图（2015 年测绘）
Front elevation of zhonggulou of Shuxiangsi (2015 survey)

殊像寺钟鼓楼横剖面图（2015 年测绘）
Cross-section of zhonggulou of Shuxiangsi (2015 survey)

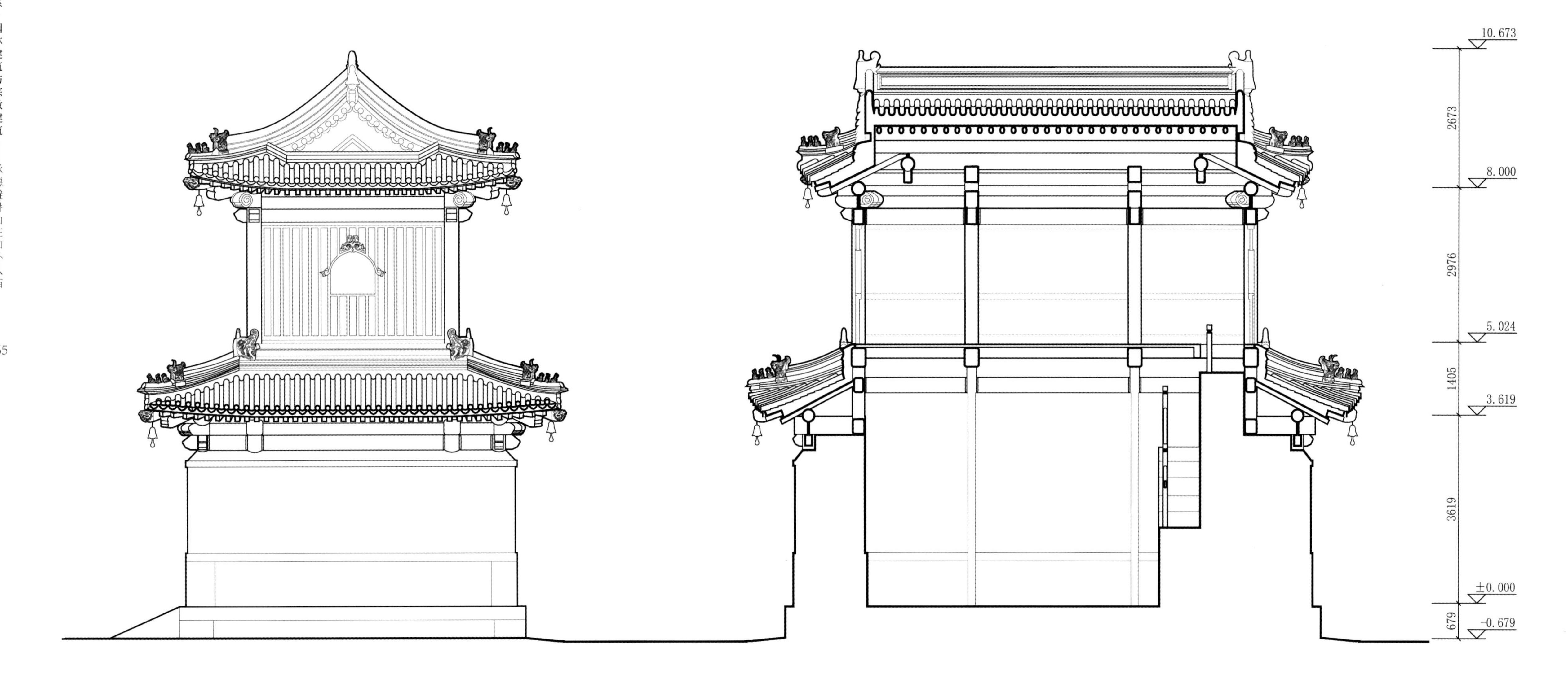

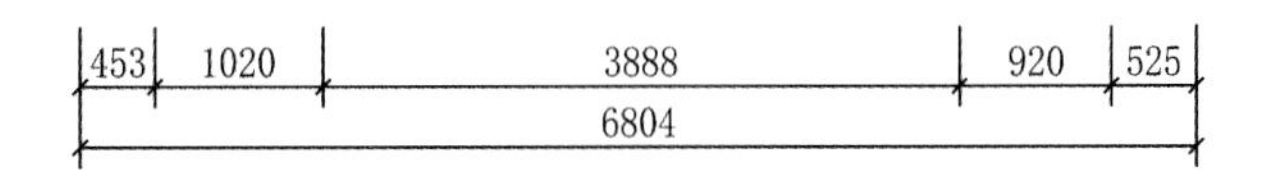

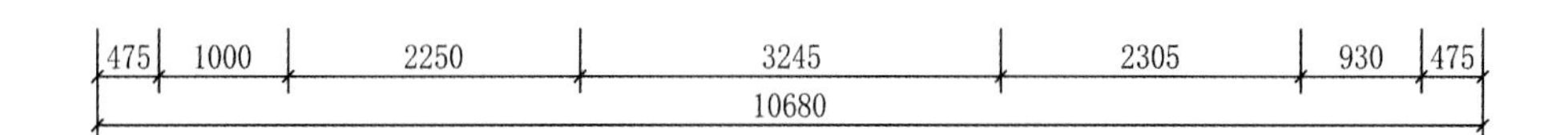

殊像寺钟鼓楼侧立平面图（2015 年测绘）
Side elevation of zhonggulou of Shuxiangsi (2015 survey)

殊像寺钟鼓楼纵剖面图（2015 年测绘）
Longitudinal section of zhonggulou of Shuxiangsi (2017 survey)

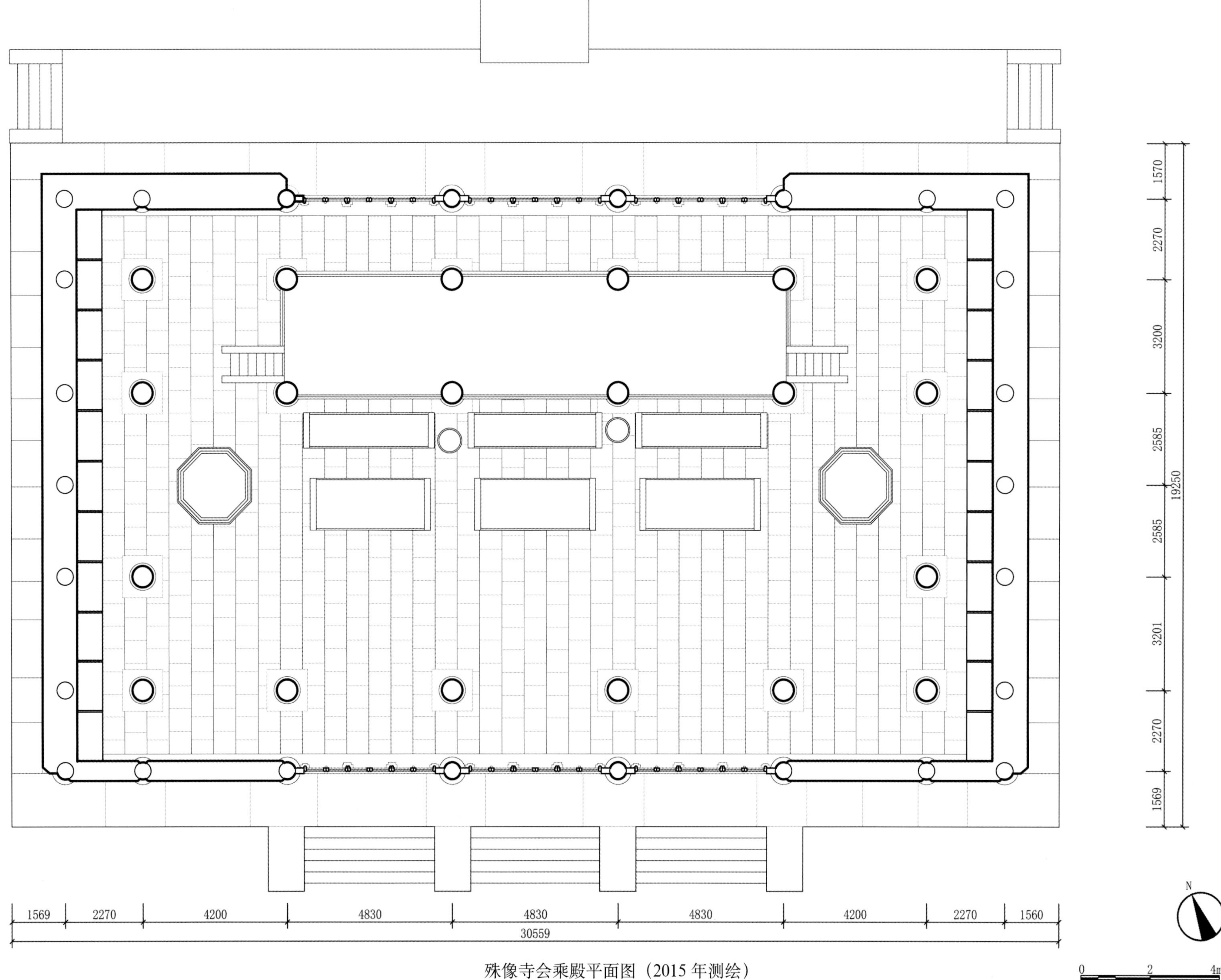

殊像寺会乘殿平面图（2015 年测绘）
Plan of Huichengdian of Shuxiangsi (2015 survey)

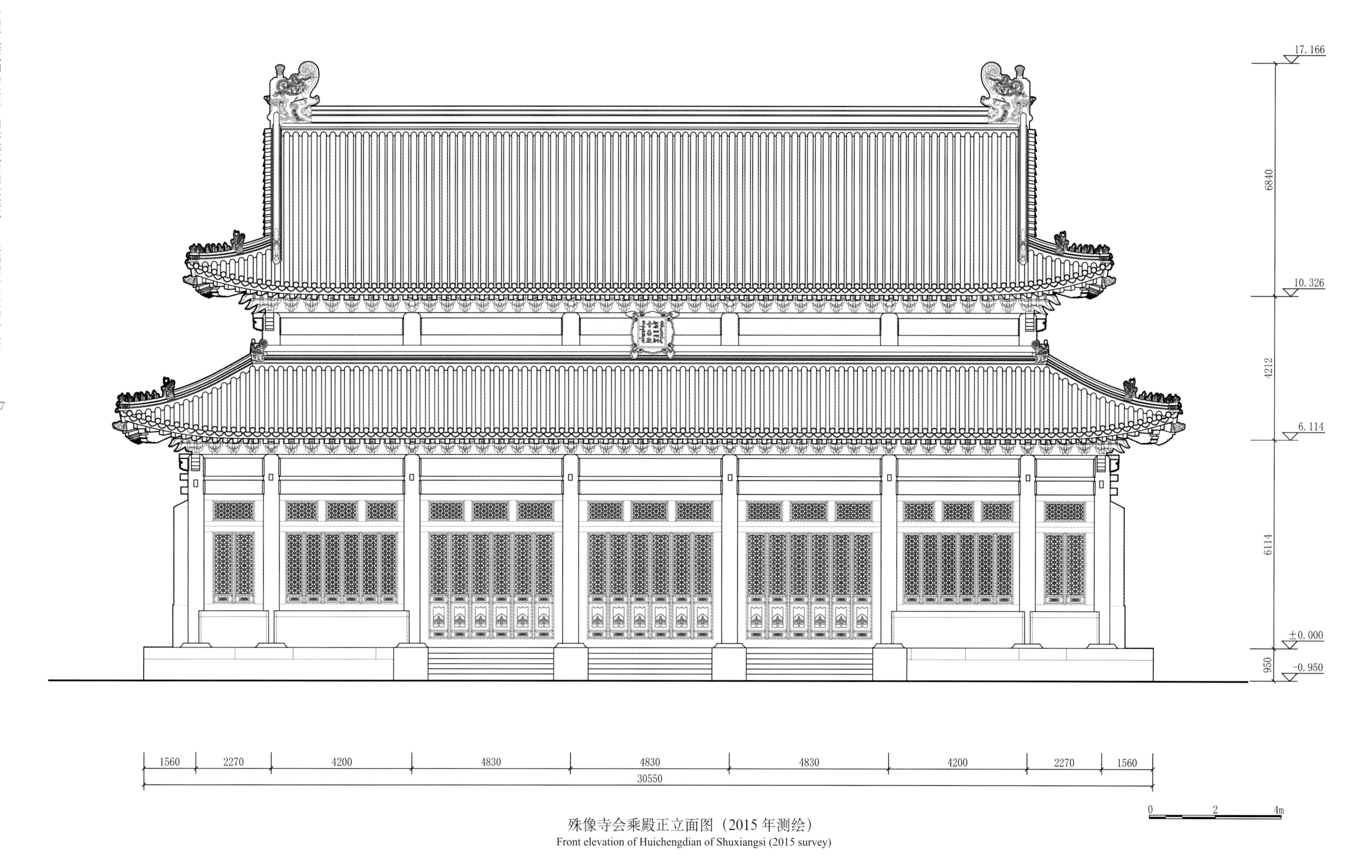

殊像寺会乘殿正立面图（2015 年测绘）
Front elevation of Huichengdian of Shuxiangsi (2015 survey)

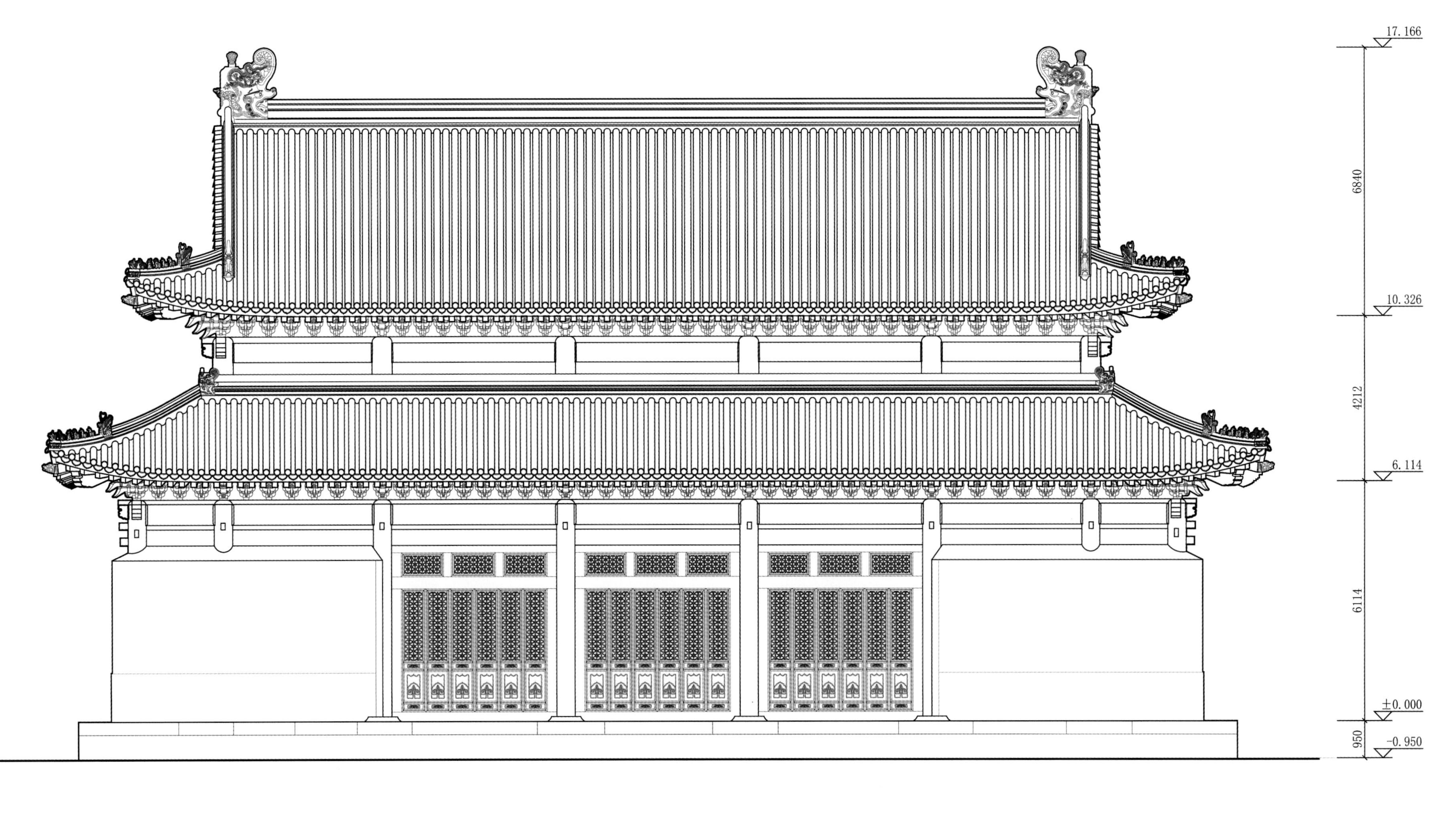

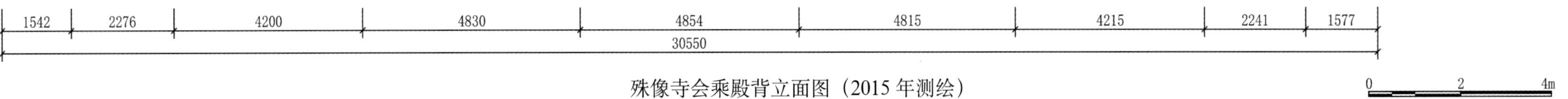

殊像寺会乘殿背立面图（2015年测绘）

Rear elevation of Huichengdian of Shuxiangsi (2015 survey)

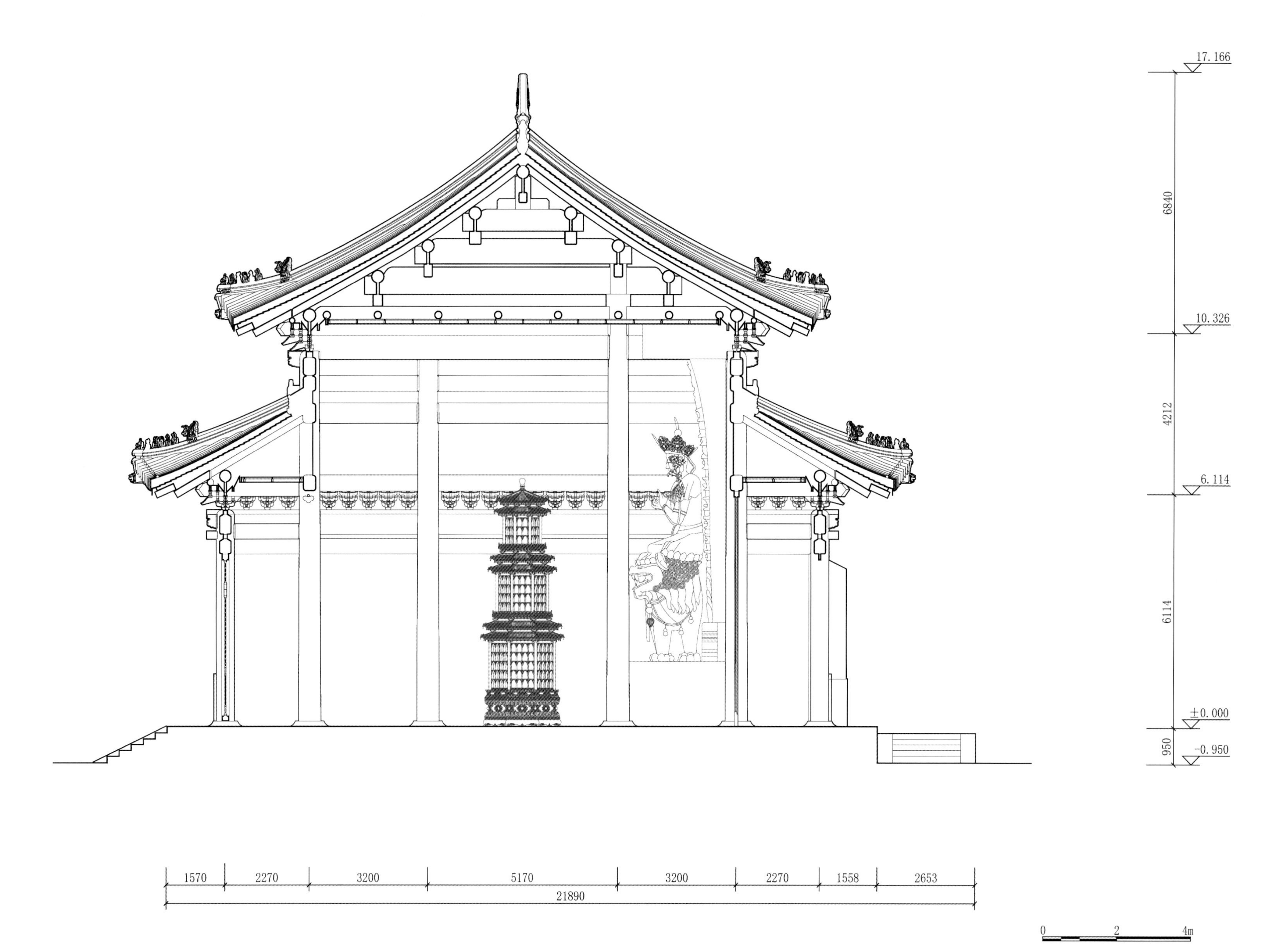

殊像寺会乘殿横剖面图（2015 年测绘）
Cross-section of Huichengdian of Shuxiangsi (2015 survey)

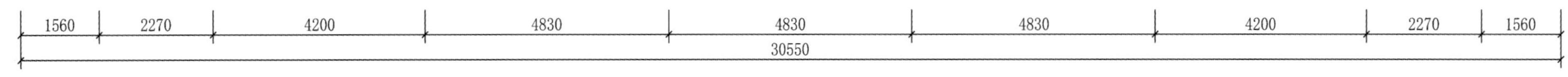

殊像寺会乘殿纵剖面图（2015 年测绘）
Longitudinal section of Huichengdian of Shuxiangsi (2015 survey)

殊像寺会乘殿观音菩萨像大样图（2015 年测绘）
Guanyin statue of Huichengdian of Shuxiangsi (2015 survey)

殊像寺会乘殿文殊菩萨像大样图（2015 年测绘）
Wenshu statue of Huichengdian of Shuxiangsi (2015 survey)

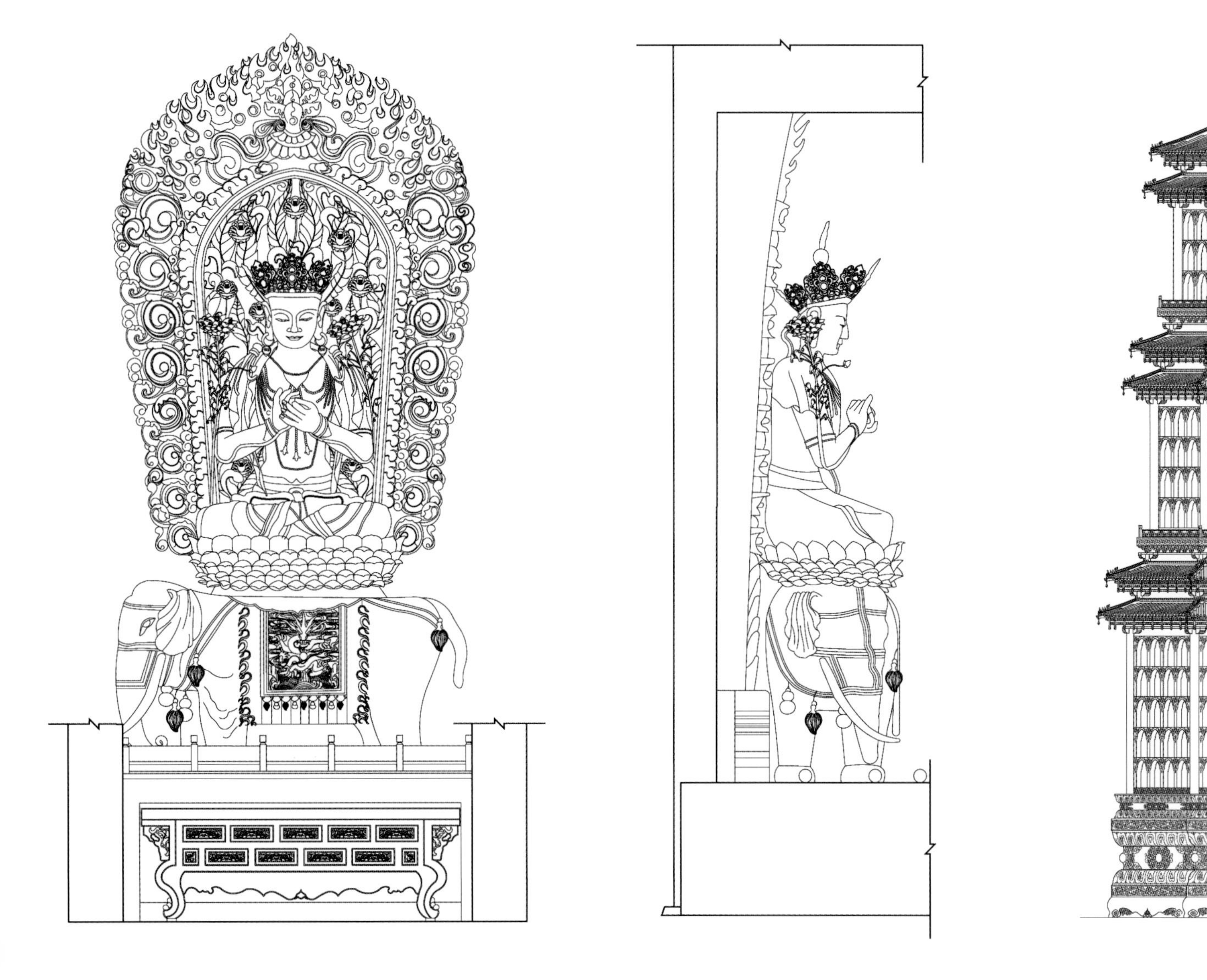

殊像寺会乘殿普贤菩萨像大样图（2015 年测绘）
Puxian statue of Huichengdian of Shuxiangsi (2015 survey)

殊像寺会乘殿楠木塔平面、立面、剖面图（2015 年测绘）
Plan, elevation and section of nanmuta of Huichengdian of Shuxiangsi (2015 survey)

参考文献

References

[一]（清）玄烨撰，清揆叙等注. 御制避暑山庄诗[Z]. 全国图书馆文献缩微中心，2001.

[二]（清）弘历. 清高宗御制诗文集[Z]. 北京：中国人民大学出版社，1995.

[三]中国第一历史档案馆，承德市文物局. 清宫热河档案[Z]. 北京：中国档案出版社，2003.

[四]（清）和珅等. 钦定热河志[Z]. 天津：天津古籍出版社，2003.

[五]石利峰校点. 热河园庭现行则例[Z]. 北京：团结出版社，2012.

[六]E. Boerschmann. Chinesische Architektur[M].Berlin:Verlag Ernst Wasuth A-G,1925.

[七]S. A. Hedin. Jehol, City of Emperors[M].London: K. Paul, Trench, Trubner, 1932.

[八]乌居龙藏. 满蒙古迹考[M]. 陈念本译. 北京：商务印书馆，1933.

[九]关野贞，竹岛卓一. 热河解说[M]. 东京：座右宝刊行会，1937.

[十]五十岚牧太. 热河古迹与西藏艺术[M]. 东京：第一书房，1942.

[十一]卢绳. 承德避暑山庄[J]. 文物参考资料，1956（9）：13-21.

[十二]卢绳. 承德外八庙建筑（一）[J]. 文物参考资料，1956（10）：59-65.

[十三]卢绳. 承德外八庙建筑（二）[J]. 文物参考资料，1956（11）：29-35.

[十四]卢绳. 承德外八庙建筑（三）[J]. 文物参考资料，1956（12）：9-13.

[十五]周维权. 避暑山庄的园林艺术[J]. 建筑学报，1960（6）：29-32.

[十六]杨伯达. 冷枚及其《避暑山庄图》[J]. 故宫博物院院刊，1979（1）：51-61.

[十七]天津大学建筑学院，承德市文物局. 承德古建筑[M]. 北京：中国建筑工业出版社，1982.

[十八]孟兆祯. 避暑山庄园林艺术[M]. 北京：紫禁城出版社，1985.

[十九]杨天在编著. 避暑山庄碑文释译[M]. 紫禁城出版社，1985.

[二十]避暑山庄研究会编. 避暑山庄论丛[M]. 紫禁城出版社，1986.

[二十一]周维权. 中国古典园林史[M]. 北京：清华大学出版社，1990.

[二十二]傅清远主编. 避暑山庄[M]. 北京：华夏出版社，1993.

[二十三]《避暑山庄七十二景》编委会. 避暑山庄七十二景[M]. 北京：地质出版社，1993.

[二十四]F. Philippe. *Mapping Chengde: the Qing landscape enterprise*[M]. Honolulu: University of Hawai`I Press, 2000.

[二十五]孙大章. 中国古代建筑史第五卷[M]. 北京：中国建筑工业出版社，2001.

[二十六]吴晓敏. 因教仿西卫，并以示中华：曼荼罗原型与清代皇家宫苑中藏传佛教建筑的创作[D]. 天津大学博士学位论文，2001.

[二十七]戴逸. 避暑山庄和康乾盛世[J]. 中国民族，2003（10）：15-17.

[二十八]师力武主编，中国承德市文物局编. 清帝与避暑山庄[M]. 北京：中国旅游出版社，2003.

[二十九]吴晓敏，史箴. 肖彼三摩耶，作此曼拿罗——清代皇家宫苑藏传佛教建筑创作的类型学方法探析[J]. 建筑师，2003（6）：89-94.

[三 十] J.Millward, R.Dunnell. *New Qing Imperial History: The Making of Inner Asian Empire at Qing Chengde* [M].Routledge, 2004.

[三十一]罗文华. 龙袍与袈裟（上、下）[M]. 北京：紫禁城出版社，2005.

[三十二]周维权. 园林·风景·建筑[M]. 天津：百花文艺出版社，2005.

[三十三]杨菁. 承德避暑山庄和周围寺庙的海外影响[A]. 中国紫禁城学会论文集（第八集）[C]. 北京：故宫出版社，2012：667-678.

[三十四]刘瑜，杨菁，王英妮. 芝加哥世博会的中国金亭——兼论早期世博会的中式建筑[J]. 世界建筑，2013（8）：P116-119.

[三十五]庄岳. 礼仪之争：马国贤《避暑山庄三十六景》铜版画与康熙《御制避暑山庄诗》木刻画的视觉差异[J]. 建筑史，2013（2）：112-121.

[三十六]Kangxi Emperor (Author), Richard E. Strassberg (Translator), Stephen H. Whiteman (Introduction), *Thirty-Six Views: The Kangxi Emperor's Mountain Estate in Poetry* and Prints [M].Washington DC: Dumbarton Oaks Research Library and Collection, 2016.

[三十七]杨菁，李声能，白成军. 文溯阁研究[M]. 天津：天津大学出版社，2017.

参与测绘及相关工作的人员名单

一、1954年测绘

指导教师：卢　绳　徐　中　沈玉麟　庄涛声　冯建逵　周祖奭　石承露
童鹤龄　赵冠洲　彭一刚　胡德君　林兆龙　宋元谨　屈浩然
张文忠（美术教师：丁莱亲　高尚廉　王学仲）

本科生（1951级）：

张　敕　杨培源　高承曾　陈忠徐　张宗文　陈　椕　陈秀兰
尚　廓　方咸孚　邵木兰　李锡钧　刘荫篁　阎万鹏　黄祥汉
于鸿图　王　立　高士苓　赵士琦　孙宝莲　张继儒　顾百刚
黄天德　孙彩文　胡松鹤　姜传宗　王有智　章竟屋　杨一伟
张晓昆　白小茹　刘稚英　谭振耀　南舜熏　王健平　刘树源
柴嵩铣　篑纪唐　周泰鎏　张培均　孙培基　李书德　徐家龙
朱志刚　马锦堂　聂守业　商志原　王绍周　曲士蕴　马　骁
张　季　王安武　蔡　青　闻凤鸣　纪瑞贤　古杏贞　王　淳
魏昌裕　赵淑静　何宁础　李崒桐　李士苓　苏家禄　刘　朴
刘济宽　周　济　郑福海　何式欧　李元德

二、1962年测绘

指导教师：卢　绳　徐　中　周祖奭　屈浩然　杨道明　荆其敏　方咸孚
（美术教师：杨化光）

本科生（1958级）：

宋宝琦　高建国　冯天意　傅庆云　桑培良　蒋　轲　闻德荣
齐元康　燕宝林　王宏明　范学信　陈锡泰　高丽莉　兆汝彪
王惠春　宋宝章　侯福安　郝顺岐　高继贤　刘增怀　邸桂生

赵平生　赵玉双　杜振远　薛来有　舒宏建　范淑光　王乃淳
王家声　王兰墀　宋霁虹　尹杰卿　宋高坪　李昌发　池学义
林宝汉　洪佛保　吴克寻　金　楣　舒松启　章世强　张建元
张幼华　张春生　张澄燕　张湛德　张汝科　张国材　刘桂馥
刘福生　陈炳玲　陈成华　董　山　郭景波　谢根瑞　沈匡德

三、1963年测绘

指导教师：卢　绳　林兆龙　潘家平　彭一刚　张文忠　屈浩然　潘家平
羌　苑　张博仁（美术教师：杨化光　高尚廉　王学仲　丁莱亲）

硕士研究生（1961级）：王乃香

本科生（1959级）：任焕章等9人

本科生（1960级）：

于　珍　王凤亮　李振中　李松林　李国华　李富南　吴凤翼
任佩珠　刘泳裳　季仁铨　金殿英　孟玉茹　唐黎明　娄维玉
庞学礼　郑小辉　齐振南　乔华涛　彭九皋　彭春维　郭水根
郭文来　杨金铭　张秀峰　赵瑞祥　简起来　滕华骅　王志航
田瑞图　李保田　林润发　周永高　宋宝玉　汪宗渝　马光明
刘金华　陈武吉　陈　喆　金长明　唐　通　徐爱珠　汤大发
施　莹　孙伯藩　张润吾　张生午　乔松年　高成春　翁钟波
郭秀琴　杨凤临　杨永祥　赵志芳　刘金德

四、2011年避暑山庄月色江声建筑群测绘

指导教师：王其亨　丁　垚　杨　菁

博士研究生（2011级）：仲丹丹（2009级）贺美芳

硕士研究生（2010级）：

朱振骅　孙立娜　曹　雪（2011级）薛　山　杜　欣　王　巍

本科生（2009级）：

余俊佳　李煜群　谈　韬　钟　山　钟渊庭　贺中超　刘鹏鹏
孟　杰　王　晨　杨骥腾　曹晓宇　靳同晖　姚　婕　韩智华
胡文涛　刘　波　马学宏　乔　峰　孙　韬　韦东方　王正通
荆　蕾　李　刚　李倩茹　柳　青　郭　昕　刘君男　辛冠华
李晶晶　张婧仪　周雅婧　张月蓉　任思宇　高　尚　姚梦佳
宋灵溪　唐　骥　李　威　阮　鹏

技术人员：蒋　根

摄影：（2008届硕士）何蓉

五、2012年避暑山庄万壑松风建筑群测绘

指导教师：王其亨　杨　菁

博士研究生：（2009级）李　江

硕士研究生：（2010级）谭　虎

本科生：（2007级）贺思琳　陈　渊　王　朝　蒋昱程　张艺明

石雪梅（河北工业大学）　王　曦

六、2012年避暑山庄如意洲、烟雨楼、文津阁建筑群测绘

指导教师：王其亨　吴　葱　杨　菁　白成军　李　哲　闫佳亮

博士研究生：（2012级）武晶

硕士研究生：（2010级）程枭翀　（2011级）杜　欣　郝　帅　侯宇楠

齐斯洋　徐龙龙　王　巍　袁　媛（2012级）陈克强　王　依

冀　凯　冯婷婷

本科生：（2010级）

曹世彪　邱　鑫　闫　瑾　班培颖　曹远行　刘　艺　杜裘伟

关春东　兰　帅　刘春孟　孙欣晔　王金晖　李　翔　张昕玮

刘未达　孙　宇　谢明轩　张星熠　从志涛　李德扬　李昊天

卢汀滢　王学浩　马宇婷　王昭宇　张浩然　郑天诚　敖子昂

王　祎　陈　恺　郭陈斐　贾梦圆　鲁世超　孙　全　王　静

张秋洋　杨　琳　陈　豪　高婉丽　亢梦荻　李梦超　李　雪

李　悦　孟令君　孙启真　孙效东　张　帆　邹春竹　周云洁

颜永国　陈嘉康　孙　超　李　洁　王业娜　邓晰元（河北工业大学）　王　曦

技术人员：张伟国　李玉龙　张志强　王　硕

摄影：（2007级本科）海　良（天津大学仁爱学院）岳意贺

新闻：（2011级本科）张艺凡

七、2014年承德殊像寺山石测绘

指导教师：杨　菁　白成军

承德市文物局：陈　东

博士研究生：（2014级）张家浩

硕士研究生：（2013级）李竞扬　张雨奇

本科生：（2010级）曹世彪

技术人员：张　珊　张志强　王　硕　吕文远　张志永　张建新　李晓燕

八、2015年承德殊像寺建筑测绘

指导教师：杨　菁　白成军

承德市文物局：陈　东

硕士研究生：（2015级）付蜜桥　范一鸣　杨洁

技术人员：张　珊　张志强　王　硕　吕文远　张志永　张建新　李晓燕

九、2016年承德殊像寺建筑复原研究

指导教师：杨　菁

硕士研究生：（2015级）付蜜桥

本科生：（2011级）郭　亮　张艺凡　冯胜村

十、2017承德测绘图整理

指导教师：王其亨　杨　菁　朱　蕾

承德市文物局：陈　东

硕士、博士研究生：高　原　张煦康　张颖娟　马胜楠　付蜜桥　郭一丹

王笑石

英文翻译：段美媛　［奥］荷雅丽　周彦邦

List of participants involved in surveying and related works

1954 Survey

Instructor: LU Sheng, XU Zhong, SHEN Yulin, ZHUANG Taosheng, FENG Jiankai, ZHOU Zushi, SHI Chenglu, TONG Heling, ZHAO Guanzhou, PENG Yigang, HU Dejun, LIN Zhaolong, SONG Yuanzhen, QU Haoran, ZHANG Wenzhong (Art Instructor: DING Laiqin, GAO Shanglian, WANG Xuezhong)

Bachelor Students (1951 Year):

ZHANG Rong, YANG Peiyuan, GAO Chenzeng, CHEN Zhongxu, ZHANG Zongwen, CHEN Mao, CHEN Xiulan, SHANG Kuo, FANG Xianfu, SHAO Mulan, LI Xijun, LIU Yinxuan, YAN Wanpeng, HUANG Xianghan, YU Hongtu, WANG Li, GAO Shiling, ZHAO Shiqi, SUN Baolian, ZHANG Xuru, GU Baigang, HUANG Tiande, SUN Caiwen, HU Songhe, JIANG Chuanzong, WANG Youzhi, ZHANG Jingwu, YANG Yiwei, ZHANG Xiaokun, BAI Xiaoru, LIU Zhiying, TAN Zhenyao, NAN Shunxun, WANG Jianping, LIU Shuyuan, CHAI Songxi, KUI Jitang, ZHOU Tailiu, ZhANG Peijun, SUN Peiji, LI Shude, XU Jialong, ZHU Zhigang, MA Jintang, NIE Shouye, SHANG Zhiyuan, WANG Shaozhou, QU Shiyun, MA Xiao, ZHANG Ji, WANG Anwu, CAI Qing, WEN Fengming, JI Ruixian, GU Xingzhen, WANG Chun, WEI Changyu, ZHAO Shujing, HE Ningchu, LI Zutong, LI Shiling, SU Jialu, LIU Po, LIU Jikuan, ZHOU Ji, ZHENG Fuhai, HE Shi'ou, LI Yuande.

1962 Survey

Instructors: LU Sheng, XU Zhong, ZHOU Zushi, QU Haoran, YANG Daoming, JING Qimin, FANG Xianfu (Art Instructor: YANG Huaguang)

Bachelor Students (1958 Year):

SONG Baoqi, GAO Jianguo, FENG Tianyi, FU Qingyun, SANG Peiliang, JIANG Ke, WEN Derong, QI Wenkang, YAN Baolin, WANG Hongming, FAN Xuexin, CHEN Xitai, GAO Lili, ZHAO Rubiao, WANG Huichun, SONG Baozhang, HOU Fu'an, HAO Shunqi, GAO Jixian, LIU Zenghuai, DI Guisheng, ZHAO Pingsheng, ZHAO Yushuang, DU Zhenyuan, XUE Laiyou, SHU Hongjian, FAN Shuguang, WANG Naichun, WANG Jiasheng, WANG Lanchi, SONG Jihong, YI Jieqing, SONG Gaoping, LI Changfa, CHI Xueyi, LIN Baohan, HONG Fobao, WU Kexun, JIN Mei, SHU Songqi, ZHANG Shiqiang, ZHANG Jianyuan, ZHAN Youhua, ZHANG Chunsheng, ZHANG Chengyan, ZHANG Zhande, ZHANG Ruke, ZHANG Guocai, LIU Guifu, LIU Fusheng, CHEN Bingling, CHEN Chenghua, DONG Shan, GUO Jingbo, XIE Genrui, CHEN Kuangde

1963 Survey

Instructor: LU Sheng, LIN Zhaolong, PAN Jiaping, PENG Yigang, ZHANG Wenzhong, QU Haoran, PAN Jiaping, QIANG Yuan, ZHANG Furen (Art Instructor: YANG Huaguang, GAO Shanglian, WANG Xuezhong, DING Laiqin

Master Student (1961 Year): WANG Naixiang

Bachelor Student (1959 Year): REN Huanzhang and 9 others. 1960 Year: YU Zhen, WANG Fengliang, LI Zhenzhong, LI Songlin, LI Guohua, LI Funan, WU Fengyi, REN Peizhu, LIU Yongchang, JI Renquan, JIN Dianying, MENG Yuru, TANG Liming, LOU Weiyu, PANG Xueli, ZHENG Xiaohui, QI Zhennan, QIAO Huatao, PENG Jiugao, PENG Chunwei, GUO Shuigen, GUO Wenlai, YANG Jinming, ZHANG Xiufeng, ZHAO Ruixiang, JIAN Qilai, TENG Huahua, WANG Zhihang, TIAN Ruitu, LI Baotian, LIN Runfa, ZHOU Yonggao, SONG Baoyu, WANG Zongyu, MA Guangming, LIU Jinhua, CHEN Wuji, CHEN Zhe, JIN Changming, TANG Tong, XU Aizhu, TANG Dafa, SHI Ying, SUN Bofan, ZHANG Runwu, ZHANG Shengwu, QIAO Songnian, GAO Chengchun, WENG Zhongbo, GUO Xiuqin, YANG Fenglin, YANG Yongxiang, ZHAO Zhifang, LIU Jinde.

2011 Bishu shanzhuang Yuese jiangsheng architecture cluster survey team

Instructor: WANG Qiheng, DING Yao, YANG Jing

Doctoral Students (2011 Year): ZHONG Dandan, (2009 Year) HE Meifang

Master Students (2010 Year): ZHU Zhenhua, SUN Lina, CAO Xue, (2011 Year) XUE Shan, DU Xin, WANG Wei

Bachelor Students (2009 Year): YU Junjia, LI Yuqun, TAN Tao, ZHONG Shan, ZHONG Yuanting, HE Zhongchao, LIU Pengpeng, MENG Jie, WANG Chen, YANG Jiteng, CAO Xiaoyu, JIN Tonghui, YAO Jie, HAN Zhihua, HU Wentao, LIU Bo, MA Xuehong, QIAO Feng, SUN Tao, WEI Dongfang, WANG Zhengtong, JING Lei, LI Gang, LI Qianru, LIU Qing, GUO Xin, LIU Ju'nan, XIN Guanhua, LI Jingjing, ZHANG Jingyi, ZHOU Yajing, ZHANG Yuerong, REN Siyu, GAO Shang, YAO Mengjia, SONG Lingxi, TANG Ji, LI Wei, RUAN Peng

2012 Bishu shanzhuang Wanhe Songfeng architecture cluster survey team

Instructor: WANG Qiheng, YANG Jing

Doctoral Student (2009 Year): LI Jiang

Master Student (2010 Year): TAN Hu

Bachelor Student (2007 Year): HE Silin, CHEN Yuan, WANG Chao, JIANG Yucheng, ZHANG Zhiming, SHI Xuemei, WANG Xi (Hebei University of Technology)

2012 Bishu shanzhuang Ruyizhou, Yanyulou, and Wenjinge architecture cluster survey team

Instructor: WANG Qiheng, WU Cong, YANG Jing, BAI Chengjun, LI Zhe, YAN Jialiang
Doctoral Student (2012 Year): WU Jing
Master Student (2010 Year): CHENG Xiaochong, (2011 Year) DU Xin, HAO Shuai, HOU Yunan, QI Siyang, XU Longlong, WANG Wei, YUAN Yuan, (2012 Year) CHEN Keqiang, WANG Yi, JI Kai, FENG Tingting
Bachelor Studet (2010 Year): CAO Shibiao, QIU Xin, YAN Jin, BAN Peiying, CAO Yuanxing, LIU Yi, DU Qiuwei, GUAN Chundong, LAN Shuai, LIU Chunmeng, SUN Xinhua, WANG Jinhui, LI Xiang, ZHANG Xinwei, LIU Weida, SUN Yu, XIE Mingxuan, ZHANG Xingyi, CONG Zhitao, LI Deyang, LI Haotian, LU Tingying, WANG Xuehao, MA Yuting, WANG Zhaoyu, ZHANG Haoran, ZHENG Tiancheng, AO Zi'ang, WANG Yi, CHEN Kai, GUO Chenfei, JIA Mengyuan, LU Shichao, SUN Quan, WANG Jing, ZHANG Qiuyang, YANG Lin, CHEN Hao, GAO Wanli, KANG Mengdi, LI Mengchao, LI Xue, LI Yue, MENG Lingjun, SUN Qizhen, SUN Xiaodong, ZHANG Fan, ZOU Chunzhu, ZHOU Yunjie, YAN Yongguo, CHEN Jiakang, SUN Chao, LI Jie, WANG Yena, DENG Xiyuan, WANG Xi (Hebei University of Technology)
Technicians: ZHANG Weiguo, LI Yulong, ZHANG Zhiqiang, WANG Shuo
Filming: Hai Liang (2007 Bachelor), YUE Yihe (Tianjin University Renai College)
News Report: ZHANG Yifan (2011 Bachelor)

2014 Chengde Shuxiang Temple shanshi survey team

Instructor: YANG Jing, BAI Chengjun
Relics Bureau of Chengde Municipality: CHEN Dong
Doctoral Student (2014 Year): ZHANG Jiahao
Master Student (2013 Year): LI Jingyang, ZHANG Yuqi
Bachelor Student (2010 Year): CAO Shibiao
Technician: ZHANG Shan, ZHANG Zhiqiang, WANG Shuo, LYU Wenyuan, ZHANG Zhiyong, ZHANG Jianxin, LI Xiaoyan

2015 Chengde Shuxiang Temple architecture survey team

Instructor: YANG Jing, BAI Chengjun
Relics Bureau of Chengde Municipality: CHEN Dong
Masters Student (2015 Year): FU Miqiao, FAN Yiming, YANG Jie
Technician: ZHANG Shan, ZHANG Zhiqiang, WANG Shuo, LYU Wenyuan, ZHANG Zhiyong, ZHANG Jianxin, LI Xiaoyan

2016 Chengde Shuxiang Temple architecture restoration team

Instructor: YANG Jing
Master Student (2015 Year): FU Miqiao
Bachelor Student (2011 Year): GUO Liang, ZHANG Yifan, FENG Shengcun

2017 Chengde Survey Drawing Organization

Instructor: WANG Qiheng, YANG Jing, ZHU Lei
Relics Bureau of Chengde Municipality: CHEN Dong
Master and Doctoral Student: GAO Yuan, ZHANG Xukang, ZHANG Yingjuan, MA Shengnan, FU Miqiao, GUO Yidan, WANG Xiaoshi

English Translator: DUAN Meiyuan, Alexandra Harrer, CHOU Yen Pang

图书在版编目(CIP)数据

承德避暑山庄和外八庙：汉英对照/杨菁，朱蕾主编；天津大学建筑学院，承德市文物局编.—北京：中国建筑工业出版社，2018.3
(中国古建筑测绘大系·园林建筑与宗教建筑)
ISBN 978-7-112-21874-5

Ⅰ.①承… Ⅱ.①杨… ②朱… ③天… ④承… Ⅲ.①承德避暑山庄-建筑艺术-图集②外八庙-建筑艺术-图集 Ⅳ.①TU-862②TU-885

中国版本图书馆CIP数据核字（2018）第036788号

丛书策划/王莉慧
责任编辑/李 鸽
英文审稿/［奥］荷雅丽（Alexandra Harrer）
书籍设计/付金红
责任校对/王 烨

中国古建筑测绘大系·园林建筑与宗教建筑
承德避暑山庄和外八庙
天津大学建筑学院
承 德 市 文 物 局 合作编写
杨 菁 朱 蕾 主编

*

中国建筑工业出版社出版、发行（北京海淀三里河路9号）
各地新华书店、建筑书店经销
北京方舟正佳图文设计有限公司制版
北京雅昌艺术印刷有限公司印刷

*

开本：787×1092毫米 横1/8 印张：37½ 字数：937千字
2019年12月第一版 2019年12月第一次印刷
定价：288.00元
ISBN 978-7-112-21874-5
(31781)